Eliciting and Analyzing Expert Judgment

ASA-SIAM Series on Statistics and Applied Probability

The ASA-SIAM Series on Statistics and Applied Probability is published jointly by the American Statistical Association and the Society for Industrial and Applied Mathematics. The series consists of a broad spectrum of books on topics in statistics and applied probability. The purpose of the series is to provide inexpensive, quality publications of interest to the intersecting membership of the two societies.

Smith, P. L., *A Primer for Sampling Solids, Liquids, and Gases: Based on the Seven Sampling Errors of Pierre Gy*
Meyer, M. A. and Booker, J. M., *Eliciting and Analyzing Expert Judgment: A Practical Guide*
Latouche, G. and Ramaswami, V., *Introduction to Matrix Analytic Methods in Stochastic Modeling*
Peck, R., Haugh, L., and Goodman, A., *Statistical Case Studies: A Collaboration Between Academe and Industry, Student Edition*
Peck, R., Haugh, L., and Goodman, A., *Statistical Case Studies: A Collaboration Between Academe and Industry*
Barlow, R., *Engineering Reliability*
Czitrom, V. and Spagon, P. D., *Statistical Case Studies for Industrial Process Improvement*

Eliciting and Analyzing Expert Judgment

A Practical Guide

Mary A. Meyer
Jane M. Booker

Statistical Sciences Group
Los Alamos National Laboratory
Los Alamos, New Mexico

siam
Society for Industrial and Applied Mathematics
Philadelphia, Pennsylvania

ASA
American Statistical Association
Alexandria, Virginia

This SIAM edition is an unabridged republication of the work first published by Academic Press Limited, London, 1991.

10 9 8 7 6 5 4 3 2 1

Library of Congress Card Number: 00-105923

ISBN 0-89871-474-5

For Dr. Thomas R. Bement,
versatile statistician,
co-developer of PREDICT,
dedicated colleague,
and wonderful friend

Contents

5 Refining the Questions 69

10 Conducting the Elicitation 163

Preface to ASA–SIAM Edition

What Is Expert Judgment?

Expert judgment consists of information and data given by qualified individuals that in turn can be used to solve problems or make decisions in a number of areas, including engineering, science, and manufacturing. The following are some applications for expert judgment:

- Determining the probability that a key event will take place
- Predicting the performance of a product or a process
- Determining the validity of assumptions used to select a model for analysis
- Selecting input and response variables for a chosen model

Why This Book?

Our aim in writing this book in 1991 was to provide for lay people a guide on expert judgment. This book is being reprinted (in its original form) because recent developments have made it useful and applicable in new ways, namely, in the creation of knowledge systems and in combination with the methodology PREDICT (Performance and Reliability Evaluation from Diverse Information, Combination, and Tracking). This innovative methodology is used to predict concept system performance and to estimate reliability (under uncertainty) before any prototyping, testing, or production take place. Rather than relying solely on sparse test data, PREDICT uses all available information to provide the best-estimated performance of a complex system in a dynamic environment. Expert judgment is a critical source of this information, and this is why PREDICT has met with success in meeting the needs of government and industry.

People in industry, government, and academia are aware that they must capture their expertise, as well as access and update it in electronic repositories—knowledge systems—if they are to preserve their chief asset and competitive edge. This book provides a shortcut to creating knowledge systems; it enables lay people to design a formal elicitation process that can form the framework for their own knowledge systems.

To this day there are few detailed sources available to lay people on how to elicit expert judgment. Sources that do exist have been specialized for particular disciplines, typically artificial intelligence, psychology, social science, and courtroom expert testimony. This book meets an emerging need of users to elicit directly and easily not only their judgments, but also the process by which they will enter and represent such judgments in their knowledge systems. We use this book to create knowledge systems both internally (e.g., at Los Alamos National Laboratory) and in our partnerships with industry.

This book also has served as the foundation for the award-winning PREDICT methodology. In 1999, PREDICT received an R&D 100 award, jointly awarded to Los Alamos National Laboratory and Delphi Automotive Systems. This award is referred to as the "Nobel Prize" of Applied Research in some circles and is given by *R&D Magazine* to the 100 most technologically significant inventions in the world, as judged by a panel of independent experts.

Prerequisites for Understanding This Book

To understand this book requires few prerequisites. The contents of Part I, *Introduction to Expert Judgment,* and Part II, *Elicitation Procedures,* are simple and procedural in orientation. Part III, *Analysis Procedures,* may require a basic understanding of probability or statistics. However, we have tried to make these procedures easier to follow by introducing techniques and notation in chapter 11 and by defining the statistical jargon as it is used in the text and also in the glossary. The data sets used in the examples are referenced where appropriate. Those data sets that are not referenced have been artificially generated using data set structures and values similar to real data.

Acknowledgments

We gratefully acknowledge the Division of Risk Analysis, Office of Nuclear Regulatory Research, Nuclear Regulatory Commission for their financial support (FIN A7225) of this effort, which culminated in the book's first printing. In particular, we are indebted to Dale Rasmussen for suggesting this work, James Johnson for overseeing the research, and P. K. Niyogi for guiding the book's development. Special thanks also go to the many scientists at Los Alamos National Laboratory, Sandia National Laboratories, Science Associates International Incorporated, U.S. Department of Defense, and Delphi Automotive Systems, who participated in our expert judgment projects. Without their contributions of time and expertise, this book would not have been possible. In addition, we extend our appreciation to Gary Tietjen and Nerses Krikorian of Los Alamos National Laboratory, Dan Brooks of the University of Arizona, and John Boose of Boeing Advanced Technology Center for their insightful reviews of the early drafts. Also, we are grateful for the help of Jeffrey Bradshaw (Boeing) in developing the sections on bias. We are also thankful for the stimulating R&D environment and encouragement provided by the Statistical Sciences Group, especially all our group leaders, including the individuals presently holding these positions, namely, Sallie Keller-McNulty and Elizabeth Kelly. We are most grateful for the desktop publishing expertise of Wilma Bunker, the composition of Patricia Leyba, and the editing of Tino Lopez and Octavio Ramos, Jr. Finally, we thank Vickie Kearn and Deborah Poulson, editors of the ASA-SIAM Series on Statistics and Applied Probability, for making this reprint possible.

Mary Meyer
Jane Booker

Preface

In this book we describe how to elicit and analyze expert judgment. This book's definition of expert judgment includes both the experts' answers to technical questions and their mental processes in reaching an answer. Expert judgment refers specifically to data obtained in a deliberate, structured manner that makes use of the research on human cognition and communication.

The book was written at the request of the Nuclear Regulatory Commission. At the time, the Risk Analysis Division of the Nuclear Regulatory Commission began to break new ground in gathering and using expert judgment in large probabilistic risk assessments. The book's content has been generalized to meet their needs and those of others using expert judgments.

Our aim in this book is to provide a guide on expert judgment for lay persons. These lay persons may be from the physical and engineering sciences, mathematics and statistics, business, or the military. Alternatively, they may work in a field that has traditionally relied on expert judgment, such as risk analysis, reliability analysis, decision analysis, operations research, or knowledge acquisition (a branch of artificial intelligence). To illustrate, people working in the sciences and the military have often told us of their desire to obtain a detailed information source on how to gather and analyze expert judgment. Earlier, no source existed that could provide the guidance they (as lay persons) needed to design and conduct their own elicitations or analyses.

The lack of usable literature on how to elicit information from experts stems from three reasons. First, the manner in which these techniques are learned does not lend itself to publication. Interviewing techniques in anthropology, psychology, and sociology are usually taught in laboratory situations. Students in these fields rarely attempt elicitation on their own; most learn the techniques by observing one of their professors. Thus, even within these specialized fields, few sources on elicitation can be found. Second, the sources that do exist have been specialized for a particular discipline or situation. As a result, attempting to generalize such sources can prove to be a most difficult task. Third, it is difficult to communicate elicitation techniques because the written medium is not well suited to conveying levels of information that are communicated through nonverbal means. Also, because most of the mechanics of elicitation have become automatic in the experienced practitioner, such mechanics become relatively inaccessible for retrieval.

To understand this book requires few prerequisites. The contents of Part I, *Introduction to Expert Judgment,* and Part II, *Elicitation Procedures,* are simple and procedural in orientation. Part III, *Analysis Procedures*, may require a basic understanding of statistics. However, we have tried to make these procedures easier to follow by introducing techniques and notation in chapter 11 and by defining the statistical jargon as it is used in the text and also in the glossary. The data sets used in the examples are referenced where appropriate. Those data sets that are not referenced have been artificially generated using data set structures and values similar to real data.

We gratefully acknowledge the Division of Risk Analysis, Office of Nuclear Regulatory Research, Nuclear Regulatory Commission for their financial support (FIN A7225) and encouragement of this effort. In particular, we are indebted to Dale Rasmussen for suggesting this work, James Johnson for overseeing the research, and P. K. Niyogi for guiding the book's development. Special thanks also go to the many scientists at Los Alamos National Laboratory, Sandia National Laboratories, and Science Associates International Incorporated who participated in our studies of expert judgment. Without their contribution of time and expertise, this book would not have been possible. In addition, we extend our appreciation to Gary Tietjen and Nerses Krikorian of Los Alamos National Laboratory, Dan Brooks of the University of Arizona, and John Boose of Boeing Advanced Technology Center for their insightful reviews of the early drafts. Also, we are grateful for the help of Jeffrey Bradshaw (Boeing) in developing the sections on bias. We are also thankful for the constant support and encouragement provided by A. Juan and our group leaders, Harry Martz and Richard Beckman. Finally, we are most grateful for the desk top publishing expertise of Wilma Bunker, the composition of Patricia Leyba, and the editing of Tino Lopez and Octavio Ramos, Jr.

Mary Meyer
Jane Booker

List of Figures

(Titles listed in italic are not labeled as figures in the text.)

Chapter

List of Tables

(Titles listed in italic are not labeled as tables in the text.)

Chapter

List of Examples

Example

Part I

Introduction to Expert Judgment

1

Introduction

In this book we provide guidance on how to gather and analyze expert judgment. The need for such a source has been great, particularly for those who are lay persons in the area of expert judgment. We have met many people working in the physical sciences, in government, or in the military who were struggling to elicit or analyze expert judgment. Their jobs required that they perform these tasks, but the information available to assist them was scarce. This book is our response to their special needs. We describe elicitation and analysis procedures, when to use them, and how to perform them in a way that allows lay readers to design methods suited to their own particular application. Additionally, we have designed this book so that the reader can start at any chapter, depending on his or her special needs. Those more experienced with expert judgment, typically those working in the fields of risk analysis, reliability analysis, operations research, decision analysis, or knowledge acquisition, may also find the book of interest because in it are mentioned techniques and options that could extend or improve their usual methods.

In this chapter we introduce expert judgment and its use in varied situations. We define expert judgment as it will be covered in this book and provide an overview of the methods that will be presented for eliciting and analyzing expert judgment. Finally, we describe our philosophy of elicitation and analysis as background for understanding the methods presented.

What Is Expert Judgment?

Expert judgment is data given by an expert in response to a technical problem. An ***expert*** is a person who has background in the subject area and is recognized by his or her peers or those conducting the study as qualified to answer questions. Questions are usually posed to the experts because they cannot be answered by other means. For example, expert judgment is needed when it would be too costly or impractical to measure the quantity of interest (e.g., the coal reserves in the United States). Expert judgment can vary from being an estimate of the number of homeless in the United States, to the probability of an occurrence of a nuclear reactor accident of a particular type, to an assessment of whether a person is likely to carry out a threatened act, to a description of the expert's thought processes in arriving at any of the above answers.

Expert judgment has also been called *expert opinion, subjective judgment, expert forecast, best estimate, educated guess,* and most recently, *expert knowledge.* Whatever it is called, expert judgment is more than a guess. It is an informed opinion based on the expert's training and experience.

When Expert Judgment Is Used

Expert judgment data has been widely used, especially in technical fields. This type of data provides information when other sources, such as measurements, observations, experimentation, or simulation, are unavailable. Furthermore, it can be employed to supplement existing data when these are sparse, questionable, or only indirectly applicable. For example, in a new reactor-risk study called NUREG-1150 (U.S. NRC, 1989), expert judgment was used where "experimental or observational data or validated computer models were not available or not widely agreed upon" (Ortiz, Wheeler, Meyer & Keeney, 1988, p. 4).

Expert judgment has been specifically gathered to meet the following needs:

- **To provide estimates on new, rare, complex, or otherwise poorly understood phenomena.** Such phenomena have also been described as being *fuzzy* or of *high uncertainty.* One example would be estimating the likelihood of rare reactor accidents. Another would be estimating the safety of new automotive fuels. For example, in the early eighties fuels such as liquid and compressed natural gas were being proposed for automotive use, but little was known about how safe they would be. To solve the problem, a group of experts estimated the relative safety of the new fuels by considering their physical properties in combination with potential accident scenarios (Krupka, Peaslee & Laquer, 1983).
- **To forecast future events.** In general, when good data are unavailable, predicting future events or actions requires use of expert judgment. To make these predictions, the experts must adjust, sometimes radically, from the status quo or past patterns. For instance, businesses often rely on expert judgment to forecast the market for their products. What the demand for various utilities will be in the United States may also come from experts' projections (Ascher, 1978). For example, by forecasting the Soviet weapon capabilities for the year 2000, experts can project the weapon needs of the United States for the same period (Meyer, Peaslee & Booker, 1982).
- **To integrate or interpret existing data.** Expert judgment is frequently needed to organize qualitative information or mixtures of qualitative and quantitative data into a framework for making decisions. ***Qualitative data*** are any nonnumeric data, such as text on the expert's reasons for giving an answer, or the expert's answer encoded in descriptive categories or preference scales like poor, moderate, and good. ***Quantitative data*** are numeric data such as estimates of probabilities, physical phenomena such as temperature, simple ***ranks*** or ***ratings*** (1-5), and error bounds on any such estimates of probability, physical phenomena, or ranks or ratings, etc. (0.75 ± 0.25).

For example, to make decisions on next year's product line, an expert might determine how to model the firm's quantitative data, such as on projected cost, and qualitative data, such as on market potential. This data might also include other experts' judgments, such as on market potential.

Similarly, expert judgment might be employed to interpret existing data, even when that data is other expert judgment. For example, decision makers interpret multiple and differing sets of expert judgment and decide which data, if any, best suit their needs.

Experts also interpret several sets of data when making a diagnosis. Medical specialists frequently must interpret differing test results in arriving at a decision.

- **To learn an expert's problem-solving process or a group's decision-making processes.** The experts often do not know how they solve a problem or reach a decision because their thoughts have become so automatic that the process is difficult to recall. However, their problem-solving techniques are needed to improve current practices, to train others, or to create systems that provide expert advice.

  One project that involved learning an expert's problem-solving processes focused on that expert's procedures for reaching a decision. In this case, the experts were police officers who specialized in resolving hostage-taking events. The experts assessed information on the situation and reached decisions on how to proceed, such as whether to negotiate or resort to an assault (Vedder & Mason, 1987).

  Another project's goal was to provide guidance on how an organization would make future decisions on the export of munitions. The experts, members of different Army offices, divided the problem into parts and specified the type of input that they wanted each office to provide in making the larger decision (Meyer & Johnson, 1985).

- **To determine what is currently known, what is not known, and what is worth learning in a field of knowledge** (Ortiz *et al.*, 1988). In the reactor risk study, NUREG-1150, the experts exchanged the most up-to-date information in preparation for giving their answers to particular questions. As a result, they identified gaps in their field's state of knowledge and determined in which areas they would most like to see research. This type of information offers several benefits: it can serve as a complement to the current state of knowledge or as motivation for further study.

Expert judgment is often used to address more than one of the above-mentioned needs. Such was the case in the new reactor risk project, NUREG-1150 (Wheeler, Hora, Cramond & Unwin, 1989), where the expert judgment met all of the above-mentioned purposes. In addition, gathering expert judgment often provides side benefits: one of the most common benefits is the facilitation of communication. The experts readily see how their judgments differ and relate to each other's views in an environment of openness and objectivity. We have noticed that the synergism of interexpert discussion stimulates results that would not have been achieved otherwise.

General Attributes of Expert Judgment

In general, expert judgment can be viewed as a representation, a snapshot, of the expert's knowledge at the time of response to the technical question (Keeney & von Winterfeldt, 1989). As Ascher (1978, p. 203) notes, "multiple-expert-opinion forecasts, which require very little time or money, do very well in terms of accuracy because they reflect the most-up-to-date consensus on core assumptions." The expert's judgment legitimately can and should change as the expert receives new information. In addition, because the judgment reflects the expert's knowledge and learning, the experts can validly differ in their judgments.

An expert frequently gives answers in quantitative form, such as probabilities, ratings, or odds. For instance, an expert's answer to the question could be respectively 0.10, 1 on a scale of 10, or 1 in 10 chances. Quantitative response modes are often requested because the numeric data are more easily analyzed than qualitative data.

Much of expert judgment is the product of high-level thought processing, also called knowledge-based cognition. By ***cognition*** is meant the mental activity that occurs when a person processes information, such as for solving a problem. ***Knowledge-based cognition*** is the high-level interpretive or analytic thinking that we do when confronted with new and uncertain decision situations (Dougherty, Fragola & Collins, 1986, p. 4-2) Thus, knowledge-based cognition is often invoked by the situations for which expert judgment is sought.

The quality of expert judgment varies according to how the data are gathered; the data can be obtained in a variety of ways, ranging from the unconscious to the deliberate. Expert judgment can be gathered unconsciously, as often occurs in technical projects. Analysts typically make decisions in defining problems, establishing boundary conditions, and screening data without being aware that they have used their own expert judgment .

Expert judgment is also gathered deliberately, although even this type of gathering varies along a continuum of informal to formal. On the informal end of the continuum, experts are asked to provide judgments *off the top of their heads.* The informal means of gathering expert judgment has been a source of current controversies involving expert judgment. The most recent controversy involves psychologists and psychiatrists serving as expert witnesses in legal proceedings. Recent articles have proclaimed that these expert witnesses are no more accurate than lay persons, particularly in predicting an individual's propensity for future violence. These situations illustrate that "without the safeguards of the scientific method, clinicians are highly vulnerable to the problematic judgment practices and cognitive limitations common to human beings" (Faust & Ziskin, 1988, p. 33). In other words, experts are subject to the same cognitive and motivational biases of other humans if certain preventative measures are not used in the elicitation. By bias is meant a skewing of the expert's judgment from some reference point. With cognitive biases, the expert judgment is considered skewed from the standpoint of mathematical or logical rules, usually because of the ways in which the expert mentally processes information. With motivational biases, the expert judgment is skewed from the standpoint of the expert's thoughts. The expressions of the expert's thinking do not match either the expert's thoughts or answers at the time of the elicitation because the expert has altered the reports of his or her thoughts or because the interviewer has misunderstood the expert's reports. (Bias will be discussed further in Chapter 3.)

Formal means of gathering expert judgment usually involve selecting experts according to particular criteria, designing elicitation methods, and specifying the mode in which the expert is to respond. The formal approach to elicitation has two advantages over its unconsciously or informally gathered counterparts. First, with the formal approach more time and care is taken in eliciting the judgments. Because the quality of expert judgment is often evaluated in terms of the methods used to gather the judgments, the greater time and effort associated with the formal approach is an advantage. Second, the formal approach lends itself more readily to documentation than those methods used unconsciously or less formally. That is, records may be kept of which elicitation methods were used and of how the experts arrived at their final judgments. Such a record allows the formal method and its results to be scrutinized. Thus, the formal approach is more likely to advance through the review process (Ortiz *et al.*, 1988).

Expert Judgment Covered in This Book

In this book, we cover how to gather expert judgment in a formal and structured manner. An additional focal point is gathering expert judgment according to those methods suggested by the research into human cognition and communication. Even the formal means of gathering expert judgment have not generally made use of the rapidly emerging body of literature on how best to obtain experts' judgments (i.e., how to avoid biases). After all, expert judgment is frequently needed, and its gatherers are not always familiar with the biases to which humans are prone. The result is that experts are asked to provide estimates without concern for bias or the benefit of methods shown to improve accuracy.

The research on how judgments should be elicited comes from three fields: psychology, decision analysis, and more recently, a branch of artificial intelligence called knowledge acquisition. Examples of works in the field are Hogarth (1980) for psychology, Spetzler and Stael von Holstein (1975) for decision analysis, and Gaines and Boose (1988) for ***knowledge acquisition***.

Using the methods suggested by the research usually enhances the quality of the expressed judgments. The following examples show just a few of the ways that research can be applied to improve the gathering of expert judgment.

The use of one such method—breaking a problem into its component parts—has been shown to yield more accurate answers (Hayes-Roth, 1980; Armstrong, Denniston & Gordon, 1975).

Other research has shown that people have difficulty correctly translating their judgments into quantities (e.g., probabilities). In the state-of-the-art elicitations of expert judgment, probabilities are either not required or the experts receive lessons in their use (U.S. NRC, 1989).

Similarly, studies show that most people cannot consider more than approximately seven things at once (Miller, 1956). To deal with this limitation, the experts may be asked to use scales that allow them to compare two, rather than seven or more, things at a time.

In addition to focusing on eliciting expert judgment as suggested by the relevant research, in this book we define expert judgment to include more than simply the expert's estimates or solutions. We have broadened the traditional meaning of expert judgment because we believe that all data associated with the expert's answer are important to understanding his or her answer. Indeed, many gatherers of expert judgment have begun to document the expert's thoughts as well as their answers, perhaps because of the influence of artificial intelligence and expert systems. From this point on, expert judgment will be defined to include the following:

1. **Any of the expert's work in selecting or defining the scope of the problem.** For example, in the reactor risk study mentioned earlier (U.S. NRC, 1989), the experts reviewed proposed problem areas (safety issues in reactor risk studies) and added or deleted issues. According to this book's definition, the selected issues that they chose and their criteria of selection are considered expert judgment.
2. **Any of the expert's work in refining the problem.** In many uses of expert judgment, the problem is depicted in some fashion, such as in scenarios, tree structures, grids of the pertinent factors, or pages of explanatory text. In this book, all these attempts to break the problem into parts and describe them are considered to be expert judgment. For example, one of the problems that experts addressed in the reactor risk study was the probability of a check valve failure—such a failure could cause a loss-of-coolant accident. To solve the problem the experts developed a series of failure scenarios by taking the issue and categorizing it into specific parts. As a result, the experts developed three scenarios: both valves fail independently, one valve fails to reclose, or the valves randomly rupture (Ortiz *et al.*, 1988). In this same study, the experts operationally defined the key variables. For instance, the experts set *independent rupture* to mean a catastrophic leak, which in turn had been defined to mean a particular flow rate per unit time.
3. **Any of the expert's mental processes in arriving at a solution.** This aspect of expert judgment often involves the expert's sources of data, definitions, assumptions, and mental procedures for processing the information. For example, in the same reactor risk study, some experts used information from event trees as their primary source of data in solving the problem, others used information from experiments, and still others relied on output from computer

models. In addition, the expert's definition of terms are considered part of his or her problem-solving processes. Experts often unconsciously assign their own meanings to terms. For this reason, the panel of experts working on the reactor risk study defined terms like *catastrophic leak*. Experts also utilize cognitive techniques to help them process the problem information. One such technique, or shortcut, is to adjust up or down from a base line. For example, an expert could evaluate the risk posed by a rare accident by setting the frequency of a related but more common risk as his or her base line.

The above-mentioned three categories of data relating to the expert's solving of the problem will be referred to as ***expert data.*** In a risk analysis application, expert data is likely to include the expert's assumptions, definitions, and decomposition of the problem into its parts. In a knowledge acquisition project, the expert data could be the expert's rules or procedures for reaching a solution. Expert data is a subset of an even more general class of information that is elicited from the experts. This more general class, called ***ancillary data*** or ***information,*** includes data gathered on the expert's educational background, work history, current job environment, or personality. Additionally, the term ***estimates*** will refer to answers such as probabilities or ratings that are given in quantified form. ***Solutions*** will refer to answers given in qualitative form, such as descriptive text or diagrams. ***Answers*** will be used as a general term for both estimates and solutions. ***Expert judgment*** will be used as a cover term that refers to a combination of the expert answers and ancillary information.

How Expert Judgment Is Elicited

The methods used to elicit expert judgment have varied widely, even within particular fields such as risk analysis and knowledge acquisition. The following factors affect how expert judgment can be best gathered in particular situations:

- The type of information the experts must provide (answers only or ancillary expert data)
- The form (***response mode***) in which the expert's answers are given
- The number of experts available
- The interaction desired among the experts
- Difficulty of preparing the problems
- The amount of time and study the experts will need to provide judgments
- The time and resources available to the study
- The methodological preferences of the data gatherers (interviewers or knowledge engineers), analysts, project sponsors, and experts.

Elicitation is the process of gathering the expert judgment through specially designed methods of verbal or written communication.

For example, in one situation, a large group of experts convenes for a week to interactively construct a problem representation and to provide the estimates.

They produce a representation of all the factors involved in deciding whether to export an Army-developed technology (Meyer & Johnson, 1985). During the week, the experts separately assign weights to the importance of these factors. To test their representation, the experts apply it to an example; they then examine the outcome, the product of the mathematical processing of their estimates through the decision framework.

In another situation, the experts are separately interviewed in depth to obtain their judgment on the performance of their computer code in modeling reactor phenomena (Meyer & Booker, 1987*b*). An interviewer asks each expert to work a selected problem and to explain in detail their thinking process as they work through it. They also rate the performance of the code's model on a linear scale.

In a third situation, the experts are interviewed separately as well as convened for discussions. Each expert first receives seismic-techtonic information which he or she must use to detail zones for the United States (Bernreuter, Savy, Mensing, Chen & Davis, 1985). They are also asked to estimate the frequency and magnitude of earthquakes by zone. After the experts independently record their judgments, they meet as a group. The group receives a combined zone map, estimates of earthquake phenomena, and the anonymous judgments of the other experts. The group discusses all the information, perhaps arriving at different judgments. As a result, each expert can then privately make adjustments to his or her independent estimates based on any new data acquired during the group discussion.

The above examples illustrate some of the diversity in elicitation processes. Elicitation processes can also differ in terms of the following:

1. The degree of interaction between experts
2. The amount of structure a group moderator or interviewer imposes on the elicitation process
3. The number of meetings
4. The time allotted for structuring the problem and eliciting the expert judgment
5. Who performs these tasks, the experts and/or the analyst
6. The response mode in which the expert estimates are elicited
7. Whether the expert's reasoning is requested or not
8. The level of detail in the expert judgment elicited
9. Whether the expert judgment undergoes some translation in a model and is returned to the experts for the next step
10. Whether all or some of the elicitation is conducted in person, by mail, or by telephone.

Despite the diversity in the elicitation processes, there are only three basic elicitation situations and a general sequence of steps. Expert judgment can be elicited through the following:

- ***Individual interviews*** in which an interviewer or knowledge engineer interviews the expert in a private, usually face-to-face, environment. This situation allows an interviewer to obtain in-depth data from the expert, such as how to solve a proposed problem, without having him or her distracted or influenced by other experts.

The individual interview is also called the *staticized* or *nominal-group* situation when the experts' estimates are mathematically combined to form one *group* answer.

- ***Interactive groups*** in which the experts and a session moderator meet in a face-to-face situation to give and discuss their data. The participants' interactions with one another can be structured to any degree: (1) a totally unstructured group resembles a typical meeting, while (2) a highly structured group is carefully choreographed as to when the experts present their views and when there is open discussion; such procedures help prevent some of the negative effects of interaction.
- ***Delphi situations*** in which the experts, in isolation from one another, give their judgments to a moderator. The moderator makes the judgments anonymous, redistributes them to the experts, and allows them to revise their previous judgments. These iterations can be continued until consensus, if it is desired, is achieved. Rand Corporation developed this elicitation situation to counter some of the biasing effects of interaction, such as when a dominant expert causes the other experts to agree to a judgment that they do not hold.

The general sequence of steps in the elicitation process follows:

1. Selecting the question areas and particular questions
2. Refining the questions
3. Selecting and motivating the experts
4. Selecting the components (building blocks) of elicitation
5. Designing and tailoring the components of elicitation to fit the application
6. Practicing the elicitation and training the in-house personnel
7. Eliciting and documenting expert judgments (answers and/or ancillary information).

Philosophy Guiding the Elicitation

The philosophy put forth in this book is that the elicitation should be designed to fit the experts and the way that humans think rather than force the experts to adapt to the methods. We propose that the research on human limitations and tendencies toward bias be taken into account in selecting the methods. For example, if the interviewer involved in selecting the elicitation methods does not consider people's limitations in comparing more than seven things at once, the resulting data will be less credible. If in the former case the expert estimates are being used to develop a model or decision process, there is the danger of *garbage in, garbage out.*

We also advocate the practice of eliciting as much of the information on the expert's ***problem-solving processes*** as possible. We believe that this data is necessary to the understanding of the experts' answers. Experts' estimates have been found to correlate to the way that they solve the problem (Booker & Meyer,

1988*a*; Meyer & Booker, 1987*b*). The expert's definitions and assumptions frequently explain how the expert arrived at one particular answer and not another. In addition, this type of data will prove valuable later if multiple experts' estimates are to be mathematically combined to form a single estimate. The expert data can guide the aggregation so that experts who construed the problem very differently will not have their answers combined inappropriately. In general, recording information on the expert's problem-solving process allows the judgments to be more easily updated as new information becomes available.

Another aspect of the elicitation philosophy is to control for the factors that can enter into the elicitation process and influence the expert's problem-solving process. For example, the phrasing of the problem, the interviewer's responses, and other participants' responses can affect the answer an expert reaches. Several elicitation methods designed to control these influences are mentioned in chapters 7 and 8. For those influences that can not be easily controlled, such as the expert's tendency to anchor to his or her first impression, we recommend gathering as much data as possible to analyze their effects.

Philosophy Guiding the Analysis

The analysis philosophy of this book complements the elicitation philosophy mentioned above. Just as the elicitation approach allows the experts' capabilities to shape the data-gathering methods, the analysis philosophy allows the data to dictate which analytic methods are used. Thus, the analyses are data driven. This analysis approach is used in the belief that it will produce the highest quality results.

As a part of the analysis philosophy, the analyses avoid assuming particular properties of expert judgment. For example, the analyses do not *a priori* assume that the expert judgment data is ***normally distributed***, that the answers of multiple experts are independent, or that the experts are perfectly calibrated (unbiased). Instead, the analysis sections offer methods that either do not require these assumptions or that can test for the existence of such properties. For example, ***nonparametric*** statistical procedures and data based ***simulation*** techniques are used because they do not depend on an assumed distribution of the data.

The analyses also avoid assuming that the experts' data are independent or dependent (i.e., conditioned on some common factor, such as the experts' education). Analysts have been driven into assuming independence in the past because they needed to combine multiple estimates into a single representative one, and most aggregation schemes have required independence. Those aggregation schemes that do not require independence are more complicated and require information on the structure of ***dependence*** in the data, information which the researcher rarely has. As part of the analysis philosophy, this book provides simple methods from recent research (Booker & Meyer, 1988*b*; Meyer & Booker, 1987*b*) to help test for independence and dependence; further methods address the problem of handling dependence.

In addition, the analysis section provides methods for investigating potential sources of bias. Many users of expert judgment have been forced to assume that the data was unbiased because they had no way of analyzing biased data. To meet this need, the analysis section also provides methods on how to determine if the expert data is biased.

A variety of methods are used to address the multivariate nature of expert judgment data. The data is multivariate because it often includes answers to multiple problems, data on the experts themselves, and mixtures of qualitative and quantitative data. ***Multivariate analysis*** techniques allow the simultaneous consideration of two or more variables of interest (Tietjen, 1986). Thus, they can be used to investigate some of the more important properties of the data, such as the dependence of experts. Several multivariate techniques, such as ***cluster analysis***, ***discriminant analysis***, and ***general linear models (GLMs)***, are used because no single one is universally applicable to the structure of expert judgment data. In addition, this book provides methods for transforming qualitative data into quantitative forms.

How To Use This Book

This book is divided into three parts—*Part I: Introduction to Expert Judgment; Part II: Elicitation Procedures;* and *Part III: Analysis Procedures.* Part I consists of three chapters. In chapter 1, *Introduction*, we present a description of what expert judgment is and our philosophy for its elicitation and analysis. Chapter 2, *Common Questions and Pitfalls Concerning Expert Judgment,* includes many questions asked about expert judgment and the hidden traps encountered in eliciting or analyzing it. Chapter 3, *Background on Human Problem Solving and Bias,* provides information on how humans think and some of the consequences of those processes, such as common biases. The background provided in chapter 3 is necessary to selecting and tailoring the elicitation methods so as to obtain the best quality data possible. Part II, chapters 4 through 10, contains sections on eliciting expert judgment, and Part III, chapters 11 through 18, provides sections on analyzing results. These two latter parts have been outlined in the flow chart that follows.

This chart gives guidance on how to best use these parts of the book, pointing out the most efficient sequencing to use in differing situations. Appendices A through D document the computer programs we have used. And finally, a glossary provides a quick reference for the terms that have been highlighted in bold italics the first time that they are mentioned in the text. We expect the glossary to be a major aid to those readers who are unfamiliar with the book's terminology and recommend that they refer to the glossary heavily at first.

Readers can use this book in several ways, depending on their needs and situation. The book proceeds sequentially from the selection of the problem (Part II) to the elicitation procedures and the analysis of results (Part III). The reader can read about each phase as he or she is ready to execute it. We assumed that readers may be at differing phases in their applications and designed the book to be entered at any chapter. As a result, some information is repeated in different chapters and much is cross referenced. Another way of using this book would be for the reader to examine sections after performing them in an application. Reading portions of the book in retrospect will allow readers to note some of the strengths and weaknesses in their performance of a phase. Most of the chapters on elicitation contain sections on common difficulties and means for resolving them. The reader can refer to these sections at any phase of gathering expert judgment.

Flow Chart for the Use of the Book

(Parts II and III)

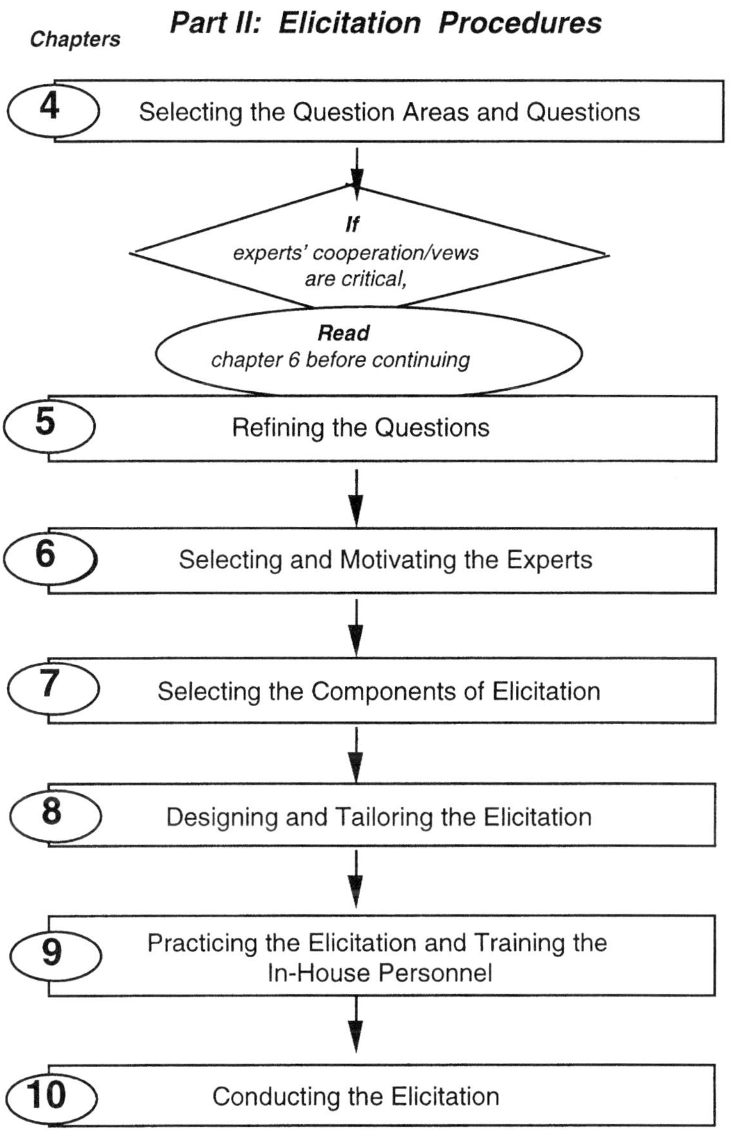

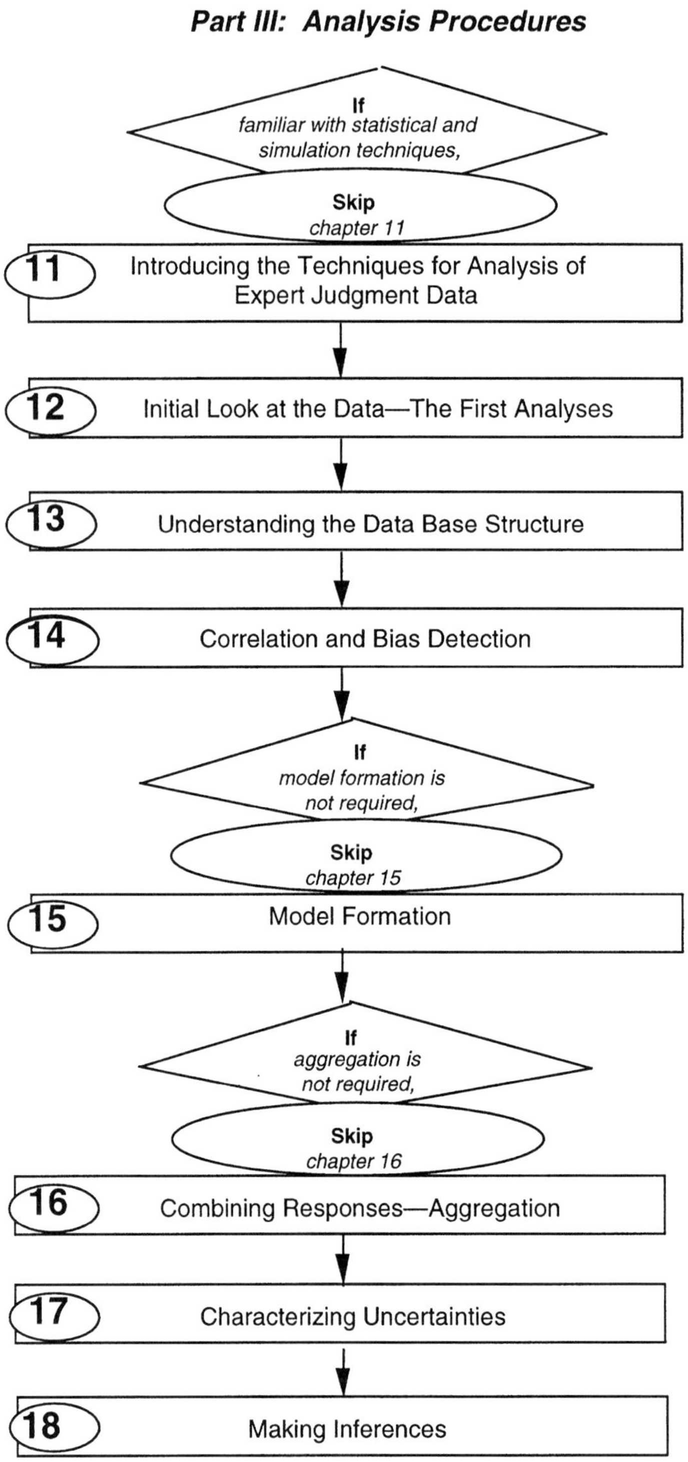
Part III: Analysis Procedures
If
familiar with statistical and
simulation techniques,
Skip
chapter 11
11
Introducing the Techniques for Analysis of
Expert Judgment Data
12
Initial Look at the Data—The First Analyses
13
Understanding the Data Base Structure
14
Correlation and Bias Detection
If
model formation is
not required,
Skip
chapter 15
15
Model Formation
If
aggregation is
not required,
Skip
chapter 16
16
Combining Responses—Aggregation
17
Characterizing Uncertainties
18
Making Inferences

2

Common Questions and Pitfalls Concerning Expert Judgment

In this chapter, we address many of the common questions that often arise from misconceptions concerning expert judgment. In addition, a section designated *Pitfalls* discusses information on those aspects of elicitation and analysis that have typically caused problems. We hope that the information on pitfalls will alert readers and prevent their falling into the difficulties described.

For the reader's convenience, the lists that follow summarize the information on both the common questions asked and the common pitfalls encountered; these questions and pitfalls are addressed at length later in this chapter.

Summary of Common Questions and Pitfalls

List of Common Questions

1. **What does it mean when the experts disagree?** When the experts disagree, as they do with irritating frequency, it can mean that they interpreted the question differently (i.e., took it to mean different things) or that they solved it using different methods (e.g., used different algorithms or data sources). If the data indicate that the expert interpreted the question differently, one option is to have the expert, who essentially answered a different question, readdress the question using the agreed upon definition. His or her previous response may be discarded as a response to a different question.
2. **Is expert judgment valid data?** Expert judgment can be considered data in the sense that it is information (cardinal, ordinal, categorical, and descriptive) separate from the theories or hypotheses of the researchers. The validity or quality of expert judgment, like any data, can vary. The quality of expert judgment depends on both the completeness of the expert's mental model of the phenomena in question and the process used to elicit, model, analyze, and interpret the data. We provide suggestions on how to select experts and gather and analyze expert judgment so as to obtain high quality data.
3. **Is the gathering and analysis of expert judgment a science?** We propose that the gathering and analyzing of expert judgment advocated in this book

aims at being scientific. Its methods are based on research that has followed the tenets of science—observation, hypothesis formation, and experimentation.

4. **Are experts Bayesian?** In general, people do not naturally follow the philosophy developed from the application of Bayes Theorem. Bayes Theorem defines the probability distribution function of the data as conditioned on its parameters. These parameters are also assumed to have a probability distribution called a prior distribution. Combining the data and prior distributions produces a posterior distribution. The expected value of this posterior distribution is the desired final estimate. The combining of distributions is often done in cases where the new or additional data need to be merged to update the existing data. The application of Bayes Theorem to expert judgment would predict that people update their existing information to account for new information as it becomes available. In laboratory and real-life settings, however, experts may fail to sufficiently adjust their estimates in view of new information.
5. **Do experts provide better data?** Whether experts provide more accurate or better quality data depends on the type of data that they are asked to give. Humans are notoriously poor in providing predictions; in addition, experts have not consistently been shown to be better than nonexperts. In contrast, experts can provide valuable and reliable information on the state of the knowledge in their field, on how to solve problems, and on the certainty of their answers. Additionally, in these areas, experts have been found to have capabilities beyond those of novices or lay persons.
6. **Can experts be fully calibrated?** Experts cannot yet be fully calibrated as some technical instruments and processes are. Calibration means to first compare unknown instruments or processes to known or correct ones then to adjust the unknown ones until they match the known. There are several reasons why the calibration of experts would be desirable. For example, experts often fail to sufficiently adjust their estimates in view of new information, to grasp the effects of sample size and variability, or to follow the axioms of probability theory in laboratory and real-life settings. To improve experts' calibration, the feedback from the known procedures must be immediate, frequent, and specific to the task (Lichtenstein, Fischhoff & Phillips, 1982), such as the feedback received in weather forecasting. This type of information is unavailable for the majority of problems requiring expert judgment, at least in the risk and reliability fields.

List of Pitfalls

1. **Interviewers, knowledge engineers, and analysts can introduce bias.** These persons can unintentionally introduce bias (i.e., alter the expert's thinking or answers) in two ways. First, interviewers or knowledge engineers are likely to misinterpret the expert's data by perceiving it to be the type of data that their training ideally equipped them to handle (training bias). Second, analysts are

likely to misrepresent the expert data by forcing it to fit the models or analytic methods with which they are most comfortable (tool bias).

2. **Experts are limited in the number of things that they can mentally juggle.** The limit to the amount of information that humans can process in their short-term memory is seven plus or minus two (Miller, 1956). This information-processing limit is something to consider when designing the elicitation situation. For this reason, we suggest that the reader carefully consider the granularity in planning the collection and analysis of the expert data.
3. **The level of detail in the data affects the analyses.** The ***granularity*** is the level of detail in a chunk of information (Waterman, 1986). An example of coarse granularity would be the basic functions of a nuclear power plant; an example of finer granularity would be the subfunctions of one such function. We have found that the granularity used in gathering the data influences the results of the analyses.
4. **The conditioning effect poses difficulties in gathering and analyzing expert data.** The data that the expert gives can be conditioned on a variety of factors, ranging from the wording of the problem to the expert's internal mood. The conditioning effect poses problems because many factors can intrude without the data gatherer's awareness, overlap with other variables, and complicate the analyses. We propose a two-step approach to handling this conditioning effect: (1) control those factors which can be controlled, and (2) gather as much data as possible on those factors that cannot be controlled so that the data may be analyzed later for their effect.

Questions

What Does It Mean When the Experts Disagree?

We are all aware of instances when the experts have disagreed. This disagreement has led some persons to question the credibility of expert judgment. The reaction, although natural, is based on a misconception about expert judgment. Many people expect the results of expert elicitation to be reproducible—they believe that experts given the same data will reach the same conclusion. If the experts' conclusions differ, they reason that their judgment is questionable. This concept comes from experiments in the *hard* sciences, where the reproducibility of results confirms their validity. However, this concept is inappropriate and misleading when applied to expert judgment for two reasons.

First, experts do not possess the same data, even if they receive the same briefings and information as background to the problem. The expert's body of knowledge, the primary reason for consulting him or her, is something which has been created over time through the expert's professional experience (education, on-the-job-training, and exposure to data bases or dramatic events in the field). Therefore, each expert's knowledge differs. In addition, what the experts do with their information (how they consider the separate pieces) is likely to differ. For example, if two experts possessed the exact same information, they would probably use the data in different ways (e.g., one might consider a datum highly relevant to the case in point and use it whereas the other might dismiss it from consideration). Because experts have different bodies of knowledge and approaches to solving problems, their answers are likely to differ. Indeed, two studies have shown this link between the expert's answer and aspects of his or her problem-solving approach (Booker & Meyer 1988*a*, 1988*b*; Meyer & Booker, 1987*b*).

Second, unlike in the physical sciences, expert judgment is frequently sought in situations in which there are no clear standards or well-developed theories. One example of such a situation is the prediction of poorly understood phenomena (e.g., the likelihood of an individual performing a threatened act, the chances of occurrence of a seismic disturbance of a particular magnitude, or the probability of occurrence of a rare sequence of reactor events). In the engineering sciences, where many standards are clearly defined, the experts have guides to follow in reaching a decision. Such standards tend to reduce expert variability in problem-solving approaches. However, in fields in which expert judgment is usually elicited, such standards are not in place and therefore do not lead to greater uniformity among the experts' solutions.

Thus, when the experts disagree, a valid interpretation would be that they have interpreted and solved the problem in differing manners. If records on the experts' problem-solving processes have been kept, it is possible to pinpoint differences on how the experts arrived at their final judgments. Based on these differences, it may be possible to justify the later inclusion or exclusion of a particular expert's answers. It may be that some of the expert's approaches are better approximations of the reality (of the problem being posed) and therefore more likely to be accurate, as proposed by the Brunswik lens model (Hogarth, 1980, p. 8). However, to choose the better approach when the correct answer or approach remains unknown can prove to be a difficult task. For this reason, differences among the experts can be interpreted in a positive light as evidence that the different perspectives on the problem are being represented. In fact, studies have shown that mathematically combining the experts' differing answers provides a better chance of covering the correct answer (when this is known) than does the use of single expert's answers.

Is Expert Judgment Valid Data?

Expert judgment has been called *expert opinion, subjective data, subjective judgments, qualitative data, expert forecasts, expert estimates, best estimates, expert knowledge,* and *expert judgment data.* If the data take the form of probabilities, then terms such as *subjective probabilities* and *probability estimates* have been used.

This book's definition of expert judgment includes the expert's answer, information on how the expert reached the answer (his or her assumptions and algorithms), and ancillary information on the expert (his or her educational background and work experience). Furthermore, expert judgment is defined as data gathered in a particular way—formally, in a structured manner, and in accordance with the research on human cognition and communication.

The term *data* has varying meanings for different people. For some, data refers exclusively to measured or observed numerical values. For others, data is used in a less restrictive context and refers to information as distinguished from hypotheses or theories. In this book, data refers to all the information that has typically been gathered in the sciences; data have been cardinal (real-numbered values), ordinal (numeric or verbal ranks), categorical (numeric or verbal classes or categories), or descriptive (words, phrases, and sentences). Thus, data refers to information either in a qualitative or quantitative form. By this definition of data, expert judgment is data.

Some people have further questioned whether expert judgment is *valid* data. Those people believe the source, the human expert, makes expert judgment *soft* data in comparison to data that has been measured or obtained from observation or instruments. Ericsson and Simon explain that data are regarded as *soft* to the degree that they incorporate the assumptions and interpretation of observers, "especially when the theoretical premises and rules of inference are themselves not completely explicit and objective" (1984, p. 4). Therefore, the *softness* of the data is not a fixed attribute of the data itself, but something that the observer or researcher can exercise control over. In their dealings with the data, the researchers or observers can strive to keep their interpretations of the data, such as in their assumptions on the properties of the data, separate from the data. They can also attempt to make the minimal amount of assumptions.

To summarize, we consider expert judgment to be valid data and comparable to other data. All data are imperfect representations of the thing that they are supposed to represent. Data taken from instruments cannot be considered perfect representations for many reasons: random noise, equipment malfunction, operator interference, data selection, or data interpretation. Just as the validity of hard data varies, so the validity of expert judgment varies. How well expert judgment represents the phenomena in question depends on the completeness of the expert's mental model of the phenomena and on the way in which the expert data was elicited, modeled, analyzed, and interpreted. It is for this reason that we emphasize carefully selecting the experts, monitoring and testing for bias (from the expert, interviewer, knowledge engineer, or analyst), using the elicitation techniques that the literature supports, and avoiding the trap of making assumptions of the properties of the expert data.

Is the Gathering and Analyzing of Expert Judgment Scientific?

To determine whether gathering and analyzing expert judgment can be considered a science, we must examine some of the views of science. A popular view of science includes the notion of reproducibility of results. Because of the dynamic nature of expertise, however, separate experts cannot be expected to give the same response, nor can a single expert at different points in time be expected to give the same response. (See *What Does It Mean When the Experts Disagree?*) Expertise is based on the expert's training and experience and, according to Gaines (1989), continuously evolves with the expert's exposure to new problems and data. Thus, no two experts will have shared the exact same experience, nor is one expert's expertise likely to remain static through time.

Therefore, let us look at another definition of science: "the process or procedure for making inquiries of our world" (Hirely, 1989, p. 25), with the basic tenets of observation, hypothesis formation, and experimentation. Expert judgment studies follow these tenets of the scientific method. The large body of research on human judgment and problem solving illustrates this point.

Tightly controlled experiments have been performed on such topics as: judgmental processes (Hogarth, 1980; Kahneman & Tversky, 1982), effects of problem decomposition (Hayes-Roth, 1980; Armstrong, Denniston & Gordon, 1975), memory functions (Ericsson & Simon, 1980; 1984), effects of group dynamics (Zimbardo, 1983), effects of question phrasing (Payne, 1951), probability estimation (Spetzler & Stael von Holstein, 1975; Hogarth, 1980), and sources of interexpert correlation (Booker & Meyer, 1988*a*).

In these studies, researchers observe aspects of human problem solving and interaction; they also postulate and test hypotheses. For example, researchers involved in the study of problem decomposition tested whether or not disaggregating a question into its component parts yields more accurate answers (Hayes-Roth, 1980; Armstrong, Denniston & Gordon, 1975). Questions are selected in which the right answer is known or can somehow be verified. The subjects are asked to solve problems which are disaggregated and those which are not, and their answers are compared to the correct answers.

The third tenet of science, experimentation, has provided expert judgment studies with challenges beyond those characteristic of the physical sciences. These challenges stem from the use of human subjects. Human subjects have the capability of manipulating the experiment while the typical subject of physical science studies does not. For example, a chemist does not have to cope with an element suddenly changing its behavior, such as its mass, because it favors a particular outcome in the experiment. In addition, with human subjects, the data is generally conditioned on multitudes of unknown and uncontrollable factors. An expert's judgment or descriptions of it can be conditioned on attributes of the expert (assumptions and algorithms used in solving the problem, training, past experiences, and so on), the interviewer's presence, the phrasing of the question, and the responses (verbal and nonverbal, real and imagined) of the interviewer or others. Some uncontrollable factors include the expert's mood before the elicitation takes place and aspects of his or her background.

In addition, the expert data gathered are often of a predictive nature which cannot be objectively verified, at least not in real time. These judgments cannot be evaluated by performing additional measurements of some physical property or by referring to some authoritative text.

These aspects of expert judgment studies pose problems in experimental design. Typically, experimental design involves the planning of a study in terms of what will be observed for measurement and how it will be observed. In expert judgment studies and applications, the investigator cannot totally control the entry of factors for observation because the subject brings a variety of unknowns to the elicitation. Also, the small sample sizes of expert judgment studies do not allow the data to rise above their effect. In addition, the type of data gathered makes testing difficult, especially in a scientific tradition that holds that "the world is objectively knowable and that deductions about it can be tested" (Denning, 1988).

In conclusion, we would argue that researchers in the field of expert judgment strive both to follow the tenets of science and to produce scientific studies. Furthermore, it is through the careful gathering, use, and examination of expert judgment that this field will make further progress as a science.

Are Experts Bayesian?

Many analysts and theorists in the mathematical community advocate the Bayesian philosophy of analysis. In the decision analysis and reliability assessment communities, this philosophy has come to mean the evaluation of gathered data as they are conditioned on other events or circumstances (variables). Given that the data are conditioned on other variables, the Bayesian philosophy implies that as these conditions change, the data change. In other words, data are updated with changing conditions. However, a major problem occurs when applying Bayesian philosophy to expert judgment—experts are not naturally Bayesian (Kahneman & Tversky, 1982). Human cognitive processes do not seem to naturally follow Bayesian philosophy.

Humans are not Bayesian for a variety of reasons demonstrated in both laboratory settings and actual applications. The studies of Kahneman and Tversky (1982) have shown that that experts fail to change or adjust their estimates in view of new information. Mathematically, the failure to update estimates means that $P(A|B)=P(A|C)$; that is, the probability of A is not altered when the conditions governing A are changed from B to C. This equation would only be true if $P(A)$ was independent of any conditioning; that is, $P(A|C)=P(A)$ and $P(A|B)=P(A)$. However, in estimating probabilities it is unlikely that any event would be totally independent of conditions.

Other characteristics of human cognition prevent experts from being Bayesian. Some of these characteristics include the inability to grasp the effects of sample size, the frequencies of truly rare events, the meaning of randomness, and the effects of variability (Hogarth, 1975).

These same failings also contribute to human difficulties in estimating probabilities in general (Kahneman & Tversky, 1982). The human brain does not follow the axioms (rules) of probability, such as all probabilities lie in the [0,1] interval and the sum of mutually exclusive event probabilities must be 1. The *probabilities* elicited from a human do not represent a true, mathematical, probability measure. We intend in this book to provide guidance on which inferences and interpretations can be applied to expert judgment data.

Do Experts Provide Better Data?

The literature differs on whether experts provide better quality data than nonexperts. Some reasons for this disparity in the literature might be the following:

- Comparative studies rarely use real populations of experts
- Expert-level questions are not usually asked because of the difficulties in evaluating the responses when there are no known answers
- A number of factors intervene in the comparison (e.g., the type of data and its form or whether the expert is being compared to a novice or person with no training in the field).

In other words, the answer to the question "Do experts provide better data?" depends on the type of data that they are asked to give. The expert can be asked: (1) to make predictions, (2) to provide information on the state of the art in the field, (3) to show how to solve the problem, or (4) to assess the accuracy of his or her responses.

In the area of predictions, experts may not significantly or consistently perform better than nonexperts. Armstrong has claimed that "there are no experts in forecasting change" (Armstrong, 1985, p. 112). Studies in the social sciences and medicine have not consistently shown experts to be better at predictions than lay persons (Armstrong, 1985). However, most of these studies have aimed at predicting human behaviors, such as patients' lengths of stay in mental institutions, or the winning scores of sports teams, or economic trends. For instance, one highly publicized study of expert witness psychiatrists and psychologists found expert prediction of future violence to be highly inaccurate and no better than that of lay persons (Faust & Ziskin, 1988). Faust and Ziskin noted (1988, p. 33) that experts are dependent on the state of their science, which in this case "lacks a formalized, general theory of human behavior that permits accurate prediction." They proposed that both lay persons and clinicians had used common cultural stereotypes and assumptions about potentially violent people to help make their predictions. Thus, their predictions were similar and inaccurate.

It should be noted that interviewers typically word prediction questions (without technical jargon) so that novices can understand and answer them even by guessing. In contrast, problem-solving questions cannot easily be understood by novices. Thus, the wording of prediction questions may minimize the difference between the performance of the expert and the novice.

In addition, experts may not perform significantly better than nonexperts because of the response modes that are used. Typically, subjects are asked to give their predictions in the form of probabilities. While the ***experts*** may be knowledgeable in the field of study, this is no guarantee that they will be expert in assigning probabilities. (Being knowledgeable in the field of study is sometimes called ***substantive expertise*** and being knowledgeable in the use of the response mode, ***normative expertise***.) In general, people do not estimate probabilities in accordance with statistical principles, as mentioned in the previous pitfall.

In general, experts know the state of the art of their field better than nonexperts. For example, novices or lay persons cannot generally describe the state of knowledge in the field; that is, what is not known and what is worth learning. Risk assessments often require and use this type of data (Ortiz, Wheeler, Meyer & Keeney, 1988, p. 3).

Experts are also better at providing data on how the problem can be solved. In more detail, this data could encompass formulating the problem, interpreting it, determining what additional information is needed to solve it, knowing whether and where this data is available, knowing how to solve the problem, providing the solution, and estimating how much confidence can be placed in the solution. Typically, all this knowledge is used in artificial intelligence applications and some of it in risk assessments, such as in a recent probabilistic risk assessment (U.S. NRC, 1989). In problem-solving uses of expert judgment, the expert's judgment is not as frequently encoded in a response mode; therefore, these uses may more directly reflect their substantive expertise.

Armstrong (1985), who has questioned the value of experts in prediction, considers problem solving to be a proper area for the use of experts. This view seems to be held even more strongly in the expert systems community, where the knowledge engineer is likely to be advised to "be sure to pick an expert highly skilled in the target domain" (Waterman, 1986, p. 92). Use of a true expert is given as the means for "extracting high quality data" and for avoiding great difficulties.

Recent evidence shows that experts think differently than nonexperts when solving problems in their area of expertise. Best documented is the expert's ability to recall greater amounts of relevant visual information. For example, skilled electronics technicians recalled more of briefly shown circuit diagrams than the novices (Egan & Schwartz, 1979). These authors proposed that experts are able to code, for memory, chunks or groups of information that are conceptually related. It is also commonly proposed that experts think in more abstract, pattern-oriented ways than nonexperts (Denning, 1986; Dougherty *et al.*, 1986). For example, the Dreyfuses (1986) characterize a novice as knowing basic facts and context-independent rules and an expert as having little conscious awareness of these rules but an ability to visualize and manipulate whole sets of objects and situations.

Experts also differ from nonexperts when assessing their own accuracy or confidence in their answers. The process of assessing or improving the accuracy of the expert judgment is sometimes called calibration. In these situations, the expert judgments are frequently given as probabilities. People in general are very poor at assessing the accuracy of their own answers and are usually overconfident. In other words, most individuals would overestimate the chances of their answer being the correct one. However, Lichtenstein and Fischhoff (1977) have found that those knowledgeable in the field are less prone to this overconfidence bias than those who are not. An individual's calibration improves with knowledge (as measured by percentage of correct answers) until the individual has over 80% of his or her answers correct. Then, those with over 80% of their answers correct become less well calibrated because they tend toward underconfidence. It may be that those who are very knowledgeable are more aware of the dangers of estimation and thus increasingly tend to underestimate their accuracy.

In sum, the quality of expert judgment may depend on the area of questioning, the wording of the question, the form in which the expert responds, and the person to whom the expert is being compared. However, the current view holds that experts provide better data in situations requiring their insights into the problem, such as in solving a problem or assessing their own accuracy. Thus, out of the four types of data that experts are asked to give—predictions, information on the state of the field, problem solving, or assessments of their own accuracy—experts seem to provide better data than nonexperts in the last three.

Note that experts can also provide benefits beyond the quality of the data. Using experts motivates other experts to participate in the study and increases the study's credibility.

Can Experts Be Fully Calibrated?

The concept of calibration is basic to the scientific method. We define *calibration* in this book to mean the comparison of an unknown (instrument or process) with a defined standard or a correct procedure to adjust the unknown until it matches the standard. Until recently, calibration was applied only to measuring devices or processes for which standards or known quantities were available. The concept of calibrating experts was attractive because people are known to violate particular standards (e.g., to be prone to motivational biases, to fail to adequately update their estimates in light of new information, to understand the effects of sample size and variability, and to follow the axioms of probability theory).

However, the conclusions from experimental studies indicate that experts cannot yet be fully calibrated. Studies by many such as Lichtenstein, Fischhoff, and Phillips (1982) show that feedback on the outcome of events can reduce but not eliminate the biases that hamper calibration. For feedback to be effective as a calibration tool, it must be immediate, frequent, and specific to the task, such as the data received by meteorologists in weather forecasting. Such feedback cannot be given in many areas of expert judgment because the outcomes lie in the future and are unknown, as is often the case in risk and reliability assessments.

While this situation of uncalibrated experts and unknown outcomes may seem problematic, many users of expert judgment (e.g., decision makers) do not worry about biases arising from their experts. Instead, they have faith in experts because they perceive them as being very knowledgeable (Morris, 1986).

This faith in the expert's judgment is not to imply that researchers ignore the calibration problem. On the contrary, calibration issues have led many decision analysts to focus on problems that arise from expert–decision maker interactions. In many applications, calibration of the expert cannot be defined independently of the decision maker (French, 1986) because the decision maker factors the expert's thinking into his or her own when reaching a final decision. (This dependent relationship between the experts and the decision maker is further discussed in the section on different application environments in chapter 16.)

Decision makers also affect calibration through the evaluation of their own as well as the expert's calibration. For example, decision makers who see themselves as miscalibrated can induce additional biases and misconceptions by over-compensating for their calibration. They may not perceive independence when it actually exists (Harrison, 1977). Thus, awareness of miscalibration and overcompensation for it, just as ignorance of it, can acerbate calibration problems.

In sum, calibration enters into many aspects of expert judgment and its use. However, the means to measure (even prevent) miscalibration require further research.

Pitfalls

Interviewers, Knowledge Engineers, and Analysts Can Introduce Bias

The interviewer, knowledge engineer, and analyst can unintentionally introduce bias into expert judgment. Bias in this case refers to motivational bias: an altering of the expert's responses because of the influence of the interviewer, knowledge engineer, or analyst. Specifically, the data gatherers and analysts can cause bias through ***misinterpretation*** or ***misrepresentation*** of the expert data.

While the interviewers and knowledge engineers are gathering the expert data, they can bias the data by misinterpreting it. Interviewers and knowledge engineers misinterpret data when they listen and record an expert's thoughts; factors like personal knowledge, training, and experience influence the interviewer's understanding of the expert's words. For example, when an engineer, economist, and decision analyst met initially with military experts on a manufacturing matter, they each interpreted the information in terms of their own training. The engineer perceived the problem as an engineering one, the economist a cost/benefit one, and the decision analyst a multiattribute decision-theory one. They each questioned the experts to obtain the additional information that they needed to apply their individual orientation in greater depth. They each believed that they had received information that confirmed the applicability of their own training to treating the matter. For this reason, we also refer to this source of bias as ***training bias***.

Data misrepresentation occurs during the later stages of an expert judgment project; specifically, misrepresentation can occur when the data is represented, modeled, or analyzed. When performing analyses, we tend to force the data to fit the models or methods with which we are most comfortable or familiar. For this reason, misrepresentation is also referred to as ***tool bias.*** A parallel example would be to use a hammer not only to drive nails into wood but screws and bolts as well. In all cases, the tool would perform its function, in some instances better than others. If the hammer continued to perform its function through time, the user would probably never realize the tool's shortcomings nor would he or she seek an alternative. The same can be said for models. For example, analysts may wish to use a model that requires either independence or a particular distribution. To use the model, they will probably assume that the data meet these requirements (or hope for robustness). Analysts may not question the validity of these assumptions because of some of the social and psychological mechanisms discussed below. Similarly, the knowledge engineers may wish to use a particular system building tool that imposes inappropriate structures on the data (Breuker & Wielinga, 1988).

Notice that training and tool bias are related. The connection exists because we have inherent in our fields values that predispose us toward particular approaches and methods. For example, artificial intelligence (Henrion & Cooley, 1987) and cultural anthropology have valued the expert's knowledge and viewed it as the *gold standard* to be extracted and emulated. By contrast, the fields of decision analysis, statistics, and operations research have viewed particular mathematical and statistical rules as the standard (Henrion & Cooley, 1987). Expert data is valued if it exhibits these standards, such as the axioms of probability and Bayesian philosophy. The methods that these two orientations use reflect their values. Knowledge acquisition and cultural anthropology favor methods designed to obtain and represent the expert's natural way of thinking. The approaches of decision analysis and statistics correct for what they consider to be limitations in human information processing.

Why are humans prone to such subtle but pervasive biases? Why do we selectively take in data that support what we already know and believe that it can be handled by the approaches, models, or methods we prefer? First, it should be noted that all human perception is selective and learned. Our perceptions of reality, of what *is*, are conditioned at a cultural, societal, and individual level.

At the cultural level, meaning and structure are imposed and then taken for reality by members of that culture. For example, members of a Western scientific culture would take the color spectrum, such as in a rainbow, and divide it into four to six colors—violet, blue, green, yellow, orange, and red. In another culture, the people would not see the segmentation that we do. Instead, they might have been conditioned to view the spectrum as consisting of wet and dry colors. The members of both of these cultures have been conditioned to see color in a particular way and to believe that it objectively exists as they perceive it.

At the societal level, our training leads us to define and structure problems in particular ways, to use our field's methods, and to value special types of data. However, we forget that these are learned values and tend to proceed as if they were simply truths that were revealed through our learning experiences. For example, many *hard* scientists believe that the only true data are the quantitative measurements gathered by instruments during physical experiments.

At the individual level, our desire to handle the problem leads us to use those tools that we know best and then to believe that they worked. A certain psychological mechanism prevents us from realizing when our beliefs and perceptions do not match, such as when the use of a favored method proves inappropriate. The psychological theory of cognitive dissonance (Festinger, 1957) predicts that when we have either two beliefs or a belief and a perception in conflict, the conflict will be resolved unconsciously. Many tests have shown that people selectively pay attention to information that confirms their beliefs and discount that which could cause conflict (Baron & Byrne, 1981). Scientists are not immune to this tendency (Armstrong, 1985; Mahoney, 1976). For example, scientists tend to notice the data that confirms their hypotheses and either miss or discount the negative evidence (e.g., the data must be noisy, the equipment probably malfunctioned, or there could have been operator interference).

How can we best prevent our own tendencies to introduce bias? First, we can strive to remain aware of this tendency. Second, we can select elicitation methods that minimize the role of, and hence the opportunity for interpretation of, the interviewer or knowledge engineer. These methods (chapter 7), used for obtaining data on the expert's answer or problem-solving processes, place the emphasis on the expert. By focusing on learning the expert's thoughts and words and using these to pursue questioning, the data gatherer is less likely to have his or her views intrude. In addition, the data gatherers can act like a blank slate to avoid translating the expert's data into their own concepts. (See the section in chapter 3, *Program for Handling Bias*.) Last, we can use analysis methods that require the making of minimal assumptions, as described in chapter 1, *Philosophy Guiding the Analysis* .

Experts Are Limited in the Number of Things That They Can Mentally Juggle

There are limits to the amount of information that we can process in solving problems. A classic paper by Miller (1956) identifies the number of things that people can accurately discriminate. In these studies, the subjects differentiated things on the basis of one attribute, such as the volume of a sound. For example, when subjects listened to a sound played at varying levels of loudness, they could accurately discern about seven levels. Other experiments included differentiating the size of drawn squares, determining the saltiness of various solutions, and distinguishing between musical notes. From many such experiments, Miller determined a limit of seven as our processing capacity because the number of errors increases greatly after that point.

The number seven does not represent a strict limit because, under particular conditions, we exceed it. We can go beyond the limit when we consider multidimensional data, when we perform relative rather than absolute comparisons, and when we make several absolute judgments in a row. Multidimensional data is the input that we simultaneously receive from our five senses, assess, and act on as functioning human beings. As in their daily judgments, the subjects also exceeded the limit of seven in experiments using multidimensional attributes. For example, in an experiment that produced combinations of six acoustical variables, subjects readily discerned without error about 150 different categories. While we are able to judge more things using multidimensional attributes, this capacity also has its limits. In particular, when the total capacity of our information processing is increased, our accuracy on any particular item is decreased. In other words, when making multidimensional judgments, "we can make relatively crude judgments of several things simultaneously" (Miller, 1956, p. 88).

We can also exceed the limit of seven when we perform relative comparisons. Relative comparisons allow individuals to judge things with respect to one another and are frequently done on two things at a time. For example, A could be compared to B, B to C, C to A, and so on.

When an individual consecutively makes several absolute judgments, the information is stored in short-term memory. Memory has its limits, such as in the number of things that can be retained for short-term consideration. Memory limits can be expanded because humans have the capability of grouping or organizing more information per thing. This capability is called *chunking*. For example, a person learning radiotelegraphic code first hears each dot and dash as separate chunks. Later, this person can organize letters, words, and even phrases into chunks. Experts have been found to be much more proficient at chunking data than novices. For example, skilled electrical technicians, in contrast to novices, can briefly view a circuit diagram and immediately reconstruct most of it from memory (Egan & Schwartz, 1979).

The information mentioned in this section has several implications for expert judgment. At the very least, an interviewer would not want to create an elicitation situation in which the experts had to mentally juggle more than seven items at a time. Miller's research suggests that limiting a rating scale to seven or less gradations produces the best results: anything larger than seven impairs an expert's ability to make fine discriminations. If the project demands that a high number of distinctions be made simultaneously, the experts will judge these more crudely than if they had considered them separately. In contrast, methods requiring subjects to compare two items at a time will avoid the limit of seven and produce more precise judgments. In addition, experts, as opposed to nonexperts, can receive and handle larger magnitudes of information because of their ability to chunk it.

The Level of Detail in the Data (Granularity) Affects the Analyses

The term *granularity* has its origins in fields such as numerical analysis and artificial intelligence. In artificial intelligence, granularity is defined as "the level of detail in a chunk of information" (Waterman, 1986). Granularity ranges from coarse (outlining the basic functions of a nuclear power plant) to fine (determining the subfunctions of a particular nuclear power plant function). In numerical analysis, granularity refers to the computational grid size used for defining the level at which the computations are made. Granularity is the level of detail at which the data is gathered, processed, and interpreted. Therefore, this level establishes the framework of operation for the problem.

The granularity is an inherent part of the experimental design of a study. In most applications, this level is dictated by some limiting aspect of the problem, such as the goals of the study or the complexity of the questions asked. Thus, in most problems, the selection of the level is done implicitly and not as a separate, conscious decision. For example, in the design of a simple voter poll, the goal of the problem defines the granularity. If the goal is to determine for which party an individual voted, it would not be necessary to gather information on the voter's property holdings. However, if one of the election issues was to determine who would support an increase in property taxes, this finer level of property information might become important. The latter goal is at a more specific level, and questions necessary to obtaining the required information must be correspondingly more detailed. Generally, providing data to answer the question *why* requires that a finer granularity of data be gathered.

The level of detail also depends upon the complexity of the problem. On simpler questions, such as those whose answers can be verified (e.g., almanac questions), the subject tends to use more structured and detailed problem-solving techniques. Thus, they are easier for the interviewer to record and for the analyst to model in full detail. On complex problems, the subject's problem-solving information tends to be more plentiful but less structured or clear. The subject and the interviewer may encounter the limitations of information processing mentioned in the previous section, *Experts Are Limited in the Number of Things That They Can Mentally Juggle.* The subject resorts to using heuristics to simplify the problem solving. The subject struggles to report these complex processes, usually simplifying them or leaving out parts in the translation. In attempting to follow the interviewee's account, the interviewer is likely to further screen and abstract the information. As a consequence, even though there is a fine granularity of data associated with solving complex questions, this level of detail is not as easy to extract or document as it is on simpler problems.

The level of granularity greatly affects the analyses, particularly the formation of models, their interpretation, and the drawing of conclusions. For example, different models can be formed, depending on the chosen level of granularity. Typically, the analyst constructs a model whose level of detail depends on the data content of the subject who has provided the least or the most general information.

Granularity is also an issue in the interpretation of the data. Analysts *see* data from their own perspective, which is not necessarily the same perspective as that of the subject from whom it was gathered. When the analyst screens, transforms, and constructs problem-solving models, the granularity becomes a function of the analyst's thinking. The analyst is led, often unconsciously, to force the data into the desired level for fitting a preconceived model or hypothesis. (See also the first pitfall under *Common Pitfalls*: *Interviewers, Knowledge Engineers, and Analysts Can Introduce Bias.*) Thus, the analyst's preconceptions can affect the way in which the data are represented. This pitfall is especially likely to occur when the data are highly qualitative, with high uncertainties, as is often the case with expert judgment.

Two studies of interexpert correlation show how granularity affects conclusions (Booker & Meyer, 1988*a*; Meyer & Booker, 1987*b*). In the first study (Booker & Meyer, 1988*a*), which dealt with the problem-solving techniques of statisticians, the experts were asked simply constructed questions. Because their data contained specific problem-solving features, the analyst was able to compare the statisticians using standard general linear models. The result from this comparison was that experts who used similar rules of thumb and assumptions reached similar solutions. Therefore, correlation among the experts appeared to exist at the detailed level of their problem-solving models.

In the second study (Meyer & Booker, 1987*b*), nuclear engineers were asked questions with a more complex structure. The specific heuristics and assumptions that the nuclear engineers used were so varied that the design matrix was prohibitively sparse for use in standard models. Thus, the problem-solving models had to be constructed at a more general level. When these models were constructed at a more general level, which mirrored the ways that the experts processed the magnitudes of information, the answers were found again to correlate with the experts' problem-solving techniques. If conclusions for the second study had been drawn at the same level as in the first study, no evidence for any interexpert correlation would have been found. This effect occurs because finding correlation depends on having the right data-to-noise ratio, something that the level of granularity determines. (Glen Shafer, originator of the Dempster-Shafer theory of belief functions and currently at the University of Kansas, called this relationship to our attention.) In sum, conclusions can differ depending on the granularity of the models chosen.

The Conditioning Effect Poses Difficulties in Gathering and Analyzing Expert Data

The data that the expert gives can be conditioned on a wide variety of factors which include the wording of the problem, the elicitation site and reference materials it contains, the expert's internal state at the time of questioning, the expert's method of solving the problem, the interviewer's or other's responses to the expert's data, and the expert's skill at articulating his or her thoughts. We believe that expert data are more highly conditioned than other kinds of data and that this trait complicates the study of expert data.

The conditioning effect creates problems in both the elicitation and analysis of expert judgment. In the elicitation, the researcher may not have control over the intrusion of factors that influence expert judgment. For example, in an elicitation session, the interviewer has little control over the state of mind that the expert brings to the session, particularly if that state has been affected by some event in the expert's private life. The conditioning effect also complicates the analyses; the factors often overlap and cannot be separated for analysis of their effects on the expert data (Meyer & Booker, 1987*a*).

The conditioning effect relates to the problem of bias in expert data. Some conditioning effects could be labeled as sources of bias. That is, they lead to an altering of the expert's responses or to judgments that do not obey mathematical and statistical standards. For example, the interviewer's negative response to some aspect of the expert's problem solving could alter or bias the expert's subsequent problem solving. In addition, the expert's use of a shortcut in problem solving, such as using the present as a baseline from which to estimate future patterns, could bias his or her answer (Hogarth, 1980).

We recommend a two-step approach for handling the conditioning effect and its offshoot, bias: (1) control those factors that can be controlled, and (2) gather as much data as possible on those factors that cannot be controlled so that the expert data may be analyzed later for their effect or, at least, annotated according to the conditions. For example, factors that relate to the question or the elicitation situation (e.g., the wording of the question, its timing, the elicitation method, response mode, and dispersion measures) are under the discretion of the project personnel and can be designed with the conditioning effect in mind. In contrast, the project personnel cannot control for other factors, such as the expert's internal state, personality attributes, and professional background. However, data can be gathered on these factors by asking a series of demographic questions before or after the expert solves the problem. In these two ways, the effects of conditioning can be examined, if not reduced.

3

Background on Human Problem Solving and Bias

In this chapter, we provide a general background on how humans solve problems. In addition, we discuss the effects of bias and propose a program for handling its occurrence.

Why Is It Necessary to Have an Understanding of Human Problem Solving?

One of this book's premises is that to optimally design an elicitation method requires an awareness of how people solve problems or the causes of bias. For example, if the interviewers expected the experts to spontaneously give accurate and in-depth descriptions of how they solved problems, they might ask the experts point blank for this information. Similarly, if the researchers were not aware of the potential for bias, they would not design measures to detect or counter it. Researchers and others have just begun to address bias as a problem affecting the quality of expert judgment.

What Is Involved in Solving Problems and in Responding?

Novice interviewers rarely understand the magnitude of the cognitive tasks that they demand of the expert in problem solving. After all, they reason, the experts solve problems every day. However, problem solving is not a simple task, and quite often the elicitation methods place additional cognitive burdens on this natural process. For example, the interviewer may request the expert to respond using difficult response modes, such as logarithms.

The Four Cognitive Tasks

The expert will perform four cognitive tasks during elicitation (Mathiowetz, 1987):

1. Understanding the question
2. Remembering the relevant information
3. Making judgments
4. Formulating and reporting an answer.

The first task involves the expert's comprehension of the wording and context of the question. The interviewers may use a different nomenclature than the expert knows or they may use terms familiar to but which hold different meanings for the expert. The expert must also determine the aims of the question and limit his or her analytic frame to focus within that context.

In the second task, the expert must remember relevant information to answer each question. To remember this information, the expert must have previously received it and stored it in memory.

However, humans do not perceive and store all the data that is available to them. Instead, they tend to selectively notice data that support information which they already possess. This failing is part of the reason why humans are not Bayesian (for further information, see *Are Experts Bayesian?* in chapter 2). For example, studies have shown that we pay attention to data that support our hypotheses but ignore data that create conflicts (Armstrong, 1985). In addition, a person does not assimilate raw data; rather, that person interprets incoming data with knowledge either learned individually or as a member of a particular culture. Thus, the expert retrieves data that has already gone through at least one level of interpretation.

Not all information is stored for later use. Also, mistakes are made in accessing memories. Often, the association used to access the memory can impact on what is remembered. For example, if the expert accessed the information through a time frame, a different reconstruction of the memory could result than if the same memory were accessed by way of a key word. In addition, the expert may not readily distinguish between similar or related events. The expert may combine memories of separate events, confusing their characteristics and the time of their occurrence.

The third task, making judgments, involves processing the information. Typically, people use mental shortcuts called heuristics to assist in integrating and processing the information (Tversky & Kahneman, 1974; Hogarth, 1980). In simplifying the processing, these heuristics often skew the final answer. For example, one heuristic is termed ***anchoring and adjustment***. This heuristic, as defined by Tversky and Kahneman (1974), occurs when an individual reaches a final answer by starting from an initial value and adjusting it as each new piece of information is received. The initial value can be supplied with the question, or it can be determined by the experts through their impressions or computations. The final answer usually reflects, or is skewed toward, the initial value. For example, if experts tried to evaluate how well their complex computer model predicted experimental results, their first impression might be that it did a *good* job. After considering the problem in more depth, they might find a number of places in which the model failed to adequately predict the experimental results. Although the experts had second thoughts, they would most likely give a final rating skewed to their first impressions (Meyer & Booker, 1987*b*).

The fourth task, formulating a response, requires that the subject report an answer. If the experts were asked to use a particular response mode, this task would include their translation of their internal answers into the response mode. The experts may use some algorithms in selecting the response option that best expresses their concept of the answer. In most cases, response modes (e.g., probabilities) are governed by logical rules. Thus, the experts may also consider these rules. For example, in putting their answers in the desired form the experts may mentally remind themselves that *my probabilities need to be values between 0.0 and 1.0.*

Two Models of Human Memory and Information Processing

These cognitive tasks can be described further by referring to models of memory and information processing. Two current models define memory: a traditional model, which describes memory as fixed images that can be retrieved from specific storage sites, and a new model, which portrays memory as a recategorization of the information in disperse groups of brain cells whose connections have received recent emphasis.

In one version of the fixed image model, Ericsson and Simon (1980, 1984) describe thought processes as involving a central processor and several types of memory that possess differing capacities and capabilities. For example, the ***short-term memory*** (STM) possesses limited capacity and intermediate duration. The ***long-term memory*** (LTM) has a large capacity and relatively permanent duration. Information recently acquired (by the central processor) relating to the problem is kept in the STM for further processing. Only the most recently heeded information is accessible in the limited storage of the STM. Thus, in solving a problem, information moves back and forth from the STM and the LTM. The individual's LTM can contain information from previous experiences or it can have by-products of his or her current efforts in solving the problem. For example, if the expert needs the solution to an equation as a step in solving the problem, the equation may be pulled from LTM. The expert solves the equation and proceeds with its product, relegating the intermediate variables and equation back to LTM. The STM has pointers to information in LTM. Some of the information in the STM can be stored in the LTM before it is lost. Once stored, the information can sometimes be retrieved from the LTM at a later date.

In contrast to this fixed image model, Rosenfield (1988) presents the recategorization model. In Rosenfield's model, the process of recollection is not depicted as a retrieval of specific images or bits of information but as a recategorization of groups of brain cells (neurons) with temporarily strengthened connections.

> Recategorization of objects or events depends on motion as well as sensation, and it is a skill acquired in the course of experience. We recollect information in different contexts; this requires the activation of different maps interacting in ways

> that differ from those of our initial encounter with the information. (Rosenfield, 1988, p. 192).

Rosenfield proposes that how we perceive stimuli depends on how these stimuli have been categorized or organized. For example, we may hear *gray chip* when *glay shrip* was said in an experimental setting because we expect the organization of vowels and consonants to make sense to us as English speakers. If we subsequently learned that one of the speakers was a sea captain, we might recollect hearing *gray ship*. Thus, memory in this model rarely functions as an exact process; instead, it helps us reconceptualize the world according to our beliefs, needs, or desires. Our knowledge evolves because we become "more richly endowed with the capacity to categorize in connected ways" (Rosenfield, 1988, p. 192).

While we believe that the recategorization model more adequately captures the diversity and creativity of human thought, we suggest that the reader consider both models and use the one which best suits the situation. (Both models should be viewed as tentative because the processes of human cognition are not yet well understood.) Specifically, the recategorization model best helps explain and understand the changes in experts' thinking processes through time (i.e., through the recategorization of their knowledge) and the differences between expert and nonexpert thinking processes (e.g., in the number and interacting of associations). The fixed image model is helpful in predicting the optimal time to question the expert about his or her problem-solving techniques. According to the fixed image model, the optimal time for eliciting the expert's thinking would be while these thoughts still occupied the STM. As time passed, most of the information in the STM would be lost, leaving only a portion fixed in the LTM; of the remaining data, only a portion can be successfully accessed and retrieved.

Bias

Another aspect of how people solve problems is bias. To understand bias requires a context; in other words, *bias* only exists in comparison to some reference point. Bias is generally defined as a *skewing,* so its definition includes the concept of its counterpart—that which is considered *unskewed* or *not biased.* Implicit in this general definition of bias is the view that what is *not* biased is *reality* or *truth.* To clarify the meaning of *bias,* a brief discussion of the nature of reality and of human perception follows.

In this book, we hold the view that humans cannot directly perceive reality because it is interpreted through the filter of their mental models. Bateson (1979, p. 30) explains that our experience of the external world is always mediated by our senses and neural pathways. He gives the example of someone stepping on his toe. He argues that what he experiences is not the stepping on of his toe but the image of this event from neural reports reaching his brain after the event has happened. He concludes that there is always "a transformation, a coding between the report, and the thing reported. . ." (1979, p. 30). Not only the quality of our emotions but also the actual degree to which we experience pain is determined by our attributions about the context of why, how, and by whom the act was performed. Our memory for the event, our perceptions of the present, and our anticipations of the future can never be *objective* because they arise from an inevitably subjective and personal model of events (Bradshaw & Boose, 1990).

Similarly, with respect to bias, we cannot determine objectively what is biased and what is not. We can only make value judgments as to what we consider desirable (e.g., more objective) in the data. Langer, in her book *Mindfullness* (1989, p. 200), demonstrates that all our decision making is ultimately based on individual values (cf., Bradshaw & Boose, 1990). To illustrate that decisions are not made solely on the basis of data, as we typically believe, she cites an example in which a group must decide whether to prolong life when a patient is in intolerable pain. Receiving additional data does not improve the decision making process because no amount of information will make one answer absolutely correct. In addition, the decision makers have no logical stopping point to cue them on when to stop collecting data. Eventually, the discussion of whether to prolong life returns to the basis of individual values. "The doctor, judge, and patient must decide between the principle of prolonging life at all costs and the 'right' to determine the quality of life" (p. 201).

Because our decision making is based on value judgments, it is appropriate that the term *bias* have a negative connotation. Bias is what we label as undesirable or less good. We compare our view of bias to dirt in Douglas's anthropological portrayal of ritual pollution.

> As we know it, dirt is essentially disorder. There is no such thing as absolute dirt: it exists in the eye of the beholder . . . Eliminating it is not a negative movement, but a positive effort to organise the environment. . . . In chasing dirt, in papering, decorating, tidying we are not governed by anxiety but are . . . positively re-ordering our environment, making it conform to an idea (1970, p. 2).

If we accept that defining bias is a value judgment, then we have the option of explicitly articulating presuppositions about the view we hold, so that the presuppositions themselves can be criticized and improved (Bateson, 1979). One way to improve these suppositions is by testing them as hypotheses, as described in Feldman, Compton, and Smythe (1989). Thus, the views that we present should be seen not as a *depiction of what bias is* but rather as a *means for obtaining a greater understanding of bias*.

There are two criteria for evaluating the adequacy of a definition of bias: its coherence within a system of related ideas and its usefulness in predicting and controlling the phenomena in question. In this case, *coherence* means "the regularity that binds the ideas together" (Bateson, 1979, p. 207). For instance, the fields that have contributed approaches to knowledge acquisition contain their own implicit and internally coherent beliefs as to what *good* data is, what bias is, and what methods can be used to achieve better data. Knowledge engineers can determine whether a particular definition of bias is consistent with these ideas. A definition of bias can also be evaluated according to "what is the most useful way of construing the problem from the point of future prediction and control"

(Adams-Webber, 1981, p. 3). For example, a definition of bias that focused on the violation of logical rules would be most useful for a researcher in charge of validating an expert's estimates against logic.

Two Views of Bias

Artificial intelligence, decision analysis, and social science literature contain two somewhat different views of ***bias***. These two views differ in terms of the reference point from which bias is judged. The reference point for one definition of bias is what the source expert actually knows or does; the reference point for the other definition is Western science's concept of *objective* criteria for reality. In the following descriptions, we have exaggerated the distinctions between the definitions for the purpose of illustration.

Those who begin from the first reference point concern themselves largely with kinds of biases that we will term *motivational.* For example, motivational bias occurs when an expert consciously or unconsciously makes accommodations to please the interviewer. The concern with motivational bias is prevalent in the soft sciences, particularly psychology and cultural anthropology. Practitioners of these disciplines generally consider the expert's way of thinking to be the *gold standard* that they wish to capture through the elicitation or the building of a knowledge-based system (Henrion & Cooley, 1987).

The second reference point leads to a definition of bias that occurs when the expert's judgment does not follow *objective* rules or standards. We group these deviations under the heading of *cognitive bias*. Based on this definition, cognitive bias occurs when the experts' predictions do not match what can be observed and measured. However, an expert's judgment often cannot be checked against accepted facts or physical measures of the phenomena. Thus, to determine cognitive bias, the researcher checks the expert's data against the mathematical and statistical standards that Western scientists apply to testing information. To illustrate, if an expert gives probability estimates on all outcomes to a problem (previously defined as being mutually exclusive) and these probabilities do not sum to 1.0, these data will be considered biased because they do not follow normative statistical rules. This view of bias comes from the fields of decision analysis (Mumpower, Phillips, Renn & Uppuluri, 1987) and statistics. Decision analysts do not deny the importance of research in describing how people actually go about making decisions but insist that research be applied to discover ways to counteract systematic sources of bias in intuitive judgment (Kahneman, Slovic & Tversky, 1982).

These views of bias are often implicit in expert judgment applications or at least in particular phases of the application. We have observed that these views coincide with particular project goals, types of elicitation techniques, selections of expert response modes, and validation strategies.

For instance, the motivational view of bias appears more prevalent in projects whose goal is the psychological modeling of the source expert. In addition, proponents of this view of bias tend to favor the use of nondirective elicitation techniques in the desire to avoid *leading* the expert. One such method involves the verbal report technique, in which the expert thinks aloud but is not questioned by the interviewer or knowledge engineer. Also, proponents of this view tend to avoid complex numeric response modes and allow the experts to respond spontaneously in their own, often nonnumeric, terms. Additionally, the motivational view of bias can be found in validations that focus on how well the knowledge base or the knowledge-based system corresponds to the source expert's knowledge.

An emphasis on the cognitive view of bias is likely to coincide with projects whose goal is to improve on the expert's judgment. In projects in which the concern with cognitive bias dominates the elicitation, the expert may be asked to respond using forms that allow greater precision of expression. For instance, researchers find it easier to determine whether experts underestimate uncertainty if the experts use probabilities than if they make qualitative statements (e.g., that some event is very likely). Thus, the formal checks on the expert's judgment can be done from the very first phases of the project. The focus on validation from the cognitive bias perspective can also occur in later validation phases. For example, Gaines (1989, p. 52) describes one validation criterion of a knowledge-based system as checking *to what extent the knowledge in the system corresponds to actuality.*

Potential Impact of Bias

Bias can degrade the quality of data, whether the bias is judged from the standard of the expert's problem-solving processes or from the standard of statistics and logic. To illustrate how motivational bias can affect data, suppose that the interviewer asks the expert if he or she mentally models the problem using a decision tree structure. Even if the expert never used this means of modeling, he or she could still be led to answer *yes*. The interviewer then faces the difficulty of resolving the discrepancies that are likely to arise between the expert's claim and his or her answers (Meyer, Mniszewski & Peaslee, 1989). In addition, cognitive bias affects expert judgment and results in solutions that are not mathematically optimal. For instance, people often use simplifying heuristics that skew their final answer (Tversky & Kahneman, 1974; Hogarth, 1980).

Because expert data is often used as input to important decisions and computer models, bias can contribute to the problem of *garbage in, garbage out.* The same is true in building expert or knowledge-based systems.

The problem of bias also needs to be addressed because of its impact on the credibility of a project. Regardless of whether or not bias presents problems, a study is open to criticism unless it addresses bias through experimental design. Several articles have recently questioned the accuracy of expert clinical judgments in psychology and psychiatry (Faust & Ziskin, 1988; Dawes, Faust & Meehl, 1989). Bias must be monitored, or controlled, and analyzed for its impact.

The topic of expert bias has recently come into vogue. In particular, many have become aware of how the interviewer or others can influence the expert's thinking.

For this reason, we have found that outside reviewers are particularly critical of methods in which there has been no attempt to control these influences. For instance, one of the first criticisms a review panel (Kouts, Cornell, Farmer, Hanauer & Rasmussen, 1987, p. 7) made of the expert judgment methods used in a Nuclear Regulatory Commission project (the 1987 draft version of NUREG-1150) was that "each expert should be free to make this characterization [of the problem areas] independently of decisions by others."

Sources of Bias

Different reasons are advanced for the presence of each kind of bias. Motivational bias is assumed to be driven by our human needs, such as for approval, whereas cognitive bias is generally seen as being driven by constraints on our information processing.

Motivational Bias

We present motivational bias as occurring when: (1) the experts do not report their actual solutions or thought processes, (2) the interviewer or knowledge engineer misinterprets the experts' reports, or (3) the analyst or knowledge engineer misrepresents the experts' knowledge.

Experts may change the descriptions of their solutions and thought processes because of concerns over the social acceptability of their responses. This concern with the social desirability of responses led us to categorize the following effects as ***social pressure biases***.

Two examples of ***social pressure from the interviewer*** follow. For instance, if the interviewer or knowledge engineer asks the experts a leading question, such as *Did you use subgoal x in solving the problem?*, the experts may feel that they should answer "yes," even if they did not solve the problem in this manner. Similarly, the interviewer's nonverbal responses can influence the expert's reporting. For instance, if the interviewer leans forward, displaying intense interest in something that the expert is saying, the expert may unconsciously respond by giving undue emphasis to statements on this topic. If an interviewer reacts in a lukewarm manner, the expert may shorten his or her response, omitting possibly relevant data. The reactions of other experts present in the room can have similar effects on what the expert says and lead to the term ***group think.*** Furthermore, experts' thinking can be affected by their imagination of how those not physically present, such as clients or supervisors, might view their responses, causing ***impression management*** bias. For example, in a study examining whether magnetic fusion projects would meet their technical milestones, some of the experts admitted that they considered their organization's possible judgments in giving their final estimates (Meyer, Booker, Cullingford & Peaslee, 1982). Similarly, the experts may report answers other than those they believe.

Additionally, the experts' hopes or involvement in the area on which they are being questioned can influence their responses. This effect, termed ***wishful thinking***, appears strongest when the experts would somehow gain from their answers. For example, an expert gives the best ratings to the technical merits of a personal project while giving poor ratings to a competitor's project. Other examples involve ethics, such as when an expert receives money to give a favorable review of some project.

There are several reasons why experts' reporting of their thoughts can be influenced by others. First, most people have an emotional need to be accepted and to receive approval (Zimbardo, 1983). Thus, in elicitation situations, experts may skew their judgments in directions they believe will meet with the approval of their managers, other experts, or the interviewer. Secondly, people are generally unaware of how they make decisions, such as in solving problems (Hogarth, 1980, p. ix). However, it is difficult for them to admit ignorance because Western scientific tradition assumes that the problem-solving process can be precisely stated (Denning, 1986, p. 345). Experts will fabricate or acquiesce to suggestions of an acceptable answer or to procedures which they might have used to solve the problem.

Another source can bias the expert's verbalizations: the interviewer's or knowledge engineer's interpretation of the expert's words. Humans tend to perceive and interpret incoming information in a selective manner that supports what they already believe. In some cases, this leads to a ***misinterpretation***, or biasing, of the information. For instance, when the expert mentions a new term, we tacitly assume a meaning based on our experience with similar sounding words. Likewise, we may think that we hear the solution that we expect. Knowledge engineers can be particularly prone to this bias because of a method that they have used for learning the expert's field. This method involves the study of the expert's domain, the building of mental constructs representing the knowledge, and an understanding of the expert's knowledge structure by comparing it to their own. This comparative means of learning the expert's knowledge structure would seem prone to bias because the knowledge engineers interpret the expert's thoughts through the filter of their own.

The expert judgment can also be altered by its ***misrepresentation***. For instance, experts' judgments can be biased if they cannot accurately encode their final judgment into the requested response mode. Some common numeric response modes include probability distributions, continuous linear scales, ratings, Saaty's pairwise comparisons, and rankings. In addition, the expert data can be misrepresented when it is modeled for analysis, such as for detecting bias or for building knowledge-based systems. The individuals performing these tasks make many tacit decisions and assumptions about the data's appearance and performance. For example, when an analyst selects a model for the data, the model assumes particular properties of the data, such as its distribution. These assumptions may not be valid. Then too, the analyst may have to aggregate multiple and differing expert judgments to provide one input value. Mathematical aggregation schemes often require assumptions; in addition, different schemes, like the mean versus the weighted average, can produce different answers. The knowledge engineers decide how to organize and represent the knowledge in the system; these decisions determine how the data will be implemented in the system.

For example, the knowledge engineers may wish to use particular system building tools that lead to "wrong, inefficient, or opaque structures being imposed on the data" (Breuker & Wielinga, 1988, p. 5).

Cognitive Bias

The human mind has limits on how much information it can process and on how much it can remember (Hogarth, 1980, p. 9). To reduce the cognitive burden, people tend to take short cuts when solving a complex problem. Thus, they start with a first impression and integrate the information in a sequential manner, making only a few minor adjustments. Later, if additional information is received, they probably will not adjust their initial impression to give a more accurate judgment. In other words, if individuals who have already reached initial solutions receive contradictory data, they will probably not take this data sufficiently into account when generating final answers. In particular, this sequential manner of integrating information handicaps us in making predictions in which large or sudden changes are involved. This limiting effect is called ***anchoring*** or ***anchoring bias***.

The human mind has limited memory capacity for information processing. As Miller (1956) noted, most individuals cannot discriminate between more than seven things at a time. This limitation in information processing causes people to be inconsistent in working through the problem. For instance, people commonly forget an assumption made earlier and contradict it, thus becoming inconsistent in their reasoning. This is one example of ***inconsistency bias***.

In addition, some data are more easily recalled than others. For instance, data involving catastrophic, familiar, concrete, or recent events may be easier to recall (Cleaves, 1986; Gold, 1987). This effect, termed ***availability bias***, can lead to the overestimation of the frequency of some events.

Another frequent effect is the ***underestimation of uncertainty***, which means the failure to account for the actual amount of uncertainty in the answers given. For example, people are frequently asked to estimate ranges around their answers to reflect their uncertainty. If people are requested to put a range around their answers such that they are 90% sure that the range encompasses the correct answer, they will underestimate the uncertainty by providing a range that is two to three times too narrow. In other words, their range will only have a 30% to 60%, not a 90%, chance of covering the correct answer (Capen, 1975).

Program for Handling Bias

Approaches to handling biases are rare (Cleaves, 1986) and in their early stages. They are perhaps as much art as they are science. The program proposed in this section is no exception. Few programs for handling bias exist because the concept is difficult to study. Studying, much less trying to counter bias, is complicated by not having a readily available reference point from which to determine the direction and magnitude of the bias. For most questions asked of experts, there are no known single *correct* answers or empirical data. Thus, the researcher cannot simply compare the expert's data to an answer in a reference book or to results obtained from modeling data. Measurement of motivational bias is further complicated by the absence of *objective* standards for comparison. At least with cognitive bias there are traits of the expert's answers that can be compared to standards. For example, do the expert's mutually exclusive and exhaustive probabilities sum to one as they should? With motivational bias, the baseline of the expert's judgment is difficult to determine, especially because expertise is not static. As Rosenfield (1988, p. 194) argues in his book on memory, higher mental functions are not fixed procedures but subject to constant reconstruction. While we recognize that both biases are difficult to detect, we believe that for progress to occur, programs like this one must be proposed and applied to expert judgment.

Our approach differs from the one presented by Cleaves (1986) in two ways. First, Cleaves' proposal focuses on cognitive bias; second, Cleaves tries to anticipate biases by the judgment processes in which they are likely to occur, namely, hypothesis and solution generation, decision rule articulation, uncertainty assessment, and hypothesis evaluation. While we agree that biases occur during these processes, we try to anticipate the biases by the elicitation components that are likely to exhibit them. We assume that many readers will be lay persons in the areas of human cognition and that this approach might be easier for them, at least as a starting point.

Another major difference in our program is its real-time emphasis. Given the evolving nature of expertise, bias is best detected when it is being elicited. This program stresses monitoring and adjusting for bias, particularly motivational bias, in real time rather than mathematically compensating for it afterwards. In particular, it is much more difficult to determine that motivational bias has occurred after the elicitation because the baseline—the expert's knowledge—is likely to have changed. For this reason, we consider each elicited datum as a snapshot that can be compared to a snapshot of the expert's state of knowledge at the time of the elicitation.

Our proposed program consists of six general steps:

Step 1. Anticipate which biases are likely to occur in the planned elicitation. We have provided lists of selected biases and the situations in which they are likely to occur (see chapter 8). These biases were selected because they represent a range of bias within the two definitions. Certainly these are not the only sources of bias, but they are ones that we have most frequently encountered.

To anticipate bias, the project personnel determine the parts of their planned elicitation and search the *Index of Selected Biases* (in chapter 8) for the biases to which the elicitation is prone. For instance, if the project personnel had planned to use the interactive group method, they would see that it was prone to the following biases: social pressure from group think, wishful thinking, and inconsistency. They then turn to the section following the table *Definitions of Selected Biases* to search for the definitions and causes of the selected biases.

As a general rule, this step should always be performed. However, step 1 becomes more important if the data gatherers or analysts are in danger of introducing bias. Reading about the biases makes the project personnel aware of their own potential for influencing the expert, misinterpreting the expert's words, or misrepresenting the knowledge. We hope this awareness lessens their tendencies to bias the expert's account.

Step 2. Redesign the planned elicitation to make it less prone to the anticipated biases. If the exercise of anticipating the biases leads the project personnel to wish to redesign the elicitation at this point, they may use the *Definitions*. The *Definitions* include information on why the biases occur; thus, project personnel can use these to design an elicitation which is less prone to the biases. Additionally, some of the *Suggestions for Countering Selected Biases* (chapter 10) could be implemented in planning the elicitation.

Step 3. Make the experts aware of the potential intrusion of particular biases and familiarize them with the elicitation procedures. The experts need to be informed about the biases that they are likely to exhibit given the elicitation situation. In particular, they need to know the definitions and causes of these biases. Without this information, the experts will be unprepared to combat their own tendencies toward bias or watch for those of data gatherers and analysts. The interviewers can use the *Index* and the *Definitions* provided for step 1 as a base for informing the experts about bias. The interviewers should also describe the elicitation procedures to the experts, as described in chapter 10. If the experts are confused about how and when they are to respond, the data gathered as well as the experts' cooperativeness will be negatively affected.

Step 4. Monitor the elicitation for the occurrence of bias. Before the elicitation session begins, a data gatherer studies the signs that warn of possible bias in *Signs of Selected Biases* (chapter 10). For instance, if group think bias was anticipated, the data gatherer would search for this bias in the *Signs* section and read about indications of its presence. One sign of group think is that the experts appear to defer to another member of the group or to each other. The interviewer, knowledge engineer, or a trained observer then watches for this sign of group think during the elicitation. In general, monitoring biases, as described here, require that the experts verbalize their thoughts and answers. Without this feedback, we have found the monitoring to be much more difficult.

Step 5. Adjust, in real time, to counter the occurrence of bias. In this step, the interviewer searches for ways to prevent a particular bias in *Suggestions for Countering Selected Biases* (chapter 10). These suggestions vary because we have used two approaches: controlling those factors contributing to a particular bias or employing the opposite bias. The first approach involves controlling the factors that contribute to the bias. For instance, fatigue is a factor that leads to increased inconsistencies in expert judgment. The interviewer can stop the elicitation sessions or schedule breaks before the experts become fatigued to prevent possible inconsistencies. An example of the second approach is to try to have experts anchor to their own judgments in an attempt to counter a group-think situation. Having the experts write their judgments encourages them to form and keep their own opinions, even when they hear the opinions of others.

Step 6. Analyze the data for the occurrence of particular biases. If step 6 is the only step of the program being followed, the analysis will be simpler than if steps 4 and 5 were also followed. If these other steps were followed, they would provide the additional data needed for performing more complex analyses. In general, adequately testing for one of the motivational biases requires this more complex testing. Mathematical tests can often determine the occurrence of a cognitive bias, such as the underestimation of uncertainty. Chapter 14 provides detailed information on how to test for the presence of bias.

Determining Which Steps to Apply

The six steps can be applied in sequence or singly, depending on the needs of the project. For example, if information on bias were not important to the project, personnel would omit all but step 6. On the other hand, if bias was a concern, the project personnel would follow more of the steps. As a general rule, we suggest that step 6 be applied to all projects because it provides a general check on the expert data.

To pick steps for use in a project, project personnel should consider the following: (1) their reason for addressing the problem of bias, (2) their view of bias, motivational or cognitive, and (3) the forms of bias in which they would be especially interested.

The Reason for Focusing on Bias. Why project personnel focus on bias can provide criteria for determining which steps of the program to implement. For example, if the project personnel's interest in bias stems from a desire to avoid having reviewers criticize the project, their selection of steps would probably differ from those whose goal is to limit the negative effects of bias. It is also possible that personnel may have no reason to delve into the problem of bias on some projects. In these instances, we recommend that bias *not* be investigated because this activity demands extra time and resources. Instead, we would suggest that the project personnel record their reasons for not investigating bias and proceed as planned with the project.

We have identified several reasons for focusing on bias:

- to become aware of bias,
- to prevent or inhibit the occurrence of bias,
- to avoid criticism on the quality of the knowledge base, and
- to analyze the data for the presence of bias.

To accomplish the above-mentioned aims, use different combinations of the six steps. The flow chart in Fig. 1 summarizes which steps to take to achieve these aims.

To become aware of bias. Wishing to learn about bias is similar to wanting to discover and articulate personal assumptions. In both cases, awareness allows individuals to consciously choose their beliefs or actions and articulate or record these for critical examination. Furthermore, we propose that awareness of bias types makes individuals less likely to unconsciously introduce them. To become aware of bias, we recommend using step 1 and step 3. If interviewers wish to learn more about biases for a particular application, they may exclusively use step 1. However, if the interviewers also want the experts to be aware of bias, both steps 1 and 3 will be needed. These two steps help the interviewers inform the experts about bias and elicitation procedures. We consider familiarizing the experts with the elicitation procedures part of the process of informing them about bias. First, the elicitation methods form a backdrop against which bias can occur. Second, if interviewers acquaint experts with the elicitation procedure, they may prevent some of the experts' confusion, a contributor to the occurrence of many biases.

Step 4 can help sensitize the data gatherers to bias. Occasionally, the experts can assist the interviewer or knowledge engineer in looking for the signs of selected biases.

To prevent or inhibit the occurrence of bias. If the reason for focusing on bias is to prevent or inhibit its occurrence, we recommend using at least steps 1, 2, 3, 5, and 6. Step 1 enables the interviewers to anticipate likely biases; step 2 allows them to use this information to redesign the elicitation to make it less susceptible to these biases. Step 3 helps reduce the likelihood of the expert introducing bias. Step 5 should be executed very carefully because the countering techniques could add bias or complicate later analyses. Step 6 acts as a check. Finally, step 6, reinforces the interviewer's claim that steps 1, 2, 3, and 5 have prevented bias. In our experience, conducting step 6 has usually allowed the interviewers to claim that no evidence was found of particular biases.

Although not necessary, step 4 can help prevent the occurrence of bias.

To avoid criticism of the quality of the data. Sometimes the reason for focusing on bias comes from forces outside the project, such as others' concerns with the quality of the data. We have noticed that those who have experienced an outside

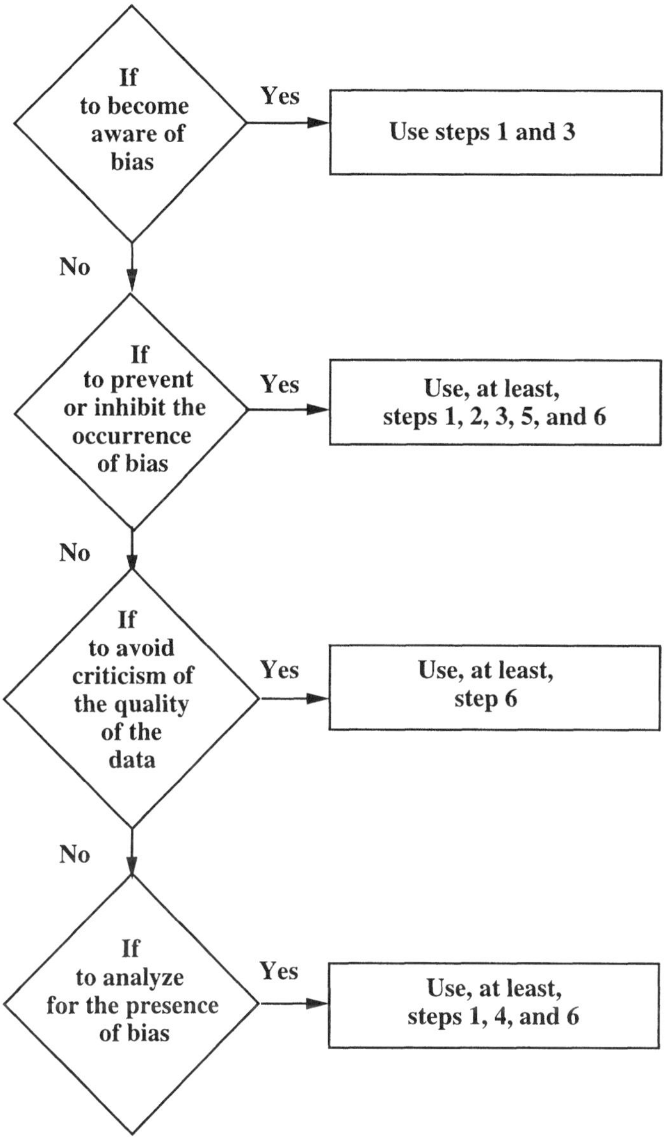

FIG. 1. Steps suggested for use with the four reasons for focusing on bias.

reviewer's criticism of their project's data are highly motivated to avoid this situation in the future. This motivation differs from the other three mentioned in this section because those who are motivated by this desire will probably not be interested in performing more than the minimum number of steps (in most cases, wishing to perform only step 6). However, to analyze the data for bias assumes some knowledge of bias, so step 1 might also be necessary. In addition, step 4 can be done to provide data for the analysis.

In our experience, external reviewers most commonly question data regarding the presence of these three types of bias: social pressure from the interviewer or knowledge engineer, social pressure from group think, and wishful thinking by the expert.

To analyze the data for the presence of bias. Interviewers analyze the data for the presence of bias for a variety of reasons. For example, interviewers may wish to study the phenomena of bias in general or they may wish to analyze a particular project's data for its presence. If the interviewers wish to study bias, particularly the techniques and situations that affect its appearance, they could use any of the steps (1, 2, 3, 4, and 5) in combination with step 6. For example, if the interviewers wished to investigate the effect of using a particular step or elicitation technique, they could use it in one study and not in the control study and then analyse and compare the results. One advantage of studying bias in this way would be the fresh look at bias that it would produce. For instance, those doing the study could determine if there was a difference in applying the step and whether they judged that difference to be positive or negative. In other words, the researchers would not automatically assume that bias led to poor results but would evaluate for themselves the peculiar properties of bias or its absence. (For this idea, we are indebted to Maurice Sharp, University of Calgary, Canada.)

If analyzing the data for bias is the aim, such as to validate data or knowledge bases, steps 1 and 6 could be used. Step 1 allows the knowledge engineers to learn about likely biases and step 6 enables them to check for the presence of these biases. Generally, step 6 is more effective in investigating sources of cognitive bias. For example, the knowledge engineers could check for underestimation of uncertainty. However, if the knowledge engineers want to analyze the data for how well it matches the expert's thinking (as would be the case with the motivational definition of bias), we would suggest adding step 4. Step 4 enables the knowledge engineers to watch for some of the motivational biases that interfere with obtaining reliable reports of the expert's thoughts.

The Selection of the Definition of Bias, Motivational or Cognitive. While the definition of motivational or cognitive bias is often implicit in a project, we recommend that the project personnel consciously select which view or views of bias to apply. By consciously selecting the view of bias, the project personnel can consider which view will best serve the application in terms of coherence and usefulness. (The application could be a project, a phase of a project, or validation of a data/knowledge base.) For example, the motivational definition of bias would be more appropriate if the goal of the project is to learn the expert's problem solving to emulate it. If the project involves estimating the likelihood of future events, the cognitive definition would be a natural choice because people tend to make inaccurate predictions, and the cognitive view of bias would help combat this weakness.

We suggest that the reader select and use only one view of bias at a time to avoid being contradictory. For example, the cognitive definition would propose the modification of a mathematically incorrect judgment. However, this act would cause a misrepresentation of the expert's data according to the motivational definition of bias. Thus, for situations in which more than one view is used, we recommend that project personnel develop rules that dictate when each view of bias will be applied. To illustrate, in a knowledge engineering project, the definition of motivational bias could be used in acquiring knowledge for building a conceptual knowledge base and the cognitive definition used thereafter.

Motivational definition. As a general rule, use steps 1, 2, and 3 when applying the motivational definition of bias. In step 1, the interviewers or knowledge engineers receive information on what biases to expect. In step 2, they refer to the *Definitions* to receive ideas on how to minimize bias in the planning or the redesigning of the elicitation. For example, if looking through the *Index* led the interviewers to expect group think, the definition of group think would provide them with ideas on how to design an elicitation that was less prone to it. Group think is more likely when ***behavioral aggregation*** is used, so the interviewers could plan to aggregate the responses mathematically. Use of step 3 would lead the interviewers to inform the experts about bias, and hopefully make the experts less prone to it. In this book, we have focused on presenting methods of elicitation and analysis which we believe minimize influencing the experts and force fitting their data. Thus, just using the methods suggested in this book, regardless of any program for handling bias, should provide some protection from motivational bias. In addition, step 4 often helps in detecting motivational biases such as group think. The *Signs of Selected Biases* tell the interviewer which behaviors might indicate particular biases.

Cognitive definition. If the cognitive definition of bias is selected, step 6 is particularly effective. Analysis is generally more definitive with cognitive than with motivational sources of bias because cognitive biases can be determined mathematically or statistically. Cognitive bias can usually be measured because it is defined as a violation of logical or statistical standards. Thus, three of the four cognitive biases mentioned in this chapter should be included in the analyses. For example, *underestimation of uncertainty* can be analyzed using the experts' ranges on their estimates. Inconsistency is analyzable through logic checks. Anchoring can be investigated by comparing experts' first impressions to their later ones, given that the problem environment has changed, such as through the experts' receipt of new information. Analysis of cognitive bias presumes that the interviewer or knowledge engineer knows which biases might pose problems. Thus, step 1 can help deal with cognitive biases. In addition, step 3 is often used with cognitive biases to lessen the expert's tendencies to introduce these.

The flow chart (Fig. 2) summarizes the steps most suited to investigating cognitive or motivational bias.

Interest in Particular Types of Bias. Frequently, those working on a project are more interested in some types of bias than others. This interest in particular types of bias can provide criteria for selecting specific steps because some steps are more effective with certain biases. The project personnel will have to identify which biases they are most interested in and then select those steps that address these biases. We describe why we picked particular steps for three of the selected biases below and simply list the steps for the rest in Table 1.

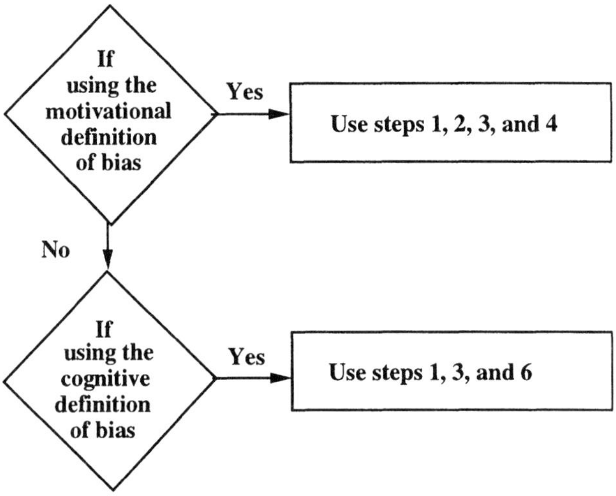

FIG. 2. Steps suggested for use with the two definitions of bias.

Social pressure from the interviewers. For instance, steps 1, 2, and 3 would be most helpful in dealing with social pressure from the interviewers. In doing step 1, the interviewers would become aware of their tendency to introduce this bias and perhaps be able to curb it. As a result of step 2, they might decide to select elicitation methods that were nondirective, like the verbal report, to minimize their tendency to influence the expert. In step 3, the experts would be made aware of this bias, so they would be able to guard against it.

Social pressure from group think. Dealing with the social pressure from group think may require use of all six steps. In step 1, the interviewers assess whether their application is susceptible to group think. If step 1 indicates this possibility, they can redesign the elicitation using information from step 2. Step 3 allows the interviewers to alert the experts to the potential occurrence of this bias, and step 4 allows both parties to monitor the sessions for its occurrence. In step 5, the interviewers can apply techniques for countering this bias. In step 6, the data may be analyzed for the presence of group think.

Wishful thinking. Use of step 1 informs the project personnel that wishful thinking can occur in many elicitation methods. Step 2 allows them to redesign the elicitation to select those experts who were less likely to exhibit wishful thinking (those who had less at stake in the judgments). In a study whose aim was to obtain probabilities of whether project deadlines will be met on time, we focused on interviewing those physically working on the project rather than their managers (Meyer, Booker, Cullingford & Peaslee, 1982). Additionally, we required that the experts disaggregate the question into its parts, give estimates for each part, and explain their reasoning for each estimate. This practice made it more difficult for the interviewees to give highly optimistic estimates.

TABLE 1
List of steps necessary to address selected biases

Types of Bias	**Steps**
Social pressure from interviewers	1, 2, and 3
Social pressure from group think	1, 2, 3, 4, 5, and 6
Social pressure, impression management	1, 2, 5, and 6
Misinterpretation	1 and 2
Wishful thinking	1 and 2
Inconsistency	1, 2, 3, 4, 5, and 6
Anchoring	1, 2, 3, 4, 5, and 6
Availability	1, 2, 3, 4, and 5
Underestimation of uncertainty	1, 2, 3, 5, and 6

Part II

Elicitation Procedures

4

Selecting the Question Areas and Questions

Selecting which questions to ask an expert is a slow and evolutionary process. As the question list evolves, however, the selection process seems clearer in retrospect. In this chapter, we illustrate the steps involved in selecting the questions: (1) definition of the project's purpose, (2) selection of the general question areas, and (3) identification of the specific questions. In particular, information is given on which persons—clients, data gatherers (interviewers and knowledge engineers), analysts, or experts—can help with each of these steps and how. In addition, this chapter covers when selecting and motivating the external experts (chapter 6) needs to be done in parallel with the steps mentioned below. The process continues in chapter 5, which deals with how to refine the questions.

Steps Involved in Selecting the Questions

Selecting which questions to ask the expert comprises only one of a sequence of steps in which the information from one step permits the completion of the next, more detailed step. The steps are summarized as follows:

Step 1. Defining the project's purpose or goals. The project's purpose is simply what the project should accomplish. For instance, the purpose of the reactor risk project, NUREG-1150, was to examine the risk of accidents in a selected group of nuclear power plants. In some cases, however, the project's purpose remains unclear. For instance, some project sponsors either have no clear concept of the project's aims or cannot express their project's purposes or goals.

Step 2. Selecting the general question areas. A ***question area*** is a specific issue for investigation. For example, the above-mentioned reactor risk project yielded nine question areas. These question areas dealt with internal events that could lead to core damage, such as the failure of the emergency core cooling system because of venting or containment failure.

To develop question areas, consider such information as the goal of the project, the client's directives, and the practicalities of gathering expert judgment on this topic (e.g., whether experts exist and whether their expert judgment would be considered proprietary information). Of the question areas initially considered, only a few may emerge as the final areas.

Step 3. Identifying the questions. Concrete, detailed points within question areas are called ***questions***. The test for whether something qualifies as a question is if the expert finds it sufficiently specific to answer. If an expert cannot address the question in its present form, it probably resembles a question area more than a question. To illustrate, the question area on an emergency core cooling system failure can be divided into different accident scenarios, each of which could lead to a cooling system failure. The experts can answer the specific questions on the probability of one of the scenarios occurring within a particular nuclear plant.

In some cases, questions take the form of technical problems. Data gatherers use such problems to examine the experts' problem-solving processes. For example, to learn how statisticians determine randomness, data gatherers could ask them to judge whether lists of numbers were random.

The reader needs to assess which of the above three steps has already been accomplished. Frequently, the defining of the project's goals has already been done by the project sponsor. On the other hand, the reader may have previously decided on the question areas. If one or more of the above steps has been completed, the reader may wish to skip ahead to the next step, *Executing the Steps with the Assistance of Sponsors, Project Personnel, and Experts.* The following section illustrates the possible variety in project goals, question areas, and questions.

Illustrations of the Variation in Project Goals, Question Areas, and Questions

The project goals, question areas, and specific questions can vary tremendously. The following three examples illustrate a variety of projects that possess project goals, question areas, and questions. In particular, these examples can provide assistance in formulating the goals, question areas, and questions for a particular application.

The first example dealt with the reactor risk project (NUREG-1150). This project's goal was to perform risk analyses of five different U.S. light water reactors to provide data on the likelihoods of severe accidents and their consequences. The data requirements for this application were large and complex. The following criteria determined which areas received expert judgment: (1) they were within the scope of the project NUREG-1150; (2) they were areas of significant importance to the estimation of risk or the uncertainty of risk; and (3) there were no other sources of data available (Wheeler, Hora, Cramond & Unwin, 1989). Questions were developed for each area. For example, one area of investigation involved the failure of a Westinghouse reactor coolant pump's shaft seals under station blackout conditions. A question asked in this area: "What is the failure probability per year for severe seal leakage?" This question included definitions of *severe leakage* and the sequence of events leading to this failure.

The second example involved a study of interexpert correlation, where the goal was to determine if experts' answers correlated, and if so, what caused the correlation. Given this general goal, the question areas, even the field of expertise from which the experts could be drawn, were completely open. The areas selected were those that would be comprehensible to the researchers, have readily available experts, and have nonproprietary data. One of the areas selected included the type of questions encountered in walk-in statistical consulting situations. For example, one question gave the sample correlation coefficient (r) between two measurements of geologic core samples as 0.70 and asked for the sample size at which this value of r would be significant at the 5% level for a one-tailed test (Booker & Meyer, 1988*a*, p. 140).

In the third example, which dealt with an Army export project, researchers wished to extract those factors experts used to decide whether or not to transfer militarily critical Army technologies, services, or data to foreign countries or persons. The purpose was to represent these factors in a structured manner that would promote better, more defensible decisions. The question areas were the perspectives of the various Army offices (intelligence, political, military, and technical) in viewing potential technology transfers. For example, one question area involved the factors that the intelligence office representatives would consider in making their decisions. The first question in the intelligence area was to evaluate whether the requesting country was vulnerable to having the particular technology compromised (Meyer & Johnson, 1985). This question included a definition of the concept of technology compromise and a description of the technology under consideration.

In a fourth example, the goal was to develop an expert system to aid police decision makers in hostage-taking incidents (Vedder & Mason, 1987). A police commander of a Hostage and Negotiation Unit acted as a source expert. The question areas were past incidents. On each incident, the expert answered such questions as: *Who is the responsible party? What information does the decision maker have at this point? What information does he or she need? Given this information, what actions would you recommend?* (1987, p. 408). These questions could be considered defined by the context of the incident in which they were being asked.

Sources of Variation

The following sources of variation in projects' purposes, question areas, and questions will allow the reader to compare his or her situation to other situations. The reader is asked to do the following:

- **Determine whether the objective is to gather the experts' answers or to gather their problem-solving processes.** While researchers frequently gather both the expert's answers and problem-solving processes, one is considered to be of first priority. For example, for most applications, especially in risk or reliability analyses, obtaining the expert's answer takes precedence. There may be some attempt to document the expert's reasoning, but this is done to support the answer and is not usually the main focus. Obtaining problem-solving data is more common to knowledge acquisition projects or to research into human cognition. While the experts' answers are usually gathered in these studies, they are considered only part of the problem-solving data.

- **Compare the complexity of the areas and the questions.** For instance, the questions of the reactor risk application mentioned above were more complicated than those classical statistical problems asked in the study of correlation because of the former's subject matter—reactor phenomenology. The reactor risk questions were extremely complex because they involved a variety of physical systems, components, and the diversity of system interaction. As a general rule, complex question areas require more honing to form questions than simpler areas do.
- **Assess the magnitude of the data required.** For example, in the reactor risk application, a tremendous amount of data was gathered because the complex questions had to be subdivided into numerous scenarios. The project personnel than had to document the questions, the experts' reasoning, and their answers. Similarly, the building of knowledge-based or expert systems frequently involves the creation of a large knowledge base.
- **Further assess the level of detail necessary in each chunk of data.** For the most part, applications that gather problem-solving data (such as for building an expert or knowledge-based system) require more detailed information than their answer-gathering counterparts. In general, gathering more detailed data coincides with fewer experts and longer elicitation sessions. In addition, the level of detail depends on how difficult the problems are to solve. For example, in the second example of the study of interexpert correlation, the solving of the statistical questions was straightforward compared to the solving of the questions in the subsequent study in which the experts were asked to evaluate how well computer-modeled results matched experimentally obtained results. The expert's *interpretation* was involved to a greater degree in answering the questions on the computer-modeled results than in answering the statistical questions of the first study. The subsequent study yielded more detailed data on the expert's problem-solving processes than the first study. The level of detail in questioning affects later analyses, especially those which involve the detection of correlation and bias (chapter 14), the aggregation of experts' answers (chapter 16), and the drawing of conclusions (chapter 18).
- **Evaluate the scope of the application based on the number of experts to be used, the personnel available to conduct elicitations, and the amount of time needed to produce the product.** For instance, the reactor risk study (NUREG-1150) had one of the largest scopes of the expert judgment applications that we have encountered. To produce a final report on risk assessment involved over 15 question areas, three or more questions per area,

50 experts, over 40 project personnel from different organizations, and more than three years.

Executing the Steps with the Assistance of Sponsors, Project Personnel, and Experts

The following list represents the generic roles of persons who work on expert judgment projects and who could help develop the questions.

Sponsors are the persons requesting the gathering of expert judgment. They may also be the persons funding the project, the decision makers who intend to utilize the project results in making decisions, or those who wish to have an expert system created for their use.

Project personnel include the in-house managers, ***data gatherers***, and ***analysts***. The data gatherers are either ***interviewers*** or ***knowledge engineers***; both types elicit information from the expert. However, knowledge engineers, in addition to interviewing, represent and enter the expert knowledge into computer systems with the goal of creating expert or knowledge-based systems.

Advisory experts are outside consultants or in-house personnel who are considered experts in the subject matter. They can assist the project personnel in the design of the elicitation methods. Generally, they do not serve as the experts for the final elicitation but assist in developing the elicitation methods by helping to select question areas, create the questions, and test each step for possible difficulties.

Experts, sometimes called *external experts* to distinguish them from advisory experts, provide the answers to the questions. They may also be called *source or domain experts* in knowledge acquisition applications. The external experts can fulfil the function of the advisory experts by selecting and refining those questions that they will later answer. Whether or when the external experts will help in developing the questions is a critical decision. For more information on making this decision, see the later section *Determining in Which Steps the Advisory or External Experts Will Assist.*

Recognize that these personnel categories are not mutually exclusive. For example, if the project were self-instigated, the sponsor could be one of the project personnel. Also, an advisory expert could be from in house, making him or her a member of the project personnel as well.

Step 1. Defining project purpose. The sponsor usually determines the focus of the investigation, what data the experts need to provide, and what resources are available. For example, the sponsor may dictate the project's products, completion dates, and funding limits. In addition, the sponsor may provide direction on the scope of the project (number of experts, question areas, time frame), the data to be gathered (primarily answers or problem solving), and the level of detail needed in the data.

Step 2. Selecting question areas. Project personnel generally work with the sponsor, advisory experts, and external experts in selecting the question areas. Occasionally, the project personnel are experts in the question areas and could forego receiving the input of the sponsor and experts. However, even if this is the case, we recommend the involvement of the sponsor and the experts for two reasons. First, with more persons working on question area selection, the less chance there is of overlooking an area or having the areas reflect one narrow viewpoint. Second, people who were involved in the selection process are more likely to support the final selection than those who were not. Indeed, for this reason the external experts should participate in question selection whenever possible. Sometimes this option is not possible because project personnel use the selected question areas to determine the areas of expertise; selection of the specific experts usually follows this process. As a result, the experts cannot participate in question area selection.

Sometimes the sponsors are not able to help in this step because they have not thought as far as the question area or have difficulty articulating their ideas for question areas. In such cases, the project personnel can interview the sponsor to determine possible question areas (e.g., by asking for the purpose of the project, its constraints in terms of time and funding, and so on).

In our experience, the project personnel have usually set the criteria for the selection of question areas either by themselves or in conjunction with the sponsor. An earlier section, *Steps Involved in Selecting the Questions*, gave an example of such criteria. Regardless of whether time is specified as a criteria, we have noticed that time limitations are often responsible for paring down the list of areas. The advisory and external experts can also assist in determining the question areas. The experts can best address the following points:

- The potential question areas possible, given the purpose of the project
- The approximate number of experts in each area
- Ideas on what would motivate the external experts to participate in the project
- The amount of subdivision necessary to make question areas answerable questions
- The determination of when a question area has been sufficiently defined to proceed to the next phase—the creation of its component questions.

While the advisory experts often address any of the above points, the external experts usually only assist on the last two. This difference exists because the external experts typically enter the project later than the advisory experts.

Step 3. Identifying the questions. Project personnel, along with either advisory or external experts, often work together to develop questions from the question areas. A sponsor qualified in the subject matter may also participate in this process.

After the experts have assisted in creating some questions, they can be asked the following:

- Are the questions answerable (e.g., will experts possess datum, reference books, or experience relevant to the question)?
- How much diversity of opinion among external experts is there likely to be?
- Would any of the expert judgment data be considered proprietary?
- How many questions can an expert answer in a specific period of time?

Determining in Which Steps the Advisory or External Experts Will Assist

Experts are needed to select the question areas, identify the questions (steps 2 and 3 in chapter 4), and refine the questions (chapter 5). Either advisory or external experts can work with the project personnel to perform these tasks. However, *when* the external experts are to become involved in the project is a critical consideration because it affects the order in which chapters 4, 5, and 6 should be applied. We recommend that the three following options be considered and one selected.

The External Experts Do Not Participate in Question Area Selection, Question Identification, or Refinement, Except Minimally Just Before Having Their Judgments Elicited

In this option, advisory experts and project personnel perform the majority of the work selecting the question areas and identifying and refining the questions. Project personnel then use the advisory experts to pilot test the questions for clarity and ease of use. (Pilot testing is discussed in chapter 9.) Surveys or questionnaires are frequently developed in this manner. When the external experts later respond to the questions, their interpretations of what the questions mean are recorded and constitute their refinement of them.

The advantages of this option are as follows:

- Project personnel control the development of the questions (e.g., either they make the choices or they direct the advisory experts in making them)
- The external experts do not have to be selected until after the questions are finalized (as described in chapter 5)

This option has three disadvantages:

- The external experts will not be as motivated to address the questions as they would have been if they had helped develop them

- External experts will have a more difficult time understanding the questions than if they had developed them. Experts often have problems encoding their answers into the response modes that the project personnel have picked, especially if they have not customarily used them in their problem solving
- Those who review the project may believe that the experts were *led* because the experts did not develop the questions.

If this option is chosen, read and apply chapters 4, 5, and 6 in sequence.

The External Experts Receive the Question Areas and Develop the Questions and Refine Them, or They Receive the Questions and Refine Them

Frequently, as in the reactor risk study, the project personnel select the question areas or questions and the experts modify them (e.g., add, delete, or reword them). If more than one expert will address a question, they will need to agree on its revision. Otherwise, the experts' responses cannot be compared because they will be answers to essentially different questions. For this reason, the project personnel usually monitor the experts' work in arriving at the final wording of the questions. This option of partial involvement by the external experts is favored when the project personnel want some control over the questions. One situation in which the project personnel would wish to retain control over the final questions is when the sponsor specifies which question areas he or she wants covered. Another situation involves cognitive studies or knowledge acquisition projects, where the project personnel vary aspects of the question to see what effect these changes might have on the expert's problem solving or answers.

The advantage of this approach follows:

- The project personnel can specify the criteria for the development of the questions and leave the specifics to the experts.

The disadvantage of this approach follows:

- Those who review the project may believe that the experts were *led* because the experts did not develop the questions.

If this option is chosen, skip to chapter 6, *Selecting and Motivating the Experts,* before finishing this chapter or reading chapter 5.

The External Experts Help Select the Question Areas and Assist in the Identification and Refinement of the Questions

This approach to developing questions works best when there are only one or two experts involved or when there are many experts who can be gathered together for a period of time. Frequent meetings require that multiple experts belong to the same geographical area or organization. This option is also applied when the experts are not located in the same place but can meet for a concentrated period of at least a week. For example, on the Army export project, experts from different Army offices met for one week to select the question areas, identify and refine the questions, and answer them for a test case. Experts can also meet to review and modify those question areas earlier proposed by the project personnel. In some cases, experts work on the question areas through mail correspondence; they may physically meet during the final question refinement.

External experts become involved early in question development only if the project personnel consider the external expert's cooperation or views critical to the success of the project. For instance, if there were very few experts in the field, the participation of each would become more important than if there was an unlimited pool of experts. Also, if only a few experts were to be interviewed intensively, as in many knowledge acquisition projects, each expert's participation would be crucial. Additionally, if the sponsor or project funder would interpret any expert's reluctance to participate as a sign that the project was a failure, expert participation becomes critical.

Two advantages of early expert involvement follow:

- Early involvement has a positive effect on the experts' willingness to participate in the study and later provide their judgment
- Experts will be more supportive of the product of the project because they helped develop it.

The disadvantages of this approach follow:

- Project personnel will not have complete or direct control over question development
- Working with a group of experts to develop the questions requires special skills (tact and the ability to keep things moving), and even then the process sometimes resembles a three-ring circus
- Having the experts meet and proceed to develop the questions requires more advance planning than the other options. (See *How to Prepare for a Delphi Situation* or *How to Prepare for an Interactive Group Situation* in chapter 10.)

If external experts will do the bulk of the question development, read chapter 6, *Selecting and Motivating the Experts*, before continuing with this chapter or reading chapter 5.

Checklist for Selected Questions

After formulating the questions, project personnel evaluate their suitability. The following is a checklist for that purpose.

1. **Will the questions provide the data necessary to meet the project's goals?** Sometimes the creation of the questions takes on a life of its own and moves in a direction that has little bearing on the goals of the project. Frequently, an outline on paper will help delimit the project goals; that is, the type of expert

judgment, answers, problem solving, or ancillary data needed for the project. An outline can also help develop and limit the questions.

2. **Will the questions be within the scope of the project?** Critical considerations include time, funding, and logistics.
   Time. A major consideration is balancing the number of questions with the time allotted. To roughly estimate how much time each question will take, consider the type of data (*answer only* or *answer plus*) needed and the required level of detail. Table 1 below illustrates the amount of time that it takes to elicit an expert's response to different types of questions. Note that the level of detail usually corresponds to whether *answers only* or *answers plus* will be gathered. When *answers plus* refers to data on the experts' problem-solving processes, the amount of data being gathered multiplies and each datum becomes more complex.

TABLE 1
Rough time estimates for eliciting expert judgment in different situations

Elicitation Situation	Type of Data	Level of Detail	Approximate Time
Experts in a group	Answer only	Very low	1-4 minutes
	Answer plus a few sentences on their rationale		5-10 minutes
Expert alone	Answer only	Medium	30 minutes
	Answer plus problem-solving data	High	1-2 hours

   Funding. The amount of personnel costs for advisory and external experts is a second consideration. For example, on one large project, the costs per external expert averaged $10,000. The experts traveled to two meetings and provided in-depth answers and problem-solving data on the questions to which they were assigned.
   Logistics. The logistics of eliciting the expert's judgments is a third consideration. For instance, in a problem-solving application in which a few experts are needed for long periods of time, are there experts located nearby? According to Waterman (1986, p. 192), the availability of experts is crucial in this situation. If the experts are not located nearby, could either the project personnel or the external experts be relocated to be together for the necessary period of time?
3. **Can the expert judgment data be obtained without extreme effort?** For example, will the data be classified or proprietary? We have found proprietary

data to be more difficult to handle than classified data. The procedures for handling classified data are clear, and although time consuming, do not deter the project. With proprietary data, however, experts may be unwilling to participate for fear of providing data to their competition. An example of an application containing proprietary data would be the failure rates of different manufacturers' pipes in nuclear reactors. If the data are likely to be proprietary, does the sponsor or the project manager have enough influence to overcome the objections to participation? Can the project personnel guarantee that access to the data will be protected and limited?

Common Difficulties—Their Signs and Solutions

Difficulty: Sponsor Cannot Provide Clear Information on the Project's Goal, the Data to Be Gathered, or the Question Areas. Some sponsors are uncertain about the project, especially during the conceptual stage, while others are unable to communicate their vision of the project. This same sort of difficulty occurs in most consulting applications—the analyst must *extract* what the sponsor wants and needs.

Solution: First, determine at which point the client's conception of the project becomes unclear. Second, elicit from the sponsor the information that is necessary for the project to proceed. For example, if the sponsor has only vague ideas on what the project should be, this basic information should be elicited, clarified, and recorded. In most instances, the sponsor will specify the project's goals but not how those goals are to be accomplished; sponsors often reason that the implementation of the goals is not their job—that is why they hired someone else. Again, the solution is to interview the sponsor to obtain as much information as possible on how to proceed within the project's scope as the sponsor has envisioned it. Two of the interviewing techniques outlined in chapter 7, the verbal probe and the ethnographic method, may prove helpful in questioning the sponsor. For instance, the sponsor could be informally questioned using these two techniques and then interviewed in more depth using the ethnographic technique (chapter 7). The ethnographic technique allows the client's own words to be used in the questioning; this process helps extract the exact meaning of his or her responses.

Difficulty: The Question Developed From the Question Area Remains Too Broad. One sign of a too-broad question is when the advisory expert or external expert breaks the question into smaller parts or requests additional information before formulating an answer. For example, experts were asked the following question about a particular nuclear plant in a reactor risk analysis study (Amos, Benjamin, Boyd, Kunsman, Lewis, Smith & Williams, 1987): "What is the frequency of ignition of the hydrogen given that there has been a station blackout causing hydrogen accumulation?" One expert felt that the frequency of ignition would depend on whether there was high or low pressure in the vessel and therefore divided the question into those two possibilities. In many instances, the experts may request an operational definition of a term mentioned in the question. In the above example, the experts may have asked for the definition of *ignition*.

Solution: Ask an expert to solve the question. If the expert states that the question is too vague, have him or her divide it into parts or subquestions. Use these parts to develop new and more narrow questions. For instance, in the above example the question could have been decomposed into high or low pressure, and ignition could have been defined as stopping short of detonation. As another check, the expert can be asked if the questions in their present form are basically answerable. If the expert replies that the questions are still too vague, ask the expert to think aloud about how to make the questions more manageable. (See chapters 7 and 10 on how to use this elicitation technique.)

Difficulty: Too Many Questions Have Been Selected for the Amount of Time Available. Selecting too many questions to be answered in the available time is a very common difficulty. An early sign of this difficulty is people's reluctance to discuss the number of questions and the amount of available time. Most persons are not aware of how time consuming in-depth elicitation is. In addition, they may estimate the amount of time on how long it would take someone to give an off-the-top-of-the-head answer. Because of wishful thinking, even persons experienced in elicitation tend to underestimate the amount of time a particular number of questions will take.

Solution: Give an advisory expert copies of the questions and describe the elicitation procedure being considered and the level of detail needed in the data. Ask for rough estimates of how long an expert would need to respond to the questions. This amount of time can be examined in light of the number of experts, the questions planned, and the total amount of time available. Several ways for reducing the number of questions follow: fewer questions can be selected from each question area, the number of question areas can be cut, fewer experts can be sampled, the elicitation method can be made simpler and faster, or less detailed data can be gathered. In addition, it may be possible to extend the project's deadlines and avoid any of the above measures.

The time-estimate chart (Table 1) shown in the previous section can also be used in approximating the amount of time that each question will take. As a general rule, elicitations of individual experts last longer than those done in groups because individual interviews are used to gather detailed data. In addition, elicitations that gather problem-solving data tend to be more lengthy than those which just gather answers. The greater the amount or detail of problem-solving data gathered, the more time the elicitation will take. For example, it takes less time to obtain the experts' general rationale (a few sentences documenting their reasoning) than it does to obtain their definitions, assumptions, heuristics, references, and calculations.

5

Refining the Questions

This chapter describes how to refine the questions selected in the previous chapter. Question refinement takes human cognitive limitations into account to create questions that the experts can readily assimilate and process. We believe that reducing the number of factors negatively affecting cognition will lead to better quality expert judgment. Chapter 5 includes suggestions for presenting the information necessary to understand the question (background, definitions, and assumptions), for ordering this information, and for dividing the question into more easily understood parts. Finally, the reader is asked to consider the wording of the question in terms of clarity and bias. In addition, this chapter describes when the experts should help refine the questions and when the stage of selecting and motivating the experts (chapter 6) needs to precede this chapter.

Reasons for Structuring the Questions

To obtain the best quality data, refine questions through structuring. Structuring questions is the process of asking them in an organized and controlled manner. Some means of structuring the question include presenting its information in an orderly way, dividing it into more easily answered parts, representing it in a pictorial or mathematical way, phrasing it in a careful, nonleading manner, and defining the key words.

Structuring the questions provides the following benefits:

- It focuses the experts' attention on what they need to provide.
- It lessens the cognitive burden of solving the question by presenting it in a more assimilable and processed form.
- It delimits the question so that the experts are not interpreting it differently (thus answering separate questions).
- It makes the question more acceptable to the experts because it has been refined to encompass their views and use their terminology.

Which structuring techniques will be used and to what degree depend on the questions, particularly their complexity. Complexity means the amount of information required to solve the question and the availability of standards that can verify the correctness of the answer. Simpler questions require less information to formulate an answer; in addition, standards can usually be found to verify the answers. An example of a simple question is as follows: "At what value would a chi-square statistic of 5.74, with 3 degrees of freedom, be significant?" (Meyer & Booker, 1987*b*, p. 40). An example of a complex question is: "What is the fraction of inventory of radionuclide group present in melt participating in pressure-driven melt expulsion that is released to containment as a result of melt expulsion?" (NUREG-1150 source-term elicitations, 4/13/88). The table below summarizes the level of need for structuring techniques with the question's complexity.

TABLE 1
Need for structuring techniques

Question Complexity	Question Information	Breaking It Into Parts	Question Representation	Question Phrasing
Simple	Low (e.g., definitions/ assumptions)	Low (e.g., textual description)	Low	High
Complex	High	High	High	High

Experts require less information to formulate answers to simpler questions. Typically, only those data, definitions, or assumptions that the experts require are provided. For example, on the simple question mentioned above, the observed values, the expected values, and the chi-square statistic were provided as question information. In addition, the options for response were provided in a form that was standard for the experts. For example, the value of 1% was defined as meaning at or greater than 0.01, 5% meaning greater than 0.01 but less than or equal to 0.05, and so on. The magnitude of this information can be judged by the space it occupied—less than half a page. With the complex example, the experts not only needed pages of background information, but information from each other concerning the question. They shared their information during briefings held before the elicitation sessions.

In addition, simple questions are less likely than complex questions to require subdivision because they are already at an answerable level. If a simple question is partitioned, it will likely yield only a few parts. Because of this simplicity, such questions usually do not require pictorial or mathematical representation. On the other hand, complex questions are frequently diagrammed with trees and charts representing their many parts and the interrelationships between those parts.

Regardless of the question's complexity, careful phrasing is always critical to the experts' understanding of the question.

Techniques for Structuring the Questions

Presentation of the Question Information

Most experts will require information beyond the current statement of the question, regardless of the question's simplicity. As a rule, the more elaborate the question is, the more information will be required, and the more time consuming the planning of its presentation will become. Planning the presentation can be divided into three considerations: (1) determining the types of information needed, such as background on the question, assumptions, and definitions, (2) determining the optimal order for their presentation, and (3) determining the roles of the project personnel and experts.

Types of Question Information Needed

Background. A crucial first step involves the selection of the types of information the experts will need. Frequently, the experts request information on the background to the question (i.e., what events have occurred, what events are supposed to occur, and what the current status is of the phenomenon being evaluated). For instance, to determine how to plan their routes, the tank-platoon leading experts needed background information on their mission (Elston, Emerson, Meyer, Osborn & Stoddard, 1986; Meyer, 1987). This background information included maps of the terrain, a starting point, the general area in which they could travel, their objective, and the probable location of enemy tanks.

Background can also be given on a physical process, if this is the focus of the question. For example, when nuclear engineers were questioned about the performance of their computer model in predicting experimental results, they were given background information on the experiment (Meyer & Booker, 1987*b*). In particular, during the elicitation sessions they received text on the experiment's procedures, its equipment (lists and schematics), and boundary and initial conditions (temperatures and pressures). For the more complex questions on severe accidents occurring in nuclear reactors (NUREG-1150), the experts were sent thick packages of the latest references before the elicitations took place.

Background information can also include representations of the question divided into parts. For example, in complex risk analyses, tree diagrams represent the possible combinations of events. The experts use these diagrams to estimate the likelihood of the occurrence of events depicted on the branches. These representations may have been developed previously by the project personnel and/or by the experts. For example, in a simple decision analysis project on the relative safety of new automotive fuels, the experts developed possible accident scenarios (Krupka *et al.*, 1983). The scenarios provided the framework in which the experts then determined how different fuels affected the likelihood of particular automobile accidents.

Assumptions. Assumptions are another type of background information provided to the expert. It is necessary to present the experts with the assumptions that they are to make in answering the question. If these assumptions are not specified, the experts are likely to make varying ones of their own, sometimes completely altering the question's meaning. (Thus, the original question may pass unaddressed and the experts may answer totally different questions.) Experts are more likely to make assumptions that conflict with the question's original meaning on complex questions. On complex questions, it is difficult to provide the experts with all the details that they believe they need. As a result, the experts may fill in the gaps by making their own assumptions. For instance, on the question of whether a particular Army technology should be exported, the experts wanted to know how many of the technologies were being requested. The experts believed that the number being requested would affect their answers. To allow the experts to proceed in answering the question and to avoid varying personal assumptions, the interviewer had them agree upon an assumption: the export request was for four to six of the technologies (Meyer & Johnson, 1985).

In some cases, the information requested cannot be provided, such as when the information concerns a rare physical process. Assumptions help the experts build a common base to continue with the problem solving.

Definitions. Definitions of terms are another type of information commonly used to refine a wide range of questions. For example, in the export study mentioned above, the experts needed to develop and define *technology transfer* in the same way, so they were asked to agree upon and use a definition: "technology transfer is the means by which technologies, goods, services, data, and munitions that are deemed militarily critical by the Department of Army are transferred to a foreign country, international organization, firm, or individual" (Meyer & Johnson, 1985, p. 7).

Ordering of Information

The next step is to determine the order in which the experts will need this information. One means of doing this is to consider the logical flow of the information. At each point in the planned elicitation (at each question) consider what information the expert needs to respond. A means for checking the flow is to have the advisory experts work through the question. If they request additional information (e.g., information not considered for inclusion or not available for inclusion), insert it at that point. For example, in the study of how tank officers planned their routes, the subjects frequently requested more information on the density of the forests than was offered on the maps (Meyer, 1987). They felt that they needed this information to assess whether the trees would offer sufficient cover.

Finally, how humans assimilate and recall information needs to be considered in sequencing the information. People generally assimilate new information more easily and quickly when it fits within the context of their previous knowledge or mental models (Waern, 1987, p. 276). Even though the subjects are experts in their field, the question information may represent a new portion or a different configuration of the data that they already possess. Because the experts need to answer the question, albeit in their own ways, they must have a thorough understanding of it. Thus, experts should be introduced to the various bits of information, such as the definitions to be used in common, in a manner that facilitates their mental categorizing and accessing of it.

In addition, in the complex and dynamic environment of problem solving, people tend to forget information. For this reason, Payne (1951) has recommended that any information critical to the interpretation of the question, such as important definitions and assumptions, be included as part of the question. For example, in a study of severe accident sequences in nuclear reactors (U.S. NRC, 1989), a few questions began with an assumption—for example, *given that x has occurred, what is the probability that y will?* If these definitions and assumptions cannot be made a part of the question because they make it confusing or too lengthy, they can be given immediately before the question. For example, in a project forecasting the weapon needs of the year 2000 (Meyer *et al.*, 1982, p. 7), the definition of the weapon was read immediately before each question.

Often the above considerations on how people best assimilate information leads to a hierarchical presentation of the information, such as from general to specific or from specific to inclusive. For example, a general-to-specific ordering would provide the experts first with the general context of the question, such as its topic and the scenario, and then the definitions or assumptions which narrow the question. An ordered presentation of the information, whether it be from general to specific or vice versa, assists the expert in assimilating the necessary information.

Roles of Project Personnel and Experts

As mentioned in chapter 4, these are the generic roles of persons who could plan the presentation of the question information. Their roles could overlap, particularly if the advisory experts doubled as project personnel. In addition, other persons such as the sponsor could have a role in presenting the question information if they had assisted earlier in selecting the questions. As mentioned in chapter 4, project personnel can include the ***data gatherers (interviewers*** or ***knowledge engineers)***, managers, and ***analysts***.

The project personnel and advisory experts can work together on determining what information needs to be presented and in what order. One approach would be to have the project personnel draft the information and the advisory experts review and pilot test it. (Pilot testing is described in chapter 9.) The project personnel qualify for this role because they know the project's aims and how they intend to perform the elicitation. The advisory experts know the field and can anticipate the external experts' information needs and responses to the proposed questions. Another approach would be to have the external experts state what background they would need to answer the question, to agree on the definitions and assumptions that they would use, and to decide on the order in which they would like to have this question information presented.

Decomposition of the Question

Another typical means of structuring the question is through ***decomposition***, also known as ***disaggregation*** (Meyer, 1986, p. 88). Frequently, questions are decomposed into parts to ease the burden of information processing and to promote accuracy (Armstrong *et al.*, 1975; Hayes-Roth, 1980). For example, Armstrong *et al.* (1975) asked straight almanac questions of half of their experts. Of the other half, they asked the same almanac questions, only these had been decomposed into logical parts. For instance, the question, "How many families lived in the U.S. in 1970?" was asked as, "What was the population of the U.S. in 1970?" and, "How many people were there in the average family then?" The persons answering the disaggregated questions gave significantly more accurate judgments. Experts often receive information on how to decompose a question as background (see the previous section).

In most instances, complex questions require decomposition. On simple questions, such as the almanac question mentioned above, decomposition is either not needed or done only slightly. A classic example of where question decomposition occurs is in risk analysis. For instance, on the study of severe accident sequences (U.S. NRC, 1989) experts disaggregated questions into a case structure. The cases were created by considering those factors that would have critical effects on the reactor phenomena under consideration. Two major factors helped develop questions that involved radionuclide release associated with pressure-driven melt expulsion in a pressurized water reactor: the reactor coolant system pressure and the state of the cavity (full, half full, or dry). Cases were formed by combining different pressures with varying cavity states (e.g., reactor coolant system pressures of 2500, 2000, and 500-1000 and a full, half full, and dry cavity, or a reactor coolant system pressure of 15-200 with a full cavity).

Considerations in Question Decomposition

Decomposition is not a simple procedure, but one that involves diverse and overlapping considerations. In addition to the question's complexity, several other factors influence the process of question decomposition. One factor is the project personnel's purpose in performing a decomposition and the aims of the project. For instance, if the question is very complex and the intent is to aid the expert in his or her information processing, a more detailed decomposition is needed. In addition, if the expert's thinking is to be documented so that someone else can track it, the decomposition will have to be taken to finer levels.

Another consideration is the amount of detail that will be needed in the data. As a general rule, as the data's level of detail gets finer, so does the level of question decomposition. For example, the level of detail tends to be greater in expert data on problem solving than on answers alone. As a general rule, the data gatherers must obtain data that are one level more detailed than that needed for analysis.

Another factor in planning the question decomposition is the external experts' involvement. How much of the decomposition will they do and at what point in the question's development? Of critical importance is the expert's acceptance of the decomposition. As a general rule, we recommend that the external experts become involved in the question decomposition as soon as possible. For example, the experts could begin refining the decomposition after it has been developed in-house and approved by the advisory experts. External experts could also decompose the questions totally on their own.

A further consideration in decomposition is the relationship between the parts of the question. For instance, the relationship could be causal, temporal, or logical (as it was in the almanac example).

In addition to the above considerations, several other concerns also guide the development of a question decomposition. First, problems can arise when project personnel combine for analysis the answers of experts who have used different decompositions. Combining such answers would be like mixing apples and oranges. A second caveat is that the decomposition be logical—that it properly model the relationship between the parts. A third problem involves having gaps or redundancies in the decomposition that lead to under- and overrepresentation of the parts when these are mathematically modeled.

Roles of Project Personnel and Experts

As mentioned above, the disaggregation can involve the project personnel, advisory experts, and external experts. One method is to have the project personnel propose the disaggregation, the advisory experts review and refine it, and the external experts use it *as is* in responding to the question. A second way is to follow the above procedure but to allow the external experts to modify the disaggregation, as was done on the reactor risk project, NUREG-1150 (U.S. NRC, 1989). Still a third way is to have the experts propose their own disaggregation. We recommend using the second or third approach because the experts are likely to reject a disaggregation that they have not in large part developed.

Representation of the Question

Most questions disaggregated in any detail require representation. Representation is the pictorial or mathematical depiction of the question showing the factors that have bearing on the question and their relationship to one another (relative likelihood, consequence, and importance). For example, in probabilistic risk assessments of nuclear reactors, experts diagram accident sequences and determine their outcomes. The accident sequences resemble a decision analytic model (Barclay, Brown, Kelly, Peterson, Phillips & Selvidge, 1977) in that the sequences consist of branches displaying the outcomes of particular events. The accident sequences can be represented in two ways:

> . . . as event trees, which depict initiating events and combinations of system successes and failures, and fault trees, which depict ways in which the system failures represented in the event tree can occur (U.S. NRC, 1983, p. 2-3).

Fig. 1 represents a simple event tree that shows events leading to a safe termination of their sequence or to a specific plant-damage state. Representations can be used to accomplish the following:

- Guide the external experts in making judgments (e.g., by providing the disaggregation that has been agreed upon).
- Document how the experts reached an answer.
- Provide future guidance on how similar questions should be solved.
- Provide guidance on how the expert judgment will be processed and analyzed.

Note that *representation* here possesses a different meaning than the term *knowledge representation* used in artificial intelligence. Representation only includes depictions or models of the question; knowledge representation refers to the means of describing the expert knowledge in some computer tractable form and of performing automatic inferences. For further references on knowledge representation in artificial intelligence, see Brachman and Levesque (1985), Sowa (1984), or Skuce and Sowa (1988).

Considerations in Representation

Arriving at a representation involves the same sort of considerations and concerns as decomposition (see *Decomposition of the Question*). The main considerations in selecting a representation scheme are that it be compatible with the question decomposition, the analysis plans, and the expert's problem-solving procedures. If one of the existing representation schemes cannot be applied, a new idiosyncratic one can be created. A few representation schemes are described below.

The accident sequence representation, used in performing probabilistic risk assessments (PRA), is based on causal and sequential relationships. That is, one event can cause another and lead to particular outcomes. The PRA accident sequence is widely used in the nuclear risk/reliability community, but it also has general applications to modeling physical phenomena. For example, a related representation scheme is used to diagram automobile accidents (Krupka, Peaslee & Laquer, 1985). This scheme depicts the events that could cause accidents among automobiles run on new alternative fuels. For details on how to create an accident sequence representation, see the PRA Procedures Guide NUREG/CR-2300 (U.S. NRC, 1983).

Another type of representation scheme involves parts that are temporally related. These schemes are often used for managing technical programs that must achieve each milestone on schedule to accomplish the end product. Such representations show whether activities or milestones should be achieved in parallel or in sequence. PERT is an early example of this type of representation. PERT was developed by governmental agencies as an aid to planning and evaluating the costs and scheduling of objective-oriented work (PERT Coordinating Group, 1963).

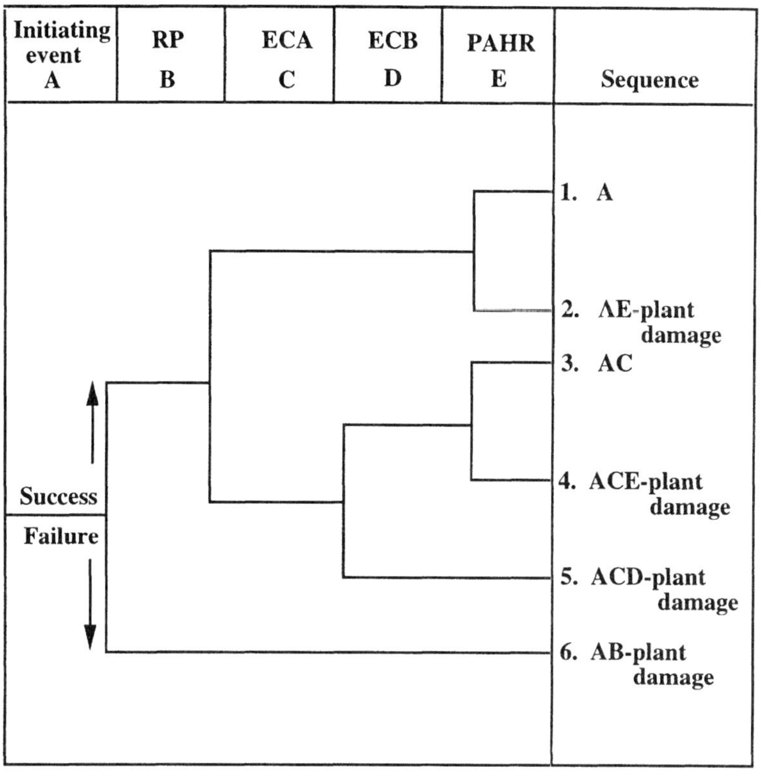

Source: NUREG CR-2300 (U.S. NRC, 1983)

FIG. 1. *An example of a simple event tree for Probabilistic Risk Assessment (PRA).*

In another type of representation scheme, the parts are alternatives which are evaluated in terms of particular attributes they are judged to possess. A common scheme of this type is Saaty's Analytic Hierarchy Process (AHP). Decision makers or experts frequently use AHP to decide which alternative actions or products will best address some agreed-upon criteria (Saaty, 1980, 1982). For example, in the export control project, the AHP representation was used to make decisions on whether a specific Army technology should be either exported, not exported, or exported with particular conditions (Meyer & Johnson, 1985).
New representations created especially for a project usually yield idiosyncratic representation schemes. Idiosyncratic schemes differ, so the relationship between their parts cannot be characterized as being one thing, such as causal. As a general rule, idiosyncratic schemes are developed and used when an existing scheme cannot be applied. Thus, idiosyncratic schemes may be very suited to one application but not easily generalizable to others. For example, an idiosyncratic scheme was created for a project that was to forecast the weapon needs of the United States for the year 2000 (Meyer *et al.*, 1982). This scheme provided the experts with a framework for thinking in a structured manner about potential

threats to the future defenses of the United States—that is, what offensive weapons other countries might develop and how the United States would need to respond with their own weapons development.

Roles of Project Personnel and Experts

Project personnel and experts have the same roles for representation as they did for disaggregation. As with disaggregation, the external experts should have a major role in creating a representation. This is particularly true if the purpose of the project is to model the experts' problem-solving processes or to create one that will serve as a guide to future decision making.

Question Phrasing

Another element of structuring is question phrasing or wording. ***Question phrasing*** refers to the wording of the question and of the selection of the response mode. Careful question phrasing maximizes the chances that the expert will understand the question and not be unduly influenced or biased by it. Biased wording can influence an expert's thoughts or answers; such influence can be considered motivational bias (refer to chapters 3 and 8). Specifically, we consider biased phrasing to be a type of social pressure because the experts' thinking is unconsciously affected by the question wording.

Payne (1951) dramatically showed the biasing effect of phrasing through his use of the split ballot technique in survey questions. The split ballot technique entails giving each half of the sample one wording of the question or response option. For example, a question can be worded in this manner: *Do you believe that X event will occur by Y time?* The same question could also be worded like so: *Do you believe that X event will occur by Y time, or not?* This second option is more balanced because it mentions both possibilities. For this reason it would likely receive a higher percentage of "no" responses. Often the difference measured by the split ballot technique is 4–15%, even when the rewording has been very slight (Meyer, 1986, p. 88).

In another example (Meyer & Booker, 1987*b*), the experts were asked to identify where two curves (one generated by a computer code, the other by an experiment) diverged. An early wording of the question asked the experts to mark the places where they felt the curves diverged. This wording led the experts (e.g., to believe that there must be divergences) and was therefore changed to "mark the places, if any, where the curves diverge." As Payne (1951), the grandfather of surveys, noted, all question phrasings are biasing—the best that one can do is aim for equal but opposite biases.

Another problem with unclear wording is that the experts are likely to interpret the question differently and give answers to essentially different questions. For example, in the study of interexpert correlation mentioned above (Meyer & Booker, 1987*b*), the experts were found to have separate interpretations of the term *diverge*. To some diverge meant where the lines were not exactly the same; to others, it meant where the distance between the lines increased with time; and to still others, it meant where the differences exceeded the error bars of 10–20%.

Another factor that has bearing on the question's clarity is its length. Payne (1951) found that people's comprehension of written sentences tends to drop off after 25 words. For this reason, we recommend that sentences be kept as short as possible.

Considerations in Question Phrasing

When phrasing the questions, consider clarity and bias. Experts and project personnel can improve clarity by reviewing and offering feedback on what the questions meant to them. (This procedure, pilot testing, will be described in detail in chapter 9.) There is no easy procedure for combating biased phrasing—the best strategy is to be sensitive to this issue and to carefully scrutinize the wording of the questions.

Wording a question that multiple experts will understand and accept presents many problems. If a phrasing (or representation or disaggregation) has been reviewed and tested by only a few experts, it may be slanted (e.g., reflect only their experiences, views, and use of terms). The chances of slanted phrasing increase greatly if only one advisory expert is used.

We encountered this problem while drafting a linear scale for experts to use in estimating the magnitude of a computer model's divergence from the experimental results (Meyer & Booker, 1987*b*). In this case, we did not assign more than one advisory expert because we did not want to reduce an already small population of experts. Thus, only one well-qualified expert served as advisory expert in helping develop the scale for the experts' responses. However, when we requested feedback on the scale, we learned that several of the experts did not accept a major premise of the scale. The advisory expert had equated magnitude of the divergences with the presence of deficiencies in the computer model (e.g., insufficient agreement was associated with serious modeling deficiencies that required immediate fixing). (See chapter 7, example 7.2, for an illustration of the linear scale.) Several of the external experts did not make this direct association. Instead, they considered the possibility that the experiment had not been conducted as reported. They reasoned that poor agreement between the model's curve and the experiment's curve could have been caused by trying to model an experiment that was conducted differently than reported. Thus they did not necessarily equate poor agreement with deficiencies in the computer model. Because of the scale's wording, we had to make adjustments to the scale much later than was desirable.

Roles of Project Personnel and Experts

Project personnel can propose the first phrasing for the advisory experts' review. This first draft and later ones should be examined by the project personnel for bias. Assigning tasks in this manner will improve the project personnel's awareness of bias. The advisory experts can review the phrasing to provide information on the experts' reaction to it. At the very least, the external experts need to have the opportunity to modify the question phrasing before their elicitation sessions. Otherwise, during the elicitations the external experts may state that they do not agree with the question and, therefore, cannot answer it. It is generally best to allow the external experts a major and early role in the development of the questions.

When the Refinement of the Questions Should Be Preceded by the Selection of the Experts

The motivation and selection of external experts (as described in chapter 6) should precede the refinement of the questions if any of the following conditions are met:

- If the purpose of the project is to capture the expert's problem solving or to serve as a guide for future decision making.
- If the experts show any indication of not accepting the questions.
- If outside reviewers are likely to be concerned about bias in the question selection or phrasing.

If any of these conditions exist, it is critical to have the external experts do the primary work on refining the questions as soon as possible. Not adequately involving the external experts in question refinement occurs frequently and leads to serious problems. For example, in a reactor study, tight time schedules led to an attempt to save time by having a panel select the questions and the response scale for the experts. The experts had the freedom to modify these, but when they met, they did not feel that they had sufficient time to do so. As a result, several experts questioned what their answers would have been otherwise and stated that they did not wish to defend the study (Benjamin, Kunsman, Williams, Boyd & Murfin, 1987, appendix F). Such statements by the experts impair the credibility of a study.

If the external experts are selected first, they will assist in creating the presentation of the information, the disaggregation, the representation, and the question phrasing. Their early involvement may lessen the need for the use of the separate and usually in-house advisory experts. The external experts' work on refining the questions can occur in different ways. For instance, the experts may do the above tasks as part of their elicitations or they may do these as a separate step, depending on the project. For example, in the export control project (Meyer & Johnson, 1985), the experts met only once to disaggregate, represent, and phrase their questions and to give their answers. By contrast, on the reactor risk study NUREG-1150 (U.S. NRC, 1989), the experts met several times to (1) receive briefings on the project and view the disaggregations proposed by the project personnel, and (2) to give their answers to their disaggregated questions (i.e., disaggregations that they had developed from those proposed earlier by the project personnel).

Common Difficulties—Their Signs and Solutions

Difficulty: The External Experts Did Not Provide Enough Input in Refining the Question. Refining includes providing input on what information is needed to answer the question or assisting with the question's disaggregation, representation, or phrasing. Because refining the question is an evolving process, at its early stages the question will not have received sufficient expert input. In addition, given the means of refining the question (particularly who is doing it and at what stage), the signs of this difficulty may vary. The earliest signs may be the responses of the advisory experts who are reviewing the question. They are likely to be confused by the question; they may state that they do not view the question in this manner or that they cannot answer the question in its current form. If advisory experts have not been used to screen the question, this difficulty may be revealed later when the external experts view the questions for the first time. External experts too may have the same reactions when they read the questions for the first time. If this problem exists, the external experts are likely to demand more information on the question or criticize its disaggregation, representation, or phrasing. Then, if they are to provide their answers as input into this structuring, they may refuse to do so. If they only use the structuring as a first cut, they may insist on extensive modifications of it. In either case, the effects present a serious problem. The project can lose its credibility or run over schedule when the external experts challenge the questions.

Insufficient input from the experts in refining the questions commonly happens because of tight time schedules. When a project has tight time constraints, project personnel may seek to save on time in several ways. Sometimes they will try to minimize the number of times that the experts meet or the number of times that the structuring of a question is redone. Another means for trying to conserve time is to have the project personnel do most of the structuring, thus keeping the length of this phase under the project management's, rather than the expert's, control.

Solution: The simplest way to avoid this serious situation is to involve the external experts as early as possible in the refining of the questions. This is particularly important if the project has any of the following conditions:

- If the purpose of the project is to capture the expert's problem solving or to serve as a guide to future decision making.
- If the experts show any indication of not accepting the questions.
- If outside reviewers are likely to be concerned about bias in the question selection or phrasing.

If one of the above conditions exists in the project, we recommend selecting and motivating the external experts (chapter 6) before beginning the refinement of the questions (chapter 5).

Difficulty: The Question Decomposition Becomes Too Complicated or Too Detailed. In this situation, the disaggregation, and hence its representation, exceeds the level of detail needed for the project's goals and analyses. While it is frequently necessary to gather data one level more detailed than needed, the level of detail discussed here is excessive. The desire to disaggregate *ad infinitum* appears to be a natural tendency. Perhaps the motivation behind it could be explained as *give us enough rope and we will hang ourselves.* Newcomers to disaggregation wish to do a good, thorough job. In addition, they may be trying to take the question to a more easily answerable point, such as where experimental data can be applied. This quest is works well in moderation and is the reason that questions are disaggregated. However, this inclination to divide things ever more finely, if unchecked, can be counterproductive.
Solution: People usually need to experience excessive disaggregation once or twice to recognize the tendency in themselves and others. If the interviewers or moderators see the experts suffering from this tendency, they can do one of two things: (1) allow the experts to continue so that they can come to their own realization about the practicalities of excessive disaggregation, or (2) show the experts through sensitivity analyses when they have more detail than they need. The first approach has the advantage of producing experts who, after their own experience, are supportive of the more general level of disaggregation proposed by the project personnel. However, this approach has the disadvantage of being more time consuming.

Difficulty: The Questions are Ill Defined or Open to Differing Interpretations. This difficulty is the other side of the one mentioned above. It tends to occur when the structuring of the questions has been rushed because of tight deadlines. Often, our expressions are not as clear as we would like to believe. To us, the question may be perfectly clear because we are reading between the lines and unaware of some of the definitions and assumptions that we are making. Anyone who has ever drafted survey questions is aware of how many ways, often other than intended, a question can be interpreted. It is to be expected that the questions will be ill defined when they are first being structured. After all, that is why refining the questions has been designated as a separate phase. However, the questions should be as specific and structured as needed before the expert judgment data are elicited.

If the advisory experts pilot test the question, they may provide the first sign of trouble. If this is the case, they are likely either to request more information or to try to fill in the question's gaps by making assumptions. The same thing can happen later with the external experts if the question has not been sufficiently defined in the meantime.
Solution: To avoid having ill-defined questions during the elicitations, solicit the input of the advisory experts on the question. For instructions on how to pilot test the question, see chapter 8. If the worst comes to pass and the external experts are being elicited with ill-defined questions, there are two options. The first and best

approach is to gather the experts together and have them refine the questions. Each refined question should be recorded and declared as the new question to be used henceforth. The other option is to record any definitions or assumptions that the expert individually uses in attempting to answer the question. The advantage of the first approach is that the experts are answering the same question and their answers can be legitimately combined (if one composite answer is later required for analysis).

6

Selecting and Motivating the Experts

In this chapter we detail how to select and motivate the experts for the two types of applications—those meant primarily to gather the experts' answers and those meant to gather data on the experts' problem-solving processes. These two applications are so different that they determine the approach to obtaining the experts' data. For example, studies that focus on experts' answers usually obtain them in quantitative form from 4 to 50 experts. On the other hand, studies that focus on the experts' problem-solving processes typically gather detailed data from a select few experts. Thus, for the first application, experts are likely to be selected for their diversity and ability to quantify their judgments in the desired form. In the second instance, however, the experts are frequently chosen for their willingness to devote a major portion of their time to being elicited and for their ability to respond to the method. For example, could the expert coherently think aloud for the verbal report method? The experts may even be screened initially by using a sample of the elicitation method.

For Applications Whose Data Will Be the Expert's Answers

For most applications, especially in risk/reliability and decision analysis, obtaining the expert's solution is the primary objective. Such applications record an expert's reasoning only to support the solution—such documentation is not the application's main goal. Usually, the expert's solution is requested in a quantified form, such as a probability. Thus, the experts must have knowledge of the subject matter as well as knowledge of the rules governing the response mode.

Who Is Considered an Expert?

An ***expert*** is anyone especially knowledgeable in the field and at the level of detail (granularity) being elicited—the individual should not be considered an expert unless he or she is knowledgeable at the level of detail being elicited. For example, an expert on different types of reactors would not possess knowledge on the probability of a specific pipe's rupture in a Westinghouse boiling water reactor (BWR). Similarly, a specialized pipe expert might not know the comparative likelihood of loss of coolant accidents in Westinghouse BWRs and pressurized water reactors (PWRs).

What Constitutes Expertise?

Two types of expertise, substantive and normative, enter into projects whose goal is obtaining answer data. Substantive expertise comes from the expert's experience in the field in question (e.g., in rupture rates of Westinghouse pipes). Normative expertise is knowledge related to the use of the response mode. The response mode is the form in which the expert is asked to give a judgment (e.g., probabilities, odds, continuous scales, ranks or ratings, and pairwise comparisons). Normative expertise is based on knowing the statistical and mathematical principles of the response mode. Several response modes, such as probability estimation, follow particular mathematical or logical rules (e.g., all probabilities are values in [0,1]). The use of individuals with expertise in only one or neither of these areas has been a serious problem in studies of expert judgment (Hogarth, 1975). Both forms of expertise contribute to the giving of a judgment, so the lack of either can affect the quality of the judgment. The lack of normative expertise may explain why the quality of substantive expert's judgments is frequently not too different from that of inexperienced lay persons (Armstrong, 1985). In particular, substantive expertise does not guarantee normative expertise as discussed in chapter 2 (sections *Are Experts Bayesian?* and *Do Experts Provide Better Data?*).

In general, a substantive expert who is experienced in the response mode (e.g., a pipe specialist with experience in probability estimation) is a better expert than one who is not (e.g., a pipe specialist without any experience in probability estimation). Two means for coping with this major pitfall are (1) to allow the experts to give their judgments in the deterministic mode that they use in solving problems at work, or (2) to try to familiarize them in use of the response mode. (See the discussion in section *What Needs to Be Practiced* of chapter 9 on training project personnel in how to familiarize the experts with the response mode.)

When Expertise Matters

The above-mentioned parameters of expertise (substantive, normative, and knowledgeable at the appropriate granularity) always play an important role in gathering the expert's solutions. However, under particular circumstances, another aspect of expertise becomes important—the notability of the experts. Selecting experts who are well known and respected among their peers and the broader public can lend the project greater credibility. For example, one study forecasting America's needs and resources was initially very well received because of the endorsements of its illustrious experts (Club of Rome, *Limits to Growth*, 1974). Therefore, if the project has the possibility of being controversial, select experts who are notable as well as qualified along the other lines of expertise (substantive, normative, and level of detail). Selecting notable experts also offers a side benefit. It often motivates other experts to participate in the project in the belief that they will be in august company. Thus, obtaining additional experts for the project becomes much easier.

Additional Considerations in Selecting Experts

Multiple and Diverse Experts

It is generally advisable to obtain diverse experts so that the problems will be thoroughly considered from many viewpoints. Diverse experts are those likely to view and solve the problem in different manners. For example, Seaver (1976) proposes that having diverse experts, particularly in face-to-face meetings, leads to better quality answers. Ascher (1978, pp. 202-203), who has evaluated the accuracy of different forecasting techniques in retrospect, states:

> multiple-expert-opinion forecasts, which require very little time or money, do very well in terms of accuracy because they reflect the most up-to-date consensus on core assumptions.

Diversity of participants is one way to minimize the influence of a single individual. For example, use of a single expert will slant results toward the contents and functioning of his or her memory. Experts will differ from one another in what they have experienced, the interpretation placed on these experiences, and the ease with which these experiences can be accessed and brought to bear on the problem (Hogarth, 1980). Experts' answers are also likely to be affected by the mental heuristics that they used to simplify and solve the problem (Hogarth, 1980; Tversky & Kahneman, 1974, 1981; Meyer & Booker, 1987*b*). The use of diverse experts allows the answers to reflect individual differences in experience, recollection, and use of problem-solving heuristics.

Diverse experts play an important role in cases in which the experts forecast future events or situations (e.g., predicting the market for nuclear power in the year 2000). In forecasting, the tendency is to anchor to the *status-quo* situation and not extrapolate sufficiently in considering the future (Ascher, 1978). Having multiple experts with different viewpoints helps the group overcome the human tendency·to anchor to one, usually conservative, reference point.

The practice of using multiple experts is being encouraged in knowledge acquisition (Boose & Gaines, 1988). One advantage of eliciting and showing different expert's judgments is that the user of the knowledge-based system can pick the expert's way of thinking that he or she finds most useful or appropriate (Gaines & Shaw, 1989).

Other studies on aggregating multiple expert estimates (such as that of Martz, Bryson, and Waller [1985]), support the practice of having multiple experts. Aggregation schemes tend to show that a combined estimate has a better chance than any single expert's estimate in getting closer to the *correct* value, when this is known.

Number of Experts

The exact number of experts that *multiple* constitutes may vary according to the elicitation method. For example, if a face-to-face meeting is involved, we recommend having from five to nine experts for each interviewer available to moderate the sessions. Having less than five experts reduces the chance of providing adequate diversity or information to make inferences (chapter 18). Nine experts in a session is usually the maximum for obtaining in-depth thinking from each expert yet having enough control to counter potential effects that can arise from group dynamics, such as the *follow-the-leader* effect.

Selection Schemes

Most selection schemes are based on having the experts name other experts. In many specialized fields (e.g., seismology, high explosives, and nuclear reactor phenomenology), the experts know one another and can supply the names of other experts. The researcher starts with a few known experts, collects additional names from them, and repeats this process until more names are gathered than are likely to be needed.

The problem with using this scheme without modification is that it leaves the project open to later questions of whether people named those with similar views. Because expert diversity is the goal, this basic scheme is often combined with additional selection criteria, such as having equal numbers of experts from academia, private industry, and the government, or from the major points of view. Some selection criteria may be used to define the meaning of expert (e.g., criteria beyond being a person who is recognized as being an expert by fellow experts). For example, only experts with particular levels of publication or experience, such as in being a plant operator, might be chosen from those named.

Logistics also play a role in the selection scheme. The scheme must respond to such concerns as whether the experts will be willing to participate, have the time to participate at the necessary level, and be allowed to do so by their employer.

The selection scheme is likely to receive close scrutiny if other aspects of the project, such as its results, are questioned. The most frequent criticism is that the scheme failed to select experts who were representative of the larger population and that their answers therefore were skewed. It is commonly believed that skewed results arise from selecting the majority of experts from one place, such as the same organization (e.g., especially from the same organization as the rest of the project personnel), a class of organizations (e.g., from academia, industry, or government), or one point of view. Our studies (Booker & Meyer, 1988*a*; Meyer & Booker, 1987*b*) have not found the expert's affiliation or education to be a significant factor in explaining similarities or differences between experts' answers. However, in the interest of trying to represent different views and to avoid criticism, we recommend selecting a balanced group of experts.

Motivating Experts to Participate

To motivate experts, first consider the proposed project from their point of view. Theories on interviewing predict that obtaining participants depend on maximizing those factors of the situation that experts would consider motivating, such as recognition, and minimizing those that they would find inhibiting, such as having to devote large amounts of their time (Gorden, 1980).

For example, two aspects that could be maximized to motivate the experts are the opportunity to affect the project or contribute to the field and the opportunity to receive recognition. If the potential experts are told that they can provide input into the methods used, they will be more likely to volunteer. Generally, if individuals have control over a process, they feel better about it and will lend support to the methods used or to the conclusions reached. This optimistic attitude stems from the belief that *if I did it, it has to be good.* Also, if the experts judge that the project will be done in a manner that will bring credit to them or that their reputations will benefit from being included in this company of experts, they will be more willing to participate.

Project personnel should remedy those aspects of the project that may inhibit the experts' participation. Generally, devoting large amounts of time to participating in a project is an inhibitor. This inhibitor can be minimized directly by reducing the time required for the project. It can also be indirectly minimized by either increasing the attractiveness of other aspects of the project, such as offering the experts a larger role and thus a greater chance for recognition, or, if all else fails, offering to pay them for their time.

Motivating the Experts Through Pay

Generally, we believe that paying the expert should be reserved for those situations in which there are no aspects of the project that can be used to motivate participation or in which participation requires major investments of the expert's time and thought. Focusing on how the intrinsic aspects of the project can be developed to encourage participation can produce more effective motivators and also improve the design of the project. Paying experts for their time should be a last resort for several reasons: it is expensive; it may attract one type of participant and slant the results (Gorden, 1980, p. 118); or it may have unexpected effects on the participants' view of the project.

Payment can affect the experts' views through a means of psychological adjustment (cognitive dissonance) illustrated below. If the experts are not paid, they must convince themselves that they are expending their time and effort for good reason, or they will feel duped. Studies have shown that the participant unconsciously resolves this dilemma by focusing on the merits of the project and on the benefits derived from participation (Baron & Byrne, 1981, p. 122). On the other hand, if the experts are paid, they are not led to consider the positive aspects of the project. In addition, the experts may view the payment as a bribe for participating in a project that could not obtain experts in any other way (Baron & Byrne, 1981, p. 124). In an unconscious effort to show that their cooperation cannot be bought, the experts may take an extremely critical stance toward the project. One exception to the policy of not paying the expert is in the area of travel and lodging expenses. Experts are not likely to interpret this practical coverage of expenses as payment for their time or cooperation: thus, they should be paid for travel and lodging as required.

Motivating the Experts Through Communication of Intrinsic Aspects of the Project

How aspects of the project are communicated to the experts will affect their desire to participate. For this reason, we recommend that a brief memo (about one page) be drafted before requesting participation. This memo can be used as a script for the telephone conversation or face-to-face meeting with the experts. Generally, more individuals will respond to a request delivered in person than by mail. For this reason, contact or call the expert before sending the memo.

Guidelines abstracted from communications theory (Stroud, 1981; Gorden, 1980) and the authors' interviewing experiences suggest that particular items of information be communicated. Typically, the potential participants will want to know this information in the following order of importance:

1. **The reason for the contact.** We suggest that the interviewer phrase the first sentence requesting the expert's participation in a manner designed to motivate. For example, *I would like you to participate in a study of Y because of your considerable knowledge of X.* This request could be considered motivating because it is a personal appeal for assistance (e.g., *I would like you to . . .*) and because it recognizes the person's expertise. It is important that this introductory sentence be motivating because many individuals decide in the first few seconds of scanning a letter whether they are interested or not and if not immediately throw away the letter. Thus, if the experts cannot be called or contacted in person, the first part of the letter is a critical point in creating interest.

   In our experience, scientists have responded well to these motivators:

   - **Recognition.** Recognition can come from the project personnel's selecting of the experts, the experts working with other experts on the project, or from the experts being identified as contributors to the project and its products.
   - **Experiencing something new and different.** Most people enjoy an occasional break in their routines, and scientists are no exception. In fact, they may have more active curiosities. (A few sources on scientists' personality traits are Mahoney, 1976; Roe, 1952, 1963; Cattell, 1963; Knapp, 1963.)
   - **Need for meaning.** Scientists are often interested in how their work fits into the larger picture. For example, computer modelers of reactor phenomena might be interested in how their data are used to assess risks and set new standards. People's work becomes more meaningful if they can see how it will be used or how it will affect others.

2. **Who is conducting/sponsoring the project.** This information would generally be given before item 1 in a conversation, as opposed to a letter. For

example: *Hello Dr. Jones. I'm John Smith of the Research Division at the NRC. Because of your expertise in dry subatmospheric containment in Westinghouse PWRs, I would like you to participate in a risk assessment study of the Surry plant.* The expert may have some impression of past projects done by this organization. If the expert's impression is likely to be negative, mention how this project differs from previous ones (e.g., it attempts to rectify particular problems or to develop better methods in some area).

3. **How much time/effort this project will likely take, over what period of time, and when it will start.** Earlier, we recommended that the reader consider the factors in the project that might inhibit or motivate potential participants. Most experts will be busy and unable to devote large portions of their time. However, if the project's time demands cannot be reduced, think about how to increase motivating factors. For example, the experts may weigh the time that the project will take against its likely contribution to the field or to their reputation. If the project's goal is to set new standards and the experts will be contributing to the creation of these, this information should be mentioned to offset the heavy time demands. If the project will include the most noted experts in the field, this factor should also be mentioned as a probable motivator.
4. **How they were selected or who referred them.** Basically, the experts will want to know how they were selected. Experts tend to express more interest in participating in projects for which they were specially, rather than randomly, selected. (Thus, if the experts were selected at random, it is best to gloss over this fact.) Even more motivating for the experts is to know that they were recommended by persons that they respect, some of whom may also be involved in the project.
5. **What tasks, in detail, would they be performing in the project.** Before committing themselves (even tentatively) to the project, the experts may wish to know their tasks or roles. Try to avoid using technical jargon to refer to the methods for eliciting or modeling the experts' responses. The following is an example of a general description of the tasks abstracted from another study (Bernreuter *et al.*, 1985): *Your role will consist of three parts: (1) helping define seismotectonic zones east of the Rocky Mountains; (2) giving your opinion on the occurrence rate and magnitude distribution of earthquakes within the zones; and (3) reviewing/refining your input and that of the other experts.*

   If the tasks seem too demanding, the expert may request background materials or training. If the details on providing these to the experts have not yet been worked out, state that background materials and training will be provided as needed.

   The experts may also want to know if they will be required to give answers on tasks or on areas beyond their scope of knowledge. Scientists often raise this issue when they are still new to the project and concerned about their ignorance. When such cases arise, tell the experts that they will not have to provide judgment until they have received training in the response mode and have become familiar with the project. [Note: If the experts express reluctance even after they have received the training and briefings, they should not be forced to provide judgment for two reasons: (1) the expert may lack knowledge in a particular area; and (2) experts may react negatively to being forced, which may detrimentally affect their view of the entire project. A reviewer's statements about a reactor risk assessment project illustrate such a reaction: "The participants were forced to provide unsubstantiated guesses as input" (Benjamin *et al.*, 1987, pp. F5-6).]

In general, the experts can be told that they will not be forced to give judgments when they are not expert because this practice would detrimentally affect the quality of the data. Emphasize that the project's goal is to collect judgments that are based on careful consideration and experience.

6. **Will the judgments be anonymous, and if so, how will confidentiality be maintained.** If the project is sensitive in nature (e.g., politically charged), potential experts will probably want to know how confidentiality will be handled before learning about the project in more detail. Thus, for a sensitive project, the information on confidentiality should be given before the details of the project listed in item 5. In the case of a sensitive or controversial project, consider making the experts' estimates and comments anonymous. However, it is best to be guided by the experts' wishes on this issue. If questions of anonymity remain at the time of contacting the experts, they could suggest guidelines on how to handle anonymity. In some cases, experts may want to have their estimates or thoughts identified: this is especially true if they perceive the project as being important and if they had a significant role in shaping it. However, expert judgment studies have traditionally followed the social or behavioral science norm of not identifying the expert's data.

   Some levels of confidentiality follow: (1) list the organizations or offices that have contributed experts; (2) list the names of the experts plus their affiliations; or (3) identify the data provided by each expert in addition to listing them and their affiliations.

   Generally, the wishes of experts who request the highest level of confidentiality should be applied. For example, the majority of experts may want their judgments kept anonymous but may want to be listed as having participated in the project. In some cases, one expert may oppose being listed as an expert or even having his or her organization named as having provided a representative. The one expert should be given the level of anonymity he or she requests, even to the point of extending this level to include all the experts.

   For the occasional highly sensitive project, the potential participants may wish to hear exactly how confidentiality will be maintained (e.g., how the data files will be stored and who will have access to them). They will use this description to judge whether they would be likely to lose the protection of confidentiality if they chose to participate.

7. **The anticipated product of the project and the experts' access to it.** Generally, one product of the project will be a written report and the experts will wish to know if they will be sent copies. For many people, knowing that there will be something tangible to show for their efforts is a major source of motivation.
8. **Whether participation will be required or voluntary.** For most studies, participation will be voluntary, and this fact should be stated. However, we suggest that this statement need not be made until after the benefits of participation have been fully elaborated.

For Applications Whose Data Will Be Problem-Solving Processes

It has become more common to obtain problem-solving data, perhaps as a result of the influence of artificial intelligence. However, this focus occurs for a variety of reasons, such as when the application is to do one of the following:
- Determine how problems are currently being solved and perhaps set new standards.
- Use the data to evaluate novice's methods of solving problems or to train novices.
- Gather data to build an expert or knowledge-based system.

While these studies usually gather the expert's answer, it is not considered the main data but rather a part of the problem-solving information.

What Is Needed in an Expert

Applications whose goal is to gather problem-solving data require more of the expert than those whose goal is primarily to gather answers. The expert needs to be extremely skillful in solving problems (Welbank, 1983, p. 8), articulate in describing his or her problem-solving processes (Waterman, 1986, p. 192), and willing to commit to this difficult, time-consuming task.

Articulate experts are rare because so many of their thought processes have become automatic or unconscious. Current theories propose that humans progress from learning and consciously manipulating rules, such as those of grammar, to more abstract thinking and less conscious use of rules or procedures (Dougherty, 1986; Denning, 1986). However, experts must somehow regain awareness of their thoughts to assist in explaining, representing, checking, and refining that process.

In general, the experts must be available for providing this information. Usually, availability requires that the interviewer or knowledge engineer and the expert occupy the same city. The expert's accessibility can be a separate concern. Frequently, even with the experts and project personnel occupying the same city, the experts become less accessible as the project slowly progresses and their interest decreases.

If the project's goal is to gather problem-solving data for building an expert or knowledge-based system, the qualities of the expert become even more critical. McGraw and Harbison-Briggs (1989, p. 99) list the following personal characteristics and attitudes as desirable:

> . . . domain experience, sense of humor, good listener, sense of commitment, patience, ability to communicate ideas and concepts, introspective of own knowledge, willingness to prepare for the session, honesty with self and others, and persistence.

Method-Driven Selection

In many cases, the interviewer or knowledge engineer does not choose the experts—the organization that funds the work simply appoints the experts. If this is the case, elicitation methods must be designed and tailored so that each expert can easily respond to them. On the other hand, if there is a plethora of experts, the experts can be selected according to the methods planned. Because of the in-depth nature of this application, the focus is usually either on one expert or one elicitation method at a time. Thus experts can be chosen for their willingness and ability to be elicited by a particular method. (See chapter 7 for a description of the elicitation methods.)

A trial for each elicitation method can be conducted on each expert to determine which combinations of expert/method work best. (To run a trial, select a sample problem and follow the instructions in chapter 10 on how to administer the verbal report, the verbal probe, or the ethnographic technique.) Of the elicitation methods, the verbal probe and ethnographic technique work well on the greatest number of people. The verbal report is more restrictive in that some experts cannot use it. We have found that about one in thirty experts becomes extremely frustrated in using this method because it interferes with his or her thinking.

Motivating the Expert

The goal in problem-solving applications is to maximize those aspects of the project that humans find motivating and minimize those which have the opposite effect (see *Motivating Experts to Participate*). The potential participant then receives the information in the same manner as detailed in *Motivating the Experts Through Communication of Intrinsic Aspects of the Project.* We describe the differences in motivating participation in a problem-solving application so that the project personnel can adjust their communication to the expert accordingly.

The main factor discouraging experts' participation in the problem-solving application is the great amount of time that it requires. Sometimes, the burden on an expert can be lessened by using several experts. However, if the number of experts is limited or if learning one expert's thinking in depth is important (as is common at the beginning of such projects), this may not be possible. A second inhibitor can be the experts' fears that the model/system could replace them or show their thinking to be faulty. For example, if management volunteered their participation, the experts may be suspicious of why management wishes to examine their thinking. These fears can be alleviated by explaining the purpose of the project and eventual capabilities of its product. To further reassure them, state that there are no *correct* answers, explain that this is not a test, and convey a nonjudgmental interest in learning their thinking. A third inhibitor is the experts' concern that they will not be able to tell how they solve problems because they do not know. Address their concerns by explaining that this lack of awareness is part of being an expert and that methods have been developed to extract the information. (For inhibitors specific to experts involved in building expert or knowledge-based systems, see McGraw and Harbison-Briggs, 1989, pp. 117-125.)

Just as the problem-solving application presents additional inhibitors, it also offers stronger inducements for participation than its answer-gathering counterpart. In addition to the motivators mentioned earlier for the answer-gathering application, the problem-solving application offers a greater opportunity for motivation through altruism and the inducement of experiencing something new and different.

One of the altruistic rewards offered by problem-solving applications is the opportunity to help different groups of people. For example, an opportunity exists to help students if the application is to provide the expert's thinking as instruction as in the study of tank platoon leaders' tactics (Elston *et al.*, 1986). Others in the expert's field can benefit if the application performs a function such as identifying likely locations for petroleum deposits (PROSPECTOR) or offering decision-making guidance in export control (Meyer & Johnson, 1985). The work can also serve the larger public if the application provides a service, such as medical diagnostics (MYCIN).

Problem-solving applications offer participants the opportunity to experience something new and different—they provide an understanding into how the participants think. While this opportunity appeals to most people, it is especially attractive to scientists. In fact, scientists will often request references on how practitioners in their field think and act. (Good sources for this information include Mahoney, 1976; Roe, 1952, 1963; Cattell, 1963; Knapp, 1963.)

Lastly, if the product of the application will be marketed, recognition and financial gain can be incentives for participation.

Keeping the expert motivated through time presents a major problem. Generally, the incentives that were used to initially capture the expert's interest in the project can be used to maintain that interest. Waterman (1986, p. 194) notes the following:

> Making the expert feel like an integral part of the system-building process will motivate him or her, as will showing the expert how the system will ultimately produce a useful tool.

Waterman adds that long periods between interviews should be avoided because they diminish the expert's interest. Welbank (1983, p. 10) recommends the following means for recharging the expert's enthusiasm: select an interesting problem (e.g., one that has received recent publicity), hold panel discussions in the belief that many experts enjoy criticizing one another, and produce a prototype of the expert's thinking for him or her to review.

Common Difficulties—Their Signs and Solutions

Difficulty: The Experts Do Not Wish to Participate. In most applications, somewhere between one-third and three-quarters of those experts called will initially agree to participate. If fewer than one-third agree to participate, expect a potential problem in motivating the experts.

Solution: While the culprit is likely the communication with the expert, we recommend gathering more specific information before reworking the presentation of the rewards of participation. For example, the interviewer can again contact an expert, this time focusing on what may be inhibiting participation. If one of these experts is an acquaintance or known to be outspoken, this expert would be a natural choice. Ask this expert to provide, in confidence, his or her real reasons for not participating. If asking an expert for this information seems awkward, consider asking the advisory experts. The advisory experts, because they are working on the project, will likely have a different view of it than the external experts. For this reason, the advisory experts are not our first choice. The advisory expert will need to be briefed on what was said to the other experts and on the experts' responses. Ask the advisory expert to place him or herself in the expert's place and to suggest reasons for the low participation rate.

If the above suggestions prove inappropriate or unsuccessful, itemize the motivators and inhibitors of the project again. The results may show that the motivators are not strong enough to overcome those things inhibiting the experts' participation. Inhibitors may be discovered that were not taken into account. In our experience, the one factor that most causes experts to balk at participation is the belief that their effort and judgment will not be used. In some such cases, the experts thought that decisions had already been made at high levels and that their judgments would not be used as input. If this perception, true or false, hinders expert participation, it needs to be addressed. The project's sponsor and the other project personnel can help suggest solutions.

Difficulty: Everyone, Including Nonexperts, Wishes to Participate. The participation of everyone may seem to be an embarrassment of riches to those encountering the difficulty mentioned above. However, it remains a difficulty, even though it can be easily resolved.

Solution: Everyone can participate, if participation is divided into classes and implemented according to a few rules. In general, the volunteer participants should be grouped according to their expertise. For example, in the Army export control project (Meyer & Johnson, 1985), most of the experts possessed knowledge in one specific question area but wished to provide their inputs to all areas. All the experts were allowed to present their views and answers in a structured manner. However, only those experts who had been previously designated as experts in the area gave their answers as votes, which were then documented for analysis. The other experts' answers were simply recorded as commentary. As a result, both groups were satisfied. The nonvoting experts expressed their views which ensured that the voting experts were not overlooking some important information. The voting experts knew that their expertise had been acknowledged and that their answers would be used to establish policy.

Frequently, the participants can be asked to sort out their participating rights and statuses. In the above example, the participants suggested the voting scheme.

Difficulty: The Real Experts Are Too Busy to Participate at the Needed Level. As a general rule, this difficulty is considered much more serious if it occurs in problem-solving applications rather than in answer applications, especially if the goal of the application is to emulate the expert. The signs of this problem are similar to the above-mentioned one of the experts not wishing to participate.
Solution: If this difficulty occurs when the experts are initially called, the problem lies in the communication. Refer to the suggestions given above. Note that the best experts *are* often more busy than those less expert or that they may need greater or different motivators to be persuaded to participate. Consider asking the experts what would convince them to participate.

Difficulty: The System for Selecting Experts Is Criticized. We have observed that selection schemes are frequently criticized. The most common complaints are that the selection is not representative or that it does not include the *real* experts.
Solution: If the credibility of the selection is questioned after the expert judgment has been elicited, there is little that can be done. One limited solution is to document the selection scheme as a means of explaining and defending it. We recommend recording the selection criteria, the reasoning behind the use of the criteria, and the number of experts who were invited to participate versus those who actually participated.

If criticism of the selection scheme occurs during the design stage, consider the following two approaches. One approach involves the rethinking of the selection scheme. We suggest targeting the top experts in the field and trying to motivate as many of them as possible to participate. If the top experts accept, the selection will not be open to the common criticism that no real experts participated. If the top experts cannot participate full time, accept their partial help. If none of the top experts will participate, at least their participation was sought. In addition, we recommend designing the selection scheme around features that are expected to make the judgments of the experts differ. For example, the experts are often expected to differ according to where they have worked or gone to school. Experts can be selected to represent these different features.

Another approach to handling criticism of the selection scheme is to put the criticizers themselves in charge of designing a selection scheme. Having to design a selection scheme will make them more appreciative of the difficulties involved. Double check their selection scheme to ensure that it is not open to one of the common criticisms, then use them to help call the experts. The experience of trying to implement their selection scheme will make them aware of how things often go awry. If they do not participate in making the calls, they may later wonder why their scheme was not implemented exactly. If they assist in the calls, they will understand that some experts could not participate and that substitutions had to be made.

Difficulty: A Conflict Exists Between Those Wanting to Identify the Expert's Data and Those Wanting to Preserve Anonymity. The conflict stems from the two views of identifying the expert's judgments. One view is that the expert data will be more credible if it is accompanied by the expert's name. Proponents of this view believe that the experts will exercise more care in giving their judgments if these judgments are attributed to the experts and that better quality data will result. The second viewpoint favors anonymity in the belief that the experts will not give their true answers if others can trace the answers back to their sources. Proponents of this second view argue that the confidences of interviewees have traditionally been protected. A sign that this difficulty is occurring is disagreement among project personnel, sponsors, or experts on anonymity.
Solution: The means for resolving this conflict of views is to have the experts decide how they wish to have their judgments identified. The experts make this choice because they provide the data and they may withhold their judgments if they feel uncomfortable. Explain to the experts that their decision on anonymity will be followed, but that they will have to reach a consensus. This puts the burden on them for swaying the members of their own party that have differing ideas. If the experts fail to reach a consensus, impose the highest level of protection requested, even if only one expert wishes his or her judgments to remain anonymous.

7

Selecting the Components of Elicitation

This chapter is designed to guide the reader through planning what is needed to obtain and later analyze expert judgment data for a particular application. We offer the checklist shown in the next section to assist the reader in determining which of the six basic components of elicitation (Fig. 1) will be needed. Subsequent sections in this chapter will help the reader select the most appropriate methods from within the selected components. For example, the reader might use the checklist below to decide that the expert's problem-solving processes need to be gathered, and therefore that ***elicitation techniques*** will be necessary. The reader may then select the verbal report and the verbal probe as the best combination of techniques for accomplishing this task.

The basic methods of elicitation are presented because people often wish to use an existing method rather than try to create a new one—such a process would be like attempting to reinvent the wheel. Also, many people prefer to use an existing and accepted method in the belief that it will enhance the credibility of their work.

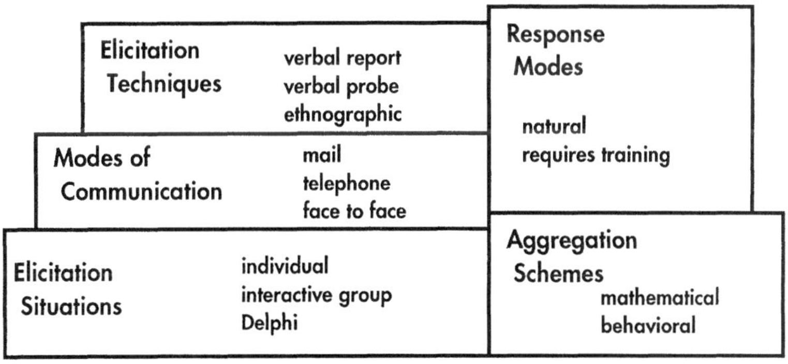

FIG. 1. Building blocks of elicitation.

Chapter 8 discusses how the methods selected in this chapter can be tailored to a specific application.

Determining Which of the Five Components Are Needed—A Checklist

Check: **When a Component Is Needed:**

____ **1. An elicitation situation is needed** if expert judgment is to be gathered. Regardless of specific project requirements, some staging is necessary for arranging how the experts and data gatherers will meet and how the expert judgment will be obtained.
Definition of component: *An elicitation situation is the setting in which the elicitation of the expert's judgment takes place. Elicitation can occur in private, in a group setting, or when the experts are alone but receiving information on the other experts' judgments. Thus, the elicitation situations differ in the amount of interaction between the experts and between the interviewer and the expert (Fig. 2). Frequently, more than one elicitation situation can be used in a project. For example, the experts could meet as a group to discuss and revise the questions. In a later session, they could be interviewed separately for their final judgments.*

____ **2. A mode of communication is needed** if expert judgment is to be gathered because the data gatherer must have some means of imparting and receiving information.
Definition of component: *A mode of communication is the means (face to face, telephone, or mail) by which the data gatherer and the expert communicate during the above-mentioned elicitation situations. Several modes can be combined in one elicitation process. For example, the interviewer could mail the expert the problem, answer the expert's questions about it over the telephone, and then elicit the expert's problem solving in person. In addition, modes of communication can sometimes be substituted for one another. For example, in the Delphi situation, the information could be communicated, if it was simple and brief, by telephone rather than by mail.*

____ **3. Elicitation techniques will be needed if**

- the project's goal is knowledge acquisition, such as in building an expert or knowledge-based system;
- there is likely to be interest in how the experts arrived at their answers; or
- the answers are to be aggregated. If the experts define the questions differently, combining their answers could be like mixing apples and oranges. Use of data on their problem solving could prevent this mistake.

Definition of component: *Elicitation techniques are the means by which the experts are led to describe aspects of their knowledge, such as their definitions for terms and the processes and procedures by which they solve* *problems. (How experts verbalize their answers is covered under the response mode component.) One elicitation technique, interviewing, is composed of different ways of asking questions, such as the verbal probe and the ethnographic technique (Meyer, Mniszewski & Peaslee, 1989). In another elicitation technique, the verbal report, the interviewer observes the experts (usually individually) and listens to them as they solve problems. A variety of elicitation techniques are available, most of which can be applied to gathering differing levels of detail.*

____ **4. A response mode/dispersion measure is needed if** the experts' answers must be in a specific form and if the experts make this conversion mentally as opposed to having someone later translate their judgments into the desired form. Project personnel often wish to use a particular model and therefore want the experts' answers in a particular form. This means that either the expert must conform to the desired mode or the analyst must translate the expert's judgments into that form. In general, we favor using a response mode rather than converting the expert's judgment into the desired form later. We believe that the former practice is less likely to lead to misinterpretation or misrepresentation of the experts' data, providing that the experts can accurately encode their thoughts into the requested response mode (either naturally or with training).
Definition of component: *A response mode is the form in which the experts are asked to encode their judgments. Some modes handled in this book include estimates of physical quantities, probability estimates, odds-ratio, probability distributions, continuous scales, pairwise comparisons, ratings or rankings, and Bayesian updating. Additionally, the experts are often asked to provide measures of dispersion on their judgments (e.g., 0.9 ± 0.1). A dispersion measure is the amount of variation or spread in the data. Dispersion measures can also indicate the amount of uncertainty in the data. The dispersion measures covered below include ranges, percentiles, and variances or standard deviations.*

____ **5. Aggregation of experts' answers is needed** if multiple experts (and, therefore, usually disagreeing experts) will be used and the project personnel require a single representation of their answers. For example, a knowledge engineer may want to combine several experts' problem-solving procedures to produce one procedure for use in an expert system. Similarly, in a risk analysis application, a researcher may aggregate several experts' estimates to enter one estimate into the model.
Definition of component: *Aggregation is a process designed to obtain a single datum from multiple and differing expert data. For example, experts can be required to agree among themselves on a single answer or they can be allowed to give different answers that are later mathematically combined (chapter 16).*

____ **6. Documentation is needed if** having a permanent or semipermanent record of the expert's answer and/or problem-solving process is desired. Documentation records include an expert's name, his or her answer, and the reasoning behind the answer. Documentation is often used to provide traceability on the expert's decision. Traceability becomes important if the judgments are later reviewed or require updating.
Definition of component: *Documentation provides a record of the expert's judgment and/or the problem-solving processes and rationale behind the judgment.*

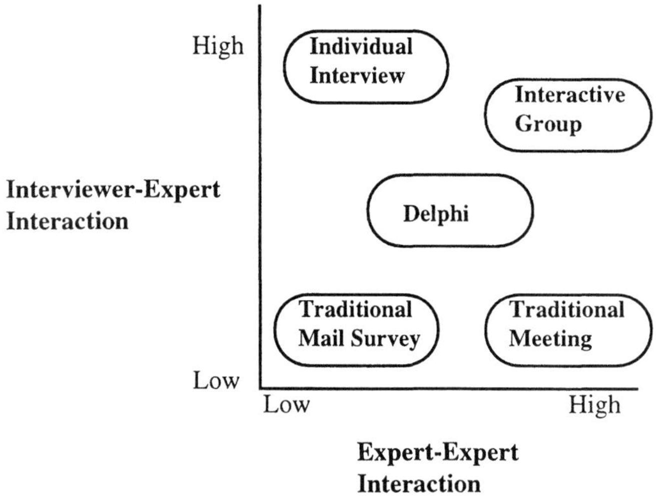

(This figure was excerpted in part from Armstrong, *Long-Range Forecasting: From Crystal Ball to Computer,* New York: Wiley-Interscience, 1985, p. 112. Available in full text at http://hops.wharton.upenn.edu/forecast.)

FIG. 2. Relative interactiveness of elicitation situations.

Selecting From Within Each Component

Selecting From Elicitation Situations

Three major situations can help elicit an expert's judgment: the interviewer can interview the expert in private; the interviewer can meet with a group of experts; and the interviewer can receive judgment from the expert, even though the expert is physically isolated from the interviewer or fellow experts. These situations can be tailored and combined to fit the application. For example, the interactive group could be structured like a technical conference, where each of the experts is scheduled to present their views before the group's discussion. The use of the interactive group method could be combined with that of the individual interview to elicit each expert's judgments apart from those of the other experts.

Interactive Group

The interactive group occurs when the experts meet in a face-to-face situation with one another and a session moderator (data gatherer). The experts' interactions with one another can be structured to any degree. An unstructured group resembles a traditional meeting; a highly structured group is carefully choreographed to prevent spontaneous interaction (to limit the negative effects of interaction, such as group think).

Advantages: *Generates more accurate data, particularly for predictions, and a greater quantity of ideas than the other two situations. (These two results are attributed to the synergism created by experts sharing their thoughts.)*

Disadvantages: *Possesses the potential for group-think bias. Poses logistical problems in scheduling and handling multiple interacting experts, particularly if there are more than four to seven experts per moderator.*

Summary of studies: *According to Seaver (1976), who performed comparative studies of these three elicitation situations, the interactive group method produces a greater quantity of ideas and a higher member satisfaction with the product than the Delphi method. He also noted that the majority of subjects in a group of diverse membership improved their accuracy following a discussion. Fogel (1967, p. 375) commends the interactive group for solving problems that require "originality and insight" and not routine tasks. He found that predictions made by groups were more often correct than those made by individuals. In general, studies comparing a structured interactive group with one or more of the other situations favor the former (Seaver, 1976; Armstrong, 1985; Gustafson, Shukla, Delbecq & Walster et al., 1973; Van de Ven & Delbecq, 1974).*

Delphi

Delphi takes place when the experts do not directly interact with one another or the moderator. The experts, isolated from one another, give their expert data. The data are collected by the moderator, made anonymous, and distributed, usually by mail, to the experts to allow them to revise their previous judgments. The experts can be allowed to revise their estimates until consensus is achieved. Rand Corporation developed this method to limit the biasing effects of interaction.

Advantages: *Designed to avoid biases arising from group dynamics. (However, some doubt exists as to whether Delphi accomplishes its design purpose.)*

Disadvantages: *Limited in the amount of data that can be gathered (e.g., not suited to gathering data on how the experts solved the problem, except for their sources of reference). Less synergism than in the interactive group. Usually the most time consuming of the three situations because of the turnaround time through the mail. (If the Delphi was done by electronic mail, it would be less time consuming.)*

Summary of studies: *For all its use, the Delphi method has not gone through an extensive empirical investigation, and its reviews range from the positive to the negative. Several researchers consider the structured interactive group to work better than the Delphi in terms of avoiding the bias that the Delphi was originally designed to avoid (Seaver, 1976; Armstrong, 1981). It would be natural* *to hope that the experts have converged on the "correct" answer when they have reached consensus in the Delphi. However, Dalkey (1969) found that the number of rounds in Delphi corresponded to increasing agreement but not to increasing accuracy.*

Individual Interview

Individual interview occurs when the expert is interviewed alone, usually in a face-to-face situation with the interviewer. This situation can be structured to any degree. An unstructured interview occurs when the interviewer has not outlined data-gathering goals or questions in advance.
Advantages: *Best method for obtaining detailed data. Main method used for obtaining data on the expert's problem-solving processes. Avoids potential bias from group dynamics and data can be combined later, usually by mathematical aggregation.*
Disadvantages: *Time consuming. No synergistic effects from interexpert discussion.*
Summary of studies: *Seaver (1976) found that mathematically combined estimates from individual interviews outperformed single estimates. When the answers from the individual interviews are mathematically combined, the individual interview is termed the staticized or nominal group method. Results from the staticized group situation have been judged poorer than those from interacting groups (Stael von Holstein, 1971). However, note that the above-mentioned studies were not using the individual interviews to elicit deep problem-solving data, the task for which it is most suited.*

Selecting From Modes of Communication

The ***modes of communication*** are as follows: face to face, telephone, and mail (postal or electronic). These modes are used in combination with the elicitation situations—in most cases, the elicitation situation determines the mode of communication. For example, use of the face-to-face mode would generally follow from the choice of the individual interview or interactive group situations. Similarly, the mail mode would be a common choice with the Delphi situation.

Face to Face

The face-to-face mode occurs when the data gatherers meet in person with the experts. This mode is usually applied in combination with the individual interview or interactive group situations and is particularly adapted to obtaining detailed data. In fact, if eliciting deep problem-solving data is the goal, this is the only suitable mode of communication. Problem-solving information is usually elicited from one expert at a time through intensive interviews. For example, one project focusing on crisis managers required two 2-hour interviews with each expert. With projects whose goal is extracting the expert's problem-solving processes to build an expert or knowledge-based system, the interviews are likely to take even longer. The latter can be a "prolonged series of intense, systematic interviews,

usually extending over the period of many months" (Waterman, 1986). The knowledge engineer usually travels to the expert's place of work and investigates the expert's problem solving *in situ*.
Advantages: *Best mode for obtaining fine granularity information, such as on the expert's problem-solving processes. The face-to-face mode is compatible with the greatest number of elicitation techniques, especially those found in the knowledge acquisition literature.*
Disadvantages: *Time consuming, labor intensive, and expensive.*

Telephone

The telephone can be used in combination with any of the elicitation situations that involve only a limited exchange of information. For instance, in the Delphi situation, the experts could give their responses (answers, a sentence or two on their reasoning, and/or the names of the references that they used) over the telephone. The coordinator could make this information anonymous and relay it by telephone to the other experts.
Advantages: *Generally less expensive than the face-to-face mode. Telephone communication has a shorter turnaround time than the traditional postal method. The data can be obtained while the interviewer is still on the telephone, rather than later by mail.*
Disadvantages: *Poor for relaying detailed or long pieces of information. The telephone would not be suited to asking the expert complex questions or to probing for the assumptions that the expert made in arriving at an answer. Similarly, the telephone mode should not be used to conduct the verbal report elicitation technique.*

Mail

Traditionally, the mail survey has been conducted by post, but electronic mail is beginning to be used for this purpose. The mail mode is frequently used in combination with the Delphi situation.
Advantages: *Good for eliciting simple data from a large sample. Mail surveys cost much less than the face-to-face mode.*
Disadvantages: *Has the lowest response rate of all the modes, although electronic mail surveys may have higher response rates than those sent by post. In addition, electronic mail surveys may have faster turnaround times than traditional postal surveys. Like the telephone mode, the mail mode is not suited to transmitting complex instructions or detailed problem-solving data. For this reason, complicated response modes that require training should not be used with either of these communication modes. Similarly, do not use the mail mode to gather in-depth problem-solving data. In particular, never use the verbal report technique when communicating with an expert through the mail.*

Selecting From Elicitation Techniques

In this book, we focus on three very simple means of eliciting the expert's problem-solving data: the verbal report, the verbal probe, and the ethnographic technique. While other techniques exist for obtaining problem-solving data, we believe that these three are among the easiest to use and least prone to bias.

For further information on interviewing methods, see Spradley (1979) or Gorden (1980). For information on elicitation techniques used for knowledge acquisition applications, see McGraw and Harbison-Briggs (1989), Boose and Shaw (1989), Shaw and Woodward (1989), Dhaliwal and Benbasat (1989), and LaFrance (1988).

Verbal Report

Verbal report involves instructing the experts to think aloud as they progress through the problem (Ericsson & Simon, 1980; 1984). This technique is used on one expert at a time.

For example, the expert receives a written copy of the problem as follows:

> What feed program would you start this colt on? The colt is 6 months, 550 lbs., has an average metabolism, and will receive light exercise through ponying. Please solve this problem as you do others that you receive in this field. Please try to think aloud as you work your way through the problem. Your thinking aloud is as important to me as the answer that you reach.

The expert's verbal report resembles someone talking to himself or herself. This technique (*protocol analysis*) comes from psychology.

Advantages: *Avoids introducing motivational bias (altering the experts' reports of their thinking) because the interviewer does not question the experts and thus is less likely to "lead" them in their thinking (Meyer et al., 1989).*

Disadvantages: *Must be used only on one expert at a time. Not suited to group situations or those in which the expert and interviewer communicate by telephone or mail. Very time consuming. It usually takes the experts at least twice the time to verbalize their thoughts as it does to simply have them. The experts may be incapable of verbalizing all of their thoughts, as is the case with all elicitation methods. These thoughts may be unconscious or they may be difficult to articulate because they must be translated from one form into another, such as from a mental picture into words.*

Summary of studies: *A main concern with this elicitation method has been that the process of verbalizing may negatively influence (bias) the expert's problem-solving process. Ericsson and Simon (1980; 1984), the authorities on this method, argue that the introduction of bias depends on the presence of the data in short-term memory and the type of information that the expert is asked to verbalize (McGraw & Harbison-Briggs, 1989, p. 335). If the information occupies the expert's short-term memory, thinking about it should not alter it. Similarly, if the experts simply report their considerations and steps, less bias is expected to occur than if the experts were to conceptualize about their reasons for these steps.*

Verbal Probe

The ***verbal probe*** is questioning done at a particular time and in a specific manner. The type of verbal probe discussed in this section is used immediately after the expert has reached a solution. The probe exclusively focuses on the problem that the expert has just solved and is indirectly phrased so as to minimize influencing the expert's thinking.

For example, immediately after the expert solves the problem and gives an answer, the verbal probe is used to learn the reasons and assumptions made to reach that particular answer.

Interviewer: Why did you give that answer—that feed program?

Expert: Well, it provides the right amount of protein, calcium, and phosphorus for a horse to grow at this age.

Advantages: *A quick means of obtaining general data on the expert's reasoning in solving the problem. Can be used on experts individually or in a group (Meyer et al., 1989).*

Disadvantages: *The verbal probe is best used when the expert can respond verbally in a face-to-face mode, such as is typically used in individual interviews and interactive group situations. Written responses to the probe are generally inadequate. We consider the verbal probe more likely to induce motivational bias than the verbal report because the probe's questioning can cause the expert to invent descriptions of this thinking (e.g., plausible reasoning).*

Ethnographic Technique

The ***ethnographic technique*** involves transposing the expert's words into questions. This technique can probe on one of the expert's responses to obtain an operational definition that could then be entered into the data or knowledge base.

In the following example, the expert has just said that the colt's feed program may need to be adjusted if the colt does not keep his weight on.

Interviewer: Not keeping his weight on?

Expert: Yes, not gaining as he should at this time.

Interviewer: At this time?

Expert: At his age, 6 months, he should be gaining between 1.5 and 2.0 lbs. per month.

Advantages: *Can be used to obtain the greatest amount of detail on the expert's problem-solving processes. The ethnographic technique is a relatively nonbiasing form of questioning (Meyer et al., 1989) because it is based on the expert's own words. In addition, this technique can act as a check on the interviewers' or knowledge engineers' tendency to assume that they know the meaning of the expert's terms (misinterpretation bias).*

Disadvantages: *Generally time consuming, except when used to elicit a few definitions. The ethnographic technique is not suited, in its usual time-consuming form, to group elicitations. Also, this technique should not be administered while the experts are still solving the problems because it can distract them.*

Selecting From Response Modes

Selecting From Response Modes The response modes that follow are organized according to the forms that experts use to answer a variety of questions. For example, a question on the number of homeless currently in the United States would require an estimate of a physical quantity. In contrast, a question on the likelihood of some occurrence would need probability estimates, odds, or distributions. Questions based on the comparison of two or more things would require pairwise comparisons, continuous scales, or ranks or ratings.

In addition, some questions require single estimates, such as one probability value, while others need multiple estimates, as in an expert's probability distribution. Of the response modes listed, only the estimate of a physical quantity, the probability estimate, the odds ratio, the ranks or ratings, and the continuous scale are utilized to obtain single estimates. Probability distributions, Bayesian updating, and pairwise comparisons are frequently employed to obtain a set of estimates from the expert, although the continuous scale and ranks or ratings can also be used for this purpose.

To help the reader find the response modes appropriate to particular questions, the modes are listed separately as well as under the more inclusive mode to which they belong. For example, odds ratios are listed separately and also as one of the ways of eliciting a probability response.

Estimate of Physical Quantity

The expert estimates a physical quantity, such as temperature, time, pressure, volume, or flow rate, in response to a technical question. For instance, the expert could be asked to provide the engineering specifications for a component, the parameters needed to achieve a particular milestone in magnetic fusion, or the amount of coal reserves currently in the United States. Estimates of physical quantities are sometimes part of the question's phrasing and thus can be used in combination with other response modes. For example, in a magnetic fusion project (Meyer *et al.*, 1982, p. 424), the parameters needed for confinement were elicited from an advisory expert, refined by an external expert, and then used to question several experts. These experts estimated the time needed (a physical quantity) and the probability (another response mode) for achieving these physical parameters. Similarly, in the reactor risk project NUREG-1150 (Ortiz *et al.*, 1988), the values of physical variables were sometimes elicited at given cumulative probabilities (e.g., *what is the inner core temperature such that the probability is 0.90 for that temperature or higher?*).

Advantages: *A convenient and flexible form for answering questions on poorly understood or difficult to measure physical processes. The experts have little or no difficulty in understanding estimates of physical quantities as response modes because they use them in their work. Thus, experts usually require no training in the use of this response mode.*

Summary of studies: *To our knowledge, there are no studies that evaluate estimating physical quantities with respect to some other response mode. However, two types of studies do address estimates of physical quantities: (1) studies that examine in retrospect the accuracy of experts' predictions, such as the size of the United States' market for petroleum in some year (e.g., Ascher, 1978); and (2) studies that examine properties of human judgment by asking* *questions with predetermined answers. There are many more studies of the second variety and most of these have used almanac questions (e.g., Armstrong et al., 1975; Martz et al., 1985), where the subjects estimate physical quantities, such as a city's population of persons over age 50. The results of these studies have shown that the accuracy of experts' estimates of physical quantities can be affected by the assumptions that the expert makes (Ascher, 1978) and by decomposition of the question—decomposition being associated with greater accuracy. (These results may apply to the other response modes as well.)*

Probability Estimate

A probability estimate is a single value given by the expert (e.g., 0.45) in response to a question. Probability estimates are usually used to predict the likelihood of some event occurring. Probability estimates can be asked for in different ways: *What is the probability that an event will occur?* (fixed value); *what are the number of occurrences in n total trials using a log scale?* (log odds); and *what are the number of occurrences or events in n total trials?* (See *Odds Ratio* below.) Multiple experts' estimates to the same question are sometimes linked to form a probability distribution, which is then used as the final answer. (See *Probability Distribution* .)

Advantages: *Commonly used in decision analysis. In fact, one advantage in using a probability-based response mode is the existence of established elicitation and analysis techniques. (For this reason we recommend the use of decision analysts and decision analysis techniques if this response mode is chosen.) In general, probability estimation is a very convenient form for modeling and analysis.*

Disadvantages: *Most experts are not good estimators of probability values and may be reluctant to use this response mode. (The use of probability wheels and training in probability estimation can mitigate these problems.) Done correctly, probability estimation is a very time-consuming process. It can fatigue the expert.*

Summary of studies: *Recent studies have shown that humans are biased in their estimation of probabilities (Tversky & Kahneman, 1974), that they do not properly interpret probabilistic phenomena, like randomness, statistical independence, and sampling variability (Hogarth, 1980), and that most do not feel capable of using this response mode (Spetzler & Stael von Holstein, 1975; Welbank, 1983). Hogarth (1980, p. 149), who has extensively studied cognitive and motivational biases, acknowledges that "probability theory itself is difficult to learn and apply" but argues that it is the best choice for expressing uncertainties. He cites the numerical precision of probabilities and their logic for structuring relationships between events as reasons for using probabilities. Welbank (1983, p. 28) notes that the precision of probabilities is not always needed or justified but states that probabilities may be useful where the knowledge is vague and where there is need for weighting of the answers.*

Odds Ratio

An odds ratio is a response that follows the form of *x* chances out of *n* total trials. For example, an expert could state that there is 1 chance in 1000 of a particular event occurring. The odds ratio is most frequently used to estimate the frequency of rare physical events. (Odds ratio is also mentioned under *Probability Estimate* and *Probability Distribution.*)

Advantages: *A convenient form for estimating the likelihood or frequency of rare events. We believe that it is easier for most people to think of rare events in terms of odds (e.g., 1 in 1000) rather than in probabilities (e.g., 0.001).*

Disadvantages: *If the researchers set the total number of trials (n) from which the experts are to estimate the occurrence of some event, too small a total can affect the experts' judgments and/or frustrate them.*

Summary of studies: *In general, humans tend to overestimate the likelihood of rare events. According to Seaver and Stillwell (1983, pp. 2-7), odds are one of the best procedures for estimating relatively unlikely events, especially if the odds appear on a logarithmic-spaced scale. In general, rare event estimates are thought to be less biased when they are elicited as odds, such as 1 in 1000, than when they are elicited as decimals, such as 0.001 (Boose & Shaw, 1989, p. 71).*

Probability Distribution

Probability distribution is used here in a very broad sense to mean a set of possible values for the estimate and the associated likelihood or probability for each value's occurrence. The set should include the absolute maximum and minimum values possible. The set is ordered into a distribution of values. Functional forms of probability distributions are commonly used to represent the probability per unit interval of values. These are ***probability distribution functions*** [f(x)], and they are equations in terms of the estimate, called a ***random variable*** [x]. (See chapter 11 for a more detailed discussion of both these terms.) The normal probability distribution function includes the *Gaussian curve* or bell-shaped *curve*.

The questions used to elicit the probabilities associated with each value in the set can be asked in different ways: What is the probability that an event will occur? (direct value); given a probability, what is the value or lower of the variable in question? (cumulative probability); what are the chances in *n* trials of an event occurring on this log scale? (log-odds); and what are the chances in *n* trials of an event occurring? (odds ratio). For example, the estimate for the probability of a pipe rupture in a specified sequence of events would be given by the following example.

EXAMPLE 7.1: ***Probability estimate of pipe rupture***

Pipe Break Estimate	Probability of That Estimate or Less
0.001	0.01
0.005	0.05
0.010	0.10
0.05	0.20
0.10	0.50
0.15	0.70
0.20	0.90
0.25	0.95
0.27	0.99

■

A *probability distribution* has the same advantages, disadvantages, and summary of studies shown in *probability estimate*.

Continuous Scales

Continuous scales have continuous number lines either with linear or log spacing of values. The end points of the scale should represent extreme values and be labeled with text or numbers. Thus, the scale could be labeled with integers, textual categories, probabilities, odds, categories, ratings, or measurements of physical quantities, such as temperature. The experts can mark their answers at or between any of the delineations on the scale. We recommend that the labels on the scale be clearly defined, especially if they are not measurements of some physical quantity. Frequently, categories or ratings require additional clarification, such as given in the linear scale in example 7.2. Experts used this scale to compare the data gathered from an experiment to the results generated from a computer model that simulated the experiment. The experts used the scale to rate both the agreement between these two sources of results and the performance of the computer model in capturing the experimentally generated reactor phenomena (Meyer & Booker, 1987*b*).

Advantages: *Requires little instruction in how to use. Easily converted to numerical, continuous variables for analysis. Most people seem to be reliable estimators when using these scales.*

Disadvantages: *Developing a continuous linear scale to fit a particular application requires time. Care must be taken to guard against biased wording of either the labels or of the definitions of these labels. Training may be required if log scales are used.*

Summary of studies: *Seaver and Stillwell (1983) compared pairwise comparisons, rankings or ratings, and continuous linear scales. Based on their experience, Seaver and Stillwell evaluated the continuous linear scale (labeled by*

EXAMPLE 7.2: ***A continuous linear scale***

EXCELLENT AGREEMENT	NO UNACCEPTABLE CODE DEFICIENCIES
MODERATE AGREEMENT	ACCEPTABLE CODE DEFICIENCIES (i.e., can still apply the code with confidence)
MINIMAL AGREEMENT	MINIMALLY ACCEPTABLE
INSUFFICIENT AGREEMENT	SERIOUS CODE DEFICIENCIES (i.e., must fix code immediately)

■

probabilities) as being an empirically tested means of estimating probabilities and as requiring little preparation for analysis.

Pairwise Comparisons

Pairwise comparisons is a process in which experts rate a set of objects, events, or criteria by comparing only two at a time. Comparisons can be made in terms of importance, likelihood of occurrence, or possession of some characteristic (e.g., cost). To perform all possible pairwise comparisons in a set of n things requires $n(n-1)/2$ comparisons. Comparing element A to B is considered the reciprocal of comparing B to A. Thus, if A is considered more likely to occur than B, B is less likely to occur than A. Using Saaty's (1980) type of pairwise comparison, the expert might be asked, *Which is more important, A or B?* and then, *How much more important?* To answer the latter, the expert would use a scale of values, such as the one listed in Table 1 that Saaty designed for pairwise comparisons (see chapter 11, Table 1 for the complete scale). A response of "3" would indicate that the expert considered A to be slightly more important than B.

TABLE 1

Saaty pairwise comparison scale

Number	Description
1	The two items are of equal importance or likely
3	A slight favoring of the first item over the second
5	A strong favoring of the first item over the second
7	A demonstrated dominance of the first over the second
9	An absolute affirmation of the first over the second

(This scale was excerpted in part from Saaty, *The Analytic Hierarchy Process: Planning, Priority Setting, Resource Allocation*, New York: McGraw-Hill, 1980, p. 54.)

Advantages: *Most people are reliable estimators using pairwise comparisons, in part because they only have to consider two things at a time. Thus, they do not exceed the capabilities of their information processing, as described by Miller (1956). Some methods, such as Saaty's (1980) Analytic Hierarchical Process, offer means for verifying the mathematical consistency of the expert's estimates. After a brief introduction, experts find pairwise comparisons an easy method to use. In addition, pairwise comparisons can provide a numerically based analysis for qualitative data.*
Disadvantages: *Time consuming to elicit all possible combinations. Pairwise comparisons provide only relative data relations. A baseline scale or value is needed to translate relative comparisons into an absolute relation (Comer, Seaver, Stillwell & Gaddy, 1984).*
Summary of studies: *Current research shows that people make better relative, indirect judgments, such as with pairwise comparisons, than direct estimates. (See Stillwell, Seaver & Schwartz, 1982, for a review.) In a study by Seaver and Stillwell (1983, pp. 2-12), pairwise comparisons were evaluated by the authors, on the basis of their experience, as being acceptable to experts, as producing a high quality of judgment, and as having a strong theoretical base. Another study examined the usefulness of paired comparisons and continuous linear scales (labeled by probabilities and matching odds ratio) for obtaining estimates of human reliability in reactor risk assessments (Comer et al., 1984). The study's conclusion was that the analyses did not dictate the selection of one response mode over another, but practical considerations could lead to a preference. The paired comparison method used required more experts and more of the expert's time than the linear scale.*

Ranks or Ratings

Ranks or ratings involve assigning numbers or descriptions to the objects, events, or values in question. These concepts are listed together because the judgments they require are thought to be based on the same underlying psychological model (Comer *et al.*, 1984). Ranks can be integers in ascending or descending order or ordinal descriptions, such as good, neutral, and poor. For example, to select the questions that would be addressed later in the elicitation sessions, the experts could rank them in importance (e.g., a "1" for the most important to address and a "5" for the least important).

Ratings are usually numbers or choices from a given scale or set of choices, such as a scale from 1 to 10 or a multiple choice set. To foster consistency in the experts' interpretation of the ranks or ratings, try to use both numerical and qualitative descriptors. Qualitative descriptors by themselves can mean different things to the experts. For example, the qualitative term *small chance* has been found to mean from 1% to 40%, depending on the expert's interpretation (Keeney & von Winterfeldt, 1989). In example 7.3, weapons' planners gave one of the numbers to rate how potential weapons related to a U.S. defense need (Meyer *et al.*, 1982). The textual descriptors on the right were provided as part of the scale to prevent inconsistent use of the numerical values.

EXAMPLE 7.3: ***A rating scale***

3	Completely related, approximately 80% related
2	Significantly related
1	Slightly related, approximately less than 20% related
0	Not at all related

■

The possibility of having experts make different interpretations of ratings was the reason that Sherman Kent developed the rating scale (example 7.4) for government use.

Advantages: *Experts find ranks and ratings easy to use. Ranks and ratings are good for applications with qualitative information or with only a limited set of possible answers. Many analysis techniques are available for rank data (Conover, 1971).*

Disadvantages: *People have difficulty keeping more than 7 (± 2) things in their minds at once (Miller, 1956); therefore, comparing and ranking many items may prove difficult and yield inaccurate results. Comparing more than 7 things requires either the use of more experts or the use of more time with fewer experts (Seaver & Stillwell, 1983).*

EXAMPLE 7.4: ***Sherman Kent rating scale***

Order Of Likelihood	Synonyms	Chances In 10	Percent (%)
			99%
Nearly Certain	Virtually (almost) We are convinced Highly probable Highly likely	9	
		8	80%
Probable	Likely We believe We estimate Chances are good It is probable	7	
		6	60%
Even Chance	Chances are slightly better than Chances are about even Chances are slightly less than	5	
		4	40%
Improbable	Probably not Unlikely We believe not	3	
		2	20%
Nearly Impossible	Almost impossible Only a slight chance Highly doubtful	1	10%

■

Summary of studies: *Seaver and Stillwell (1983, pp. 2-12) compared pairwise comparisons, rankings or ratings, and continuous linear scales. Based on their experience, they evaluated ranks or ratings as being relatively easy to collect and acceptable to the experts. They also believed that the use of ranks or ratings carried sound theoretical justification.*

Bayesian Updating

Bayesian updating is a process of revising estimates by combining different sources of information. Bayesian updating can be done by combining measured data with expert judgment data, by combining data previously supplied by one expert with that of the same expert at a different time, or by combining expert judgment from different experts. The expert judgment can be elicited using any of the previously mentioned response modes. This technique is also a means of aggregation because it is used to combine data sources. (Chapter 11 provides some background on Bayesian methods, chapter 16 deals with aggregating estimates, and chapter 17 discusses handling uncertainties.) For example, suppose a component was being tested for failure and 0 failures were found in 10 tests. Imagine also that an expert provided an estimate of 1 failure in 100. Using a binomial process for the data and assuming a beta prior distribution for the expert's data, Bayes updating would combine the test data and the expert's estimate as

$$\frac{0+1}{10+100} = 0.009.$$

Advantages: *Provides a convenient way to combine various information sources and accounts for the conditional nature of the data.*
Disadvantages: *Requires assumptions about the distributions of each source of data. May also require additional estimates by the experts for the parameters of the assumed distributions.*

Selecting From Dispersion Measures

Experts give a dispersion measure when they are asked to provide some measurement of the amount of variation or uncertainty in the data (e.g., error bars on experimental measurements). In addition, the expert's *answer* can be the datum on which a dispersion measure is requested. For instance, the experts could be asked to provide the absolute maximum and minimum possible values on their estimates of a physical quantity or a probability. The dispersion measures covered below are ranges, percentiles, and variances or standard deviations.

Ranges

Range is the difference between two values that represent a likely interval within which the estimated value lies. Experts are usually asked to provide an absolute maximum and absolute minimum possible value. The experts may be asked to provide two values that represent their version of what is likely. Two examples include error bars and uncertainty ranges.

For example, imagine that an expert estimates the probability of an event as 0.001. The expert would then be asked to estimate a minimum and maximum on that event; he or she gives a minimum of 0.0001 and a maximum of 0.005.
Advantages: *Ranges are easily elicited. Ranges are an acceptable measure of dispersion for analytical purposes.*
Disadvantages: *Humans are not very good estimators of absolute maxima and minima. They tend to underestimate maxima and overestimate minima. Ranges may not be sufficiently defined to analyze. For example, if the expert is asked to give a range and supplies two numbers, the analyst will not know what interpretation to place on these values. Similarly, if the expert must mark specified ranges, such as ± 0.10, the analyst cannot assume that the expert's values represent ± 0.10 because people tend to generally underestimate such intervals.*

Volunteered Ranges

Volunteered ranges are the same as ranges except that the experts are not asked to provide any ranges of dispersions during the elicitation.

For example, in a project evaluating the performance of a computer model (Meyer & Booker, 1987*b*), the experts used a linear scale, marked their best estimate with a tick mark, and then voluntarily made two marks on either side.

Advantages: *Can be used to represent additional values of those elicited. Thus, the volunteered ranges can be used to fill out the sample of expert estimates. The values indicate the experts' uncertainty in their original estimates. Obtaining volunteered ranges requires no special efforts, except perhaps to use a response mode like continuous scales that encourages the experts to mark their ranges.*

Disadvantages: *Interpreting the meaning and therefore analyzing the range can prove a most difficult task.*

EXAMPLE 7.5: ***Volunteered ranges on a best estimate***

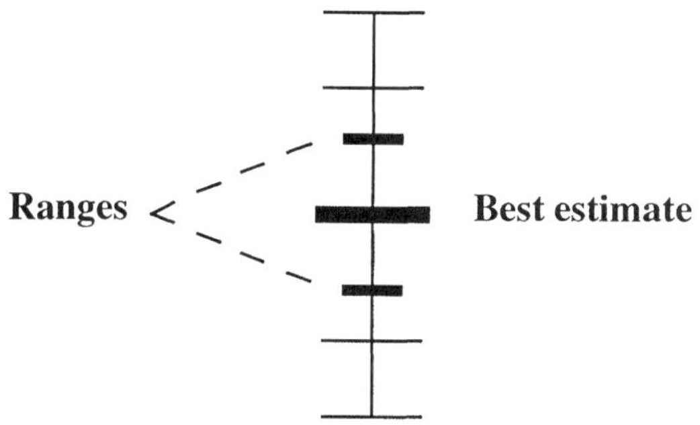

■

Percentiles

Percentiles are where the expert estimates are specified values of a distribution such that the distribution is cut into the predesignated pieces. The expert may be asked to provide an estimate such that 5% of the values are smaller than that estimate, or such that 95% of the values are larger than that estimate. Confidence interval estimation is the same as estimating two different percentiles.

For example, the expert estimates the failure probability for an event as 0.001. He or she is asked to give an estimate of the probability of failure such that there is a 5% chance or less or that the probability will be less than the estimate. The expert estimates this 5th percentile to be 0.0001. He or she then gives an estimate of the probability of failure such that there is a 95% chance or more that the probability will be greater than the estimate. The expert estimates this 95th percentile to be 0.01. This interval of (0.0001, 0.01) is the expert's estimate for a 90% coverage interval.

Advantages: *Convenient for many analyses and modeling techniques.*
Disadvantages: *Humans are not very good estimators of specified percentiles. Experts may require training to understand percentiles.*

Variances and Standard Deviations

Variances and ***standard deviations*** are statistics estimated by the expert. The variance measures the average squared deviations of all possible values of the estimates from the arithmetic mean. The standard deviation is the square root of the variance.

For example, the expert could give a best estimate for the probability of an event as 0.001. He or she might estimate that the variance of this value is 0.000001.

Advantages: *Convenient for many analyses and modeling techniques. Standard deviations are easier to estimate than variances because they are the same order of magnitude as the estimates themselves, whereas variances are squared quantities.*
Disadvantages: *Humans are not good estimators of variances or standard deviations. Experts must receive training in the concepts of variance or standard deviation to use this response mode.*

Selecting the Type of Aggregation

In many cases, the project will require a single answer from multiple and differing expert answers. Project personnel can use two basic methods to obtain a single response: (1) have the experts work until they reach a single consensus response, or (2) have the experts' answers mathematically aggregated into one answer.

Behavioral Aggregation

Behavioral aggregation relies on the experts reaching a consensus (Seaver, 1976). The aggregation occurs during, rather than after, the elicitation session. In interactive group situations, the group moderator informs the experts that they must reach a consensus; the experts usually use persuasion and compromise to do so. In the Delphi method, successive iterations are used to reach one datum. (Chapter 8 discusses how to prepare for behavioral aggregation.)
Advantages: *Produces an aggregated result during the session. Behavioral aggregation protects anonymity because no individual can be linked to the consensus response. Encourages the experts to support the product of their labors and view it as a group effort (e.g., perhaps from the thinking of "if we do not hang together, we will all surely hang separately").*
Disadvantages: *Needs advance planning because it should be used in conjunction with particular elicitation situations and measures for countering bias. In particular, it can foster a group-think situation in which no one truly thinks but unconsciously acquiesces. Behavioral aggregation can be very time consuming if group think does not facilitate unconscious agreement. Employing behavioral*

aggregation can suppress expressions of difference and thus the chances of discovering the "correct" answer. This means of aggregation obscures the differences between the experts' answers and the reasons for the differences, both of which can be critical to the understanding, analysis, and use of this data.

Mathematical Aggregation

Mathematical aggregation is the use of mathematical means to combine multiple experts' data into a single estimate or single distribution of estimates. Some mathematical methods weight the experts' data equally, such as the mean; others weight the experts' data differently in an attempt to give higher weights to the *more expert* or more valued data (These methods are not included in chapter 8 but in chapter 16 because they can be considered after the elicitation.)

Advantages: *Does not have to be planned as early or as closely in conjunction with the elicitation situations as the behavioral aggregation. (However, the choice of the response mode may limit which aggregation methods can be applied.) Also, different mathematical schemes can be applied in succession to the individual's data, whereas with the behavioral aggregation the process can usually only be done once.*

Disadvantages: *Like any type of aggregation, mathematical aggregation obscures the differences between the experts' answers and the reasons for the differences. It is easy to do mathematical aggregation incorrectly, such as by combining the estimates of experts who have made such different assumptions in answering the question that they have essentially solved different questions. In addition, mathematical aggregation can lead to the creation of a single answer that all of the experts would reject. For example, if the experts had given different distributions for describing a physical process, the aggregation of these distributions might describe a phenomenon that could not occur physically. (For further discussion of the difficulties of mathematical aggregation, consult chapter 16.)*

Summary of studies: *The problem of how to aggregate experts' estimates has received much attention recently (Morris, 1986; Clemen, 1986; Schervish, 1986; French, 1986; Winkler, 1986). Most schemes require that the experts' estimates be independent, even though experts' judgments are not known to be (Booker & Meyer, 1988a, p. 135). The few exceptions are those discussed by Winkler (1981) and Lindley and Singpurwalla (1984), but the latter two schemes assume that the structure of the correlation in the experts' estimates be somehow known or estimable (Booker & Meyer, 1988a, p. 136). Comparisons of different aggregation schemes indicated that the equal weighting of experts' estimates performs the best in covering the correct answer (Martz et al., 1985), when this answer is known.*

Selecting From Methods for Documentation

There are two major options for recording expert judgments: (1) recording the expert's answers (***answer-only documentation***), and (2) recording both the expert's answers and any information on the expert's problem-solving processes (***answer plus problem-solving documentation***), including the expert's background. The second option can be done to any level of detail. We describe three levels of documentation schemes for the second option: ***summary documentation***, ***detailed verbatim documentation***, and ***detailed structured documentation***.

Answer-Only Documentation

Answer-only documentation is a written record of only the answers or solutions.
Advantages: *Is the quickest and easiest method of documentation.*
Disadvantages: *The details of how answers were reached cannot be reconstructed later, if this becomes necessary. As a result, these details cannot be critically reviewed. At first this appears to be a blessing, but upon closer inspection not being able to review the expert's reasoning proves to be a disadvantage (i.e., what cannot be reviewed cannot be improved).*

Answer Plus Problem-Solving Documentation

Answer plus problem-solving documentation are written records of both the experts' answers and how they arrived at these answers. Such records can vary in their degree of detail. Three levels of detail are listed in *Advantages*.
Advantages: *Allows for the defense of the judgments or the processes of elicitation. Provides the data for revising or updating the judgments. Provides the data for conducting detailed analyses on factors (e.g., of the expert's problem solving or background) that may correlate to the answer.*
Disadvantages: *On the other hand, reviewers can scrutinize and criticize the documented data. We believe that it is better to document how the experts arrived at their answers and receive possible criticism of specifics than it is to not document this information and to be criticized for inscrutability. There seems to be a growing trend among reviewers to lambast studies for missing or confusing documentation (e.g., the early drafts of the NUREG-1150 reactor risk study).*

Summary Documentation. Summary documentation is when experts or project personnel provide a few sentences or paragraphs on their thinking processes (e.g., references used, major assumptions, and reasoning). This type of documentation is sometimes used to annotate the expert's answers or to compare them to each other or to some other baseline. One version of summary documentation gives a few sentences of explanation for each answer.

For instance, experts from Army offices rated the importance of particular factors to the export decisions. The importance of the factor *Triggers US Conventional Arms Transfer Restrictions* received the high rating of "3" because "these restrictions are set by Congress and are outside of the Department of Army's control" (Meyer & Johnson, 1985, p. 50).
Advantages: *Less labor intensive or time consuming than the other means of documenting problem solving.*
Disadvantages: *Generally does not provide enough detail for tracing an expert's thinking.*

Detailed Verbatim Documentation. Detailed verbatim documentation involves transcribing the expert's elicitation session, usually from an audio or video recording. This method is more common to knowledge acquisition applications in artificial intelligence.
Advantages: *Requires minimal advance planning of what should be recorded and how.*
Disadvantages: *Transcribing tape recordings is very time consuming and labor intensive. Frequently, verbatim documentation provides too much and too undifferentiated data to be useful in tracing the expert's thoughts. We have also noticed that project personnel often prefer to recontact the expert to clarify some question rather than search through tapes or transcribed records of that expert's session. Thus, even when the sessions are tape recorded and/or transcribed, this record may not be examined by the project personnel.*

Detailed Structured Documentation. Detailed structured documentation usually involves providing the person tasked with documentation with a format of what to record. The format lists those aspects considered the most important (e.g., answers and uncertainty levels, assumptions, and rules-of-thumb) in the level of detail desired.

For example, in the large reactor risk study NUREG-1150 (U.S. NRC, 1989), the experts and project personnel recorded the following:

1. The issue name (e.g., the Temperature Induced PWR Hot Leg Failure)
2. The sources of information (e.g., analysis of results of running computer codes RELAPs/SCDAP)
3. Subissues into which the issue was divided (e.g., likelihood of hot leg circulation cell and ballooning occurring)
4. Assumptions relating to the subissues (e.g., that hot leg nozzle and pipe and surge pipe are maintained as designed)
5. Answers (e.g., plot of probability of failure, given temperature).

Advantages: *Provides the most traceable and analyzable problem-solving data.*
Disadvantages: *Requires the most planning and coordination. Very labor and time intensive.*

A Common Difficulty—Its Signs and Solutions

Difficulty: The Literature on the Different Methods Is Either Scarce or Conflicting. When the project personnel select the components of their elicitation, they would like the literature to provide them with guidance. Specifically, they would like to be able to determine which method would be best for their particular application or situation. Unfortunately, there is little comparative literature on the methods, and the few sources that exist may provide conflicting advice.

Several factors create this state of the literature on elicitation components. For instance, it is very difficult to evaluate which methods provide the best results if the experts' answers to the question cannot be externally verified, such as through measurement. This is a problem with evaluating the individual interview, Delphi, and interactive group methods, as well as the response modes/dispersion measures and aggregation schemes. For this reason, elicitation components are more often compared on practical considerations such as how acceptable they are to the experts or how much time they take to administer (e.g., Seaver & Stillwell, 1983). The results of one comparative study sometimes conflict with another. One reason for this difference may be that authors define methods differently. For example, the term *Delphi* has been used to refer to a variety of elicitation situations—some where the experts are kept physically separate, their data made anonymous and redistributed to the experts, and others where the experts are together but restricted as to when they may interact. In addition, the authors may have different views about what is valuable in a method. For instance, response modes/dispersion measures can be compared in terms of how easy they are for the expert to use or how tractable they are computationally. While pairwise comparisons are easier for many experts to use than probabilities, they are not as easy to analyze. Depending on the importance that different authors attribute to these almost opposite considerations, different methods could be recommended for the same type of application.

Solution: To select the best elicitation components for a given situation, we recommend the following steps. First, review the relevant literature on methods. The references mentioned in the above text, particularly in the *Summary of studies* paragraphs found throughout the section *Selecting from Within Each Component,* should provide a starting point. In addition, there are a few new attempts in the field of knowledge acquisition to compare and evaluate different elicitation techniques (see Dhaliwal & Benbasat, 1989; Shaw & Woodward, 1989).

Second, list the possible elicitation components and their strengths and weaknesses for the situation under consideration. This list will serve as a decision aid in selecting the components of the elicitation for the application. The information on the method's strengths and weakness can be gleaned from the literature and also from the views of the project personnel. Try to use the definitions of the methods rather than the author's names for them as the basis of comparison. Points where the different sources of information truly differ (i.e., support the choice of different methods) should be noted and as much explanatory information should be provided as possible.

This list can be used later to justify the selection of methods, such as in the final report for the project. We suggest that the report give the theoretical and logistical reasons for selecting the particular method. In addition, if the standard methods were tailored in any way, the documentation should explain why.

8

Designing and Tailoring the Elicitation

In the previous chapter the reader was guided through the selection of the building blocks or components of elicitation for gathering expert data. This chapter will help the reader to tailor these components to a particular application. Five considerations are presented in the following sections to guide the tailoring of the elicitation: (1) *Logistics and Costs of Different Elicitation Situations and Modes of Communication*; (2) *Structuring the Elicitation Process*; (3) *Handling Bias—A Starting Point*; (4) *Documentation During and After the Elicitation Sessions*; and (5) *Presentation of the Question—A Quick Check.*

To arrive at the appropriate ***elicitation design*** requires detailed planning on paper because there are so many possible combinations, given the components of elicitation and the tailoring considerations mentioned above. In general, people often neglect this planning phase, only to regret it later. Indeed, the design is not done even after following all the guides in this chapter. Chapter 9, *Practicing the Elicitation and Training of Project Personnel,* describes how to find any remaining glitches by testing the elicitation design.

In addition, this chapter contains information on how to conduct the first and second steps of the program for handling bias: (1) *Anticipate Which Biases Are Likely To Occur in the Planned Elicitation*; and (2) *Redesign the Planned Elicitation To Make It Less Prone To the Anticipated Biases.*

Considerations in Designing the Elicitation

A combination of the considerations mentioned above and project-specific constraints (e.g., schedule) frequently drive the elicitation design. However, if this is not the case, the reader can selectively read those sections pertaining to his or her areas of concern.

Logistics and Costs of Different Elicitation Situations and Modes of Communication

Even though the basic elicitation situation and mode of communication for extracting the expert's data have been selected (chapter 7), these can be tailored by adding or substituting other methods. Elicitation situations and modes of communication are frequently combined to create the quickest and least expensive means of elicitation. In particular, if the interactive group and the face-to-face mode have been chosen as the basic elicitation situation and mode of communication, these could be combined with less expensive situations, such as the Delphi and the individual interview, and modes of communication, such as the mail and telephone. The expense of the interactive group situation lies in having the experts and project personnel meet together. It is usually cheaper for the interviewer to communicate by mail or telephone with the experts, as in the Delphi, or to meet with each expert separately, as in the individual interview. In addition, it is logistically difficult to schedule more than a few, usually busy, persons to meet for a week or for a series of shorter periods. Given the high expense and difficulty of gathering experts together, these meetings are often reserved for when they are absolutely necessary. Other situations, such as the Delphi and individual interview, and modes of communication are substituted for the group meetings.

Tailoring the Elicitation Situations

Frequently, other means of eliciting besides the main elicitation situation are used for the first, second, and fourth stages of the larger elicitation process. The stages are as follows:

Stage 1. Question area selection
Stage 2. Question refinement
Stage 3. Expert data elicitation
Stage 4. Data documentation

For example, in the NUREG-1150 reactor risk study (Ortiz *et al.*, 1988), individual interviews had been selected for the eliciting of the expert data (stage 3), but other means were used for the rest of the stages. A preliminary set of questions was selected by the project staff and then sent, with background information, to the experts (stage 1). The experts were allowed several months to review this information and to modify the questions. The experts then met to be briefed on the use of the response modes and on techniques for refining the questions through decomposition. The question refinement (stage 2) continued as follows: the experts separately prepared their preliminary disaggregations; they met a second time to present their preliminary results so that they could benefit from the exchange of information; and they met as an interactive group for a third time to refine the questions and to discuss their problem-solving approaches. Interviewers used individual interviews to elicit the experts' answers and aspects of their problem-solving processes (stage 3). The interviewers then contacted each expert separately after their elicitations to review the documentation of their data (stage 4).

The only exception to using different situations for the other stages occurs when the individual interview has been selected for eliciting in-depth, problem-solving data. Obtaining in-depth data in stage 3 must be preceded by working toward that level in the other stages. Group situations (the interactive group) or situations in which the expert is physically isolated from the interviewer (the Delphi) are not suited to obtaining detailed data. For this reason, if the individual interview is being used for stage 3, we would recommend that the other two situations not be used in the remaining stages.

Tailoring the Modes of Communication

Different modes of communication can also be employed to create the optimal means of elicitation. The three modes of communication that can be used are face to face, telephone, and mail. Although some modes are natural choices with a particular elicitation situation (e.g., face to face with interactive group and individual interview and mail with Delphi), these sometimes can be paired differently in the interest of logistics, time, and expense. For example, the mail mode is often used in combination with an interactive group situation to send the experts the questions before the experts are convened for the actual elicitation. Then too, if more than one elicitation situation is planned, such as for the different stages of elicitation, multiple modes of communication may be employed.

Each mode has its strengths and weaknesses. These are reviewed below to assist the reader in tailoring the different modes to his or her application. (For further discussion, see *Selecting from Modes of Communication* in chapter 7.)

Face to Face. The face-to-face mode works best for obtaining detailed data. Thus, if eliciting deep problem-solving data is the goal, the other modes cannot be substituted. However, the face-to-face mode is the most time consuming and expensive of the modes and thus should only be used if necessary. It may be that the telephone or mail mode can be used for elicitation stages 1 through 3 or for 4 even if information on the experts' problem solving will be gathered.

Telephone. The telephone mode could be substituted for the mail mode if only limited bits of information were being communicated. For example, in a Delphi situation, the experts could give their responses (answers, a sentence or two on their reasoning, and/or the names of the references that they used) over the phone. The coordinator could make this information anonymous and relay it by telephone to the other experts. If there were many experts and thus magnitudes of data had to be relayed, the above-mentioned information could be sent by mail. After the experts had received the mailed information, they could be interviewed by telephone for their revised responses.

The telephone mode fails to work well in situations that involve relaying detailed or long pieces of information. Thus, this mode would not be suited to probing for the assumptions that the expert made in arriving at an answer. Nor would it be suited to eliciting expert judgment when these judgments were requested in a complex response mode that required the experts to be trained.

Telephone communication is generally less expensive than either the face-to-face or mail mode. The telephone communication also has a shorter turnaround time than the traditional postal method, which makes it a good substitute for it.

Mail. The traditional mail survey works well for sending and receiving simple data from a large sample of experts. Mail surveys cost less than face-to-face interviews or meetings of the experts. Thus, the mail mode may be used in stage 1 to have the experts select the question areas or in stage 4 to record their answers and return them.

However, the mail mode does not work well for the transmission of complex instructions or detailed problem-solving data. For this reason, the mail mode should not be used with complicated response modes that require the experts' training. Similarly, individual interviews for eliciting in-depth problem-solving data should not be conducted by mail or telephone. In particular, the verbal report, or thinking aloud method, should never be used in mail communications with an expert.

The mail survey has the greatest problem of all modes of communication in its response rate, probably because it is easier to ignore a mailed request than to refuse one made in person or over the phone. In addition, most people find it more difficult to record their thoughts and mail them than to just state them; the added effort of writing the response leads many to abandon the attempt. For this reason, a mail survey is not recommended if it is critical to receive a response from a high proportion (i.e., over half) of the targeted population. The response rate can be boosted by calling the sample to request their response, but these attempts usually generate only half again as many as that of the earlier rate.

Reasons for Gathering the Experts Together

Another consideration involved in tailoring the elicitation is whether the experts should meet, as in an interactive group situation. In the final decision, the reasons for convening the experts will need to be balanced against the possible costs. These reasons and costs are listed below to help the reader make this decision. For additional information on the relative costs and speed of the elicitation situations, see Armstrong (1985, pp. 108-132).

1. **If complex response modes and dispersion measures (e.g., probability distributions or percentiles) have been selected for use with more than a few experts.** The experts will require training in these complex modes and measures; in cases in which more than a few experts will participate, training should take place in a group setting. In addition, a less complex response mode, Saaty's pairwise comparison, should also be introduced in an interactive setting. Experts sometimes experience initial confusion in using this mode (e.g., *Does a three indicate that A is slightly more important than B, or the reverse?*). For this reason, they need the extra clarification that a face-to-face situation offers.
2. **If the synergistic effect of group discussion is necessary to the elicitation process.** For example, if the experts in this field have not previously or recently met, if the field continues to evolve, or if the questions require knowledge of the field's state of the art, then meeting together as a group

would be advisable. In addition, if it is critical that the group of experts identify with the outcome of their problem solving (i.e., see it positively as the product of their labors), meeting as a group is recommended. According to Seaver (1976), members of interactive groups express higher satisfaction with their product than members of Delphi situations.

3. **If the experts will be unable give their uninterrupted attention to the project because of other demands on their time.** Having the experts meet in a place of the interviewer's designation allows the interviewer to place controls on others' demands of the experts' time. For instance, the interviewer can place controls on when and how the experts' telephone messages are delivered to the meeting room. Occasionally, the experts themselves have requested that the meetings be held away from where they could be easily reached by their offices so that they could focus on the project (Meyer & Johnson, 1985). We believe that gathering the experts together works especially well if their management has not given the project top priority.

Expenses for Gathering the Experts Together

Possible expenses for gathering the experts together follow:

1. **Payment for the experts' travel and lodging,** if they do not reside in the same geographical location.
2. **Payment for the meeting room and any refreshments** (e.g., coffee to help keep the experts awake).
3. **Payment for having the sessions videotaped.** This record not only helps document the data but allows experts who miss a session the opportunity to view the tape later. For example, in the NUREG-1150 reactor risk study, the experts gave presentations to the group on issues related to the technical question. These presentations and their subsequent discussions were taped. These tapes can then be sent to the experts who were unable to make the presentation meetings so that they will receive the same information as the other experts.
4. **Payment for miscellaneous administrative costs,** such as for typing, copying, and mailing any background material that the experts need to see before forming a group. Copying may take place throughout the meetings. For example, after the experts have met, often they have materials that they wish to share with the other experts. In addition, toward the end of the process, the copying machine can be used to provide copies of the project personnels' documentation of the experts' data for the experts' review.
5. **Payment for the expert's time**, if this was part of the project budget. (See chapter 6, *Motivating the Experts Through Pay,* for a discussion of when payment may act as a detriment to expert participation.)

Structuring the Elicitation Process

Structuring means imposing controls on the elicitation process. Structuring with respect to presenting the expert with a clear and assimilable statement of the question was discussed earlier in chapter 5. When the concept of structuring is applied to the larger elicitation, however, it can include using a predesigned set of questions to guide the elicitation, allowing only particular kinds of communication between the experts and requiring that the experts answer using one of the response modes.

Why Structuring Is Done

The reasons for structuring the elicitation include helping the data gatherers with the interviewing process, making the elicitation easier for the expert, or limiting the introduction of bias. For example, the structuring may be imposed by the use of a specially designed interview instrument—one that prompts the interviewer to ask particular questions at specific times (e.g., as described in Meyer, 1987). The structuring may be put on the experts' group interactions to prevent some experts from acquiescing, without thinking, to other experts, as occurs in group think. For example, the Delphi was designed to counter biases arising from group interactions and thus is structured in this manner.

In general, structuring the elicitation limits the intrusion of extraneous factors, such as bias. It seems to keep the field of observation clearer and thus eases the task of gathering and analyzing the expert data.

Degrees of Structuring

Structuring can be done to varying degrees to different aspects of the elicitation process. As a general rule, using one of the following components imposes structure: a response mode and dispersion measure, an elicitation technique, the use of behavioral aggregation, or a documentation scheme. Using one of these components means that there is a plan in place and a specific procedure that will be followed. For example, using the pairwise comparison response mode means that experts' judgments will be elicited by the interviewer or moderator asking particular questions and by the experts responding with answers on the appropriate scale. The response mode or dispersion measure, elicitation technique, and behavioral aggregation are typically used when the expert data are being extracted (stage 3). The documentation component is often used in stage 4 to record the data or the data-gathering methods.

Examples of structuring applied to each elicitation stage follow. The more that these options are applied, the more highly structured the elicitation situation becomes.

Structuring Options Applied to the Stages of Elicitation

Stage 1: Question selection

For most elicitation situations, any one of the three structuring options could be applied to the selection of the questions.

- Advisory experts could help decide which questions to ask of the external experts. This option is frequently used when the project personnel are unfamiliar with the field or if the project sponsor has not requested specific questions.
- The project personnel or the external experts can rank and select the questions according to some criteria. For example, on the reactor risk study NUREG-

1150 (U.S. NRC, 1989), project personnel initially selected the questions based on those that held the greatest potential to produce uncertainty in risk. The experts then reviewed the proposed set of questions and added, deleted, or modified these using their own criteria.

- Alternatively, the external experts could be polled to learn, in advance, which problems they would consider themselves qualified to address or which they would like to work on.

Stage 2: Question refinement

- In interactive group situations, the experts and project personnel can work together in disaggregating the question. The experts or project personnel can present the other experts and the group moderator with their interpretation of the question, means of disaggregating it, or any other relevant information.
- In any face-to-face situation, the experts can be required to review the definitions, the assumptions, or any other information that defines the question; the experts can then refine the question one last time.
- In individual interviews, the external or advisory experts could work on refining the questions that will be asked of them or other experts. For example, the expert can assist the interviewer in determining how to word the question and how to define particular variables.
- In a Delphi situation or a mail survey, the external experts can be sent information on the question and be asked to provide information on how to prepare it. For example, in a seismic study (Bernreuter *et al.*, 1985), the experts were asked to divide the eastern United States into various earthquake zones in preparation for defining the question. In some cases, qualified project personnel, such as advisory experts, can do most of the work in refining the question and then have the external experts review and modify it. For example, in the NUREG-1150 reactor risk study, the project personnel created sample disaggregations of the question, while the experts had the option of using these disaggregations as the starting point for their own.

Stage 3: Expert data elicitation

- During interactive group elicitations, the external experts can record their own data on a documentation format. The experts can also be asked to verbalize their judgments or thinking to the group or to the group moderator. For instance, the experts could state their names and their answers in the desired response mode. Only some of the experts need give data on particular questions—namely, those who were earlier judged (by themselves or the project staff) as the most qualified. On the export control project, all the experts voiced their opinions in the early discussions, but only those experts who were assigned to particular questions were allowed to *vote* on the answers (Meyer & Johnson, 1985). If a group of experts are verbally giving their responses, the group moderator can ask the group to remain silent while each expert in turn gives his or her response. Natural and official leaders in the group can give their responses last or privately to avoid group think. The verbal probe or ethnographic technique can be used briefly to question the experts on aspects of their problem-solving procedures.

- In the Delphi situation, the experts can receive questions and send their data (e.g., their answers, a few lines of explanation for each answer, and a footnote of the references used). This data can be made anonymous and redistributed to the experts to allow them to revise their earlier answers. This process can be repeated as long as necessary, until consensus (behavioral aggregation) is achieved. If the experts must provide their answers in a particular response mode, they can be sent a format, such as a copy of a continuous linear scale, and instructions on how to use it.
- In individual interviews, the interviewer prepares a list of topics to guide the questioning process. Alternatively, the interviewer could be guided by the format on which the data are to be recorded. Additionally, the interviewer can simply use one of the preexisting techniques for eliciting problem-solving data (e.g., verbal report, verbal probe, or ethnographic technique) to guide his or her gathering of the data.

Stage 4: Data documentation

- In interactive group situations, the experts can use one of the documentation schemes to guide them in recording their answers or problem-solving data. Similarly, the group moderators can use one of these to help them gather and record all the required data.
- In a Delphi, the documentation format (along with instructions) can be sent to the experts.
- In individual interviews, the expert or data gatherer can use the documentation scheme to record the desired information. A computer program that prompts the experts for their inputs and/or reasoning would serve the same purpose as a written documentation scheme (Meyer, 1987).

General Rules in Structuring the Elicitation

As a general rule, the more structuring that is imposed on the elicitation, the greater the time needed to plan and conduct it. However, these greater demands are balanced by the greater effectiveness of the structured techniques. For example, unstructured individual interviews are described as being less effective than their structured counterparts, at least in knowledge acquisition (Hoffman, 1987; McGraw & Harbison-Briggs, 1989, p. 73).

A higher degree of structuring also seems to parallel the gathering of more detailed data. The fact that a structured approach allows the data gatherers to focus the questioning to a finer level makes sense. An unstructured interview allows the expert to jump from one topic or level to another, much to the interviewer's dismay.

We would also emphasize that conducting a structured elicitation often demands more tact than administering an unstructured one. This need arises because the experts do not always follow the structuring. They may be confused or ignorant about what they should do, or they may simply decide that following instructions takes too much effort. The interviewer should exercise tact in getting the experts to follow the plan because the interviewer depends on the experts' good will for obtaining data. For example, imagine that the interactive group method has been structured to minimize the occurrence of group think bias. The experts have been asked to present their views to the group before entering into a discussion of all the presentations Additionally, the natural leader of this group has been requested to go last in stating his or her views. However, the plan goes awry. The leader interrupts and criticizes other experts' presentations. To make this session run again according to the plan, the interviewer must employ diplomacy. One way of handling this problem would be to take the leader aside and tell him or her about the group think bias and how it negatively affects people's thinking. In addition, the experts could be briefed again, as a group, on the elicitation procedure and the reasons for structuring it in this manner.

Note that the greater demands of time and effort for structuring the elicitation may not always be justified by the finer precision or detail in the expert data that they provide. We suggest that the goals and needs of the project be considered in deciding whether or how much structuring to impose on the elicitation. (For a discussion of some of the specific tradeoffs involved in structuring the elicitation of probabilities, see Welbank (1983, pp. 28-29).

Handling Bias—A Starting Point

Designing elicitation with regards to bias, as we advocate, is a new approach. The information presented in this section was largely developed from our own experiences. For this reason, it is intended to be a starting point for those who wish, as we did, to become more aware of bias and how to handle it. There is a small but growing body of literature on human biases that can be applied to handling bias in expert judgment—Payne, 1951; Hogarth, 1980; Tversky & Kahneman, 1974, 1981; Kahneman & Tversky, 1982; Cleaves, 1986; Meyer *et al.*, 1989; and Meyer & Booker, 1989.

There are two reasons for paying attention to bias: First, the occurrence of some biases can degrade the quality of the results (Hogarth, 1980; Kahneman & Tversky, 1982). Second, concerns about its presence, such as among reviewers of a project, affect the project's credibility.

Chapter 3 defines two types of bias—motivational and cognitive.

Motivational bias occurs when the elicitation process alters the experts' reports of their thoughts or answers. (For an explanation of why this altering occurs, see chapter 3, *Motivational Bias* under *Sources of Bias.*) For example, bias occurs if the experts give answers that are different from what they believe because of the interviewer or knowledge engineer's comments. Using the motivational definition of bias, if the expert's estimates or problem-solving data do not represent the expert's thinking, these are not quality data. In addition, reviewers or sponsors are liable to question the data's validity if they suspect the interviewer has *led* the experts. Also, many people understand that bias can arise from a conflict of interest. These biases need not actually occur to impair the credibility of the work; they need only to be suspected of having occurred.

Cognitive bias occurs when the expert's estimates fail to follow normative, statistical, or logical rules. Cognitive bias is frequently determined by checking the experts' data against the mathematical and statistical standards that apply to the chosen response mode. To illustrate, if an expert tends to remember failures with disproportional frequency to successes in estimating the functioning of some technology, these estimates would be considered biased, according to the cognitive definition. While the majority of people are not as aware of cognitive bias as they are of motivational bias, this situation is not likely to continue. Authors like Hogarth, Kahneman, and Tversky have led the field in revealing the cognitive limitations to which the human mind is prone. The old view that the brain acts like a computer (i.e., in being mathematically correct) is rapidly being debunked.

The approach proposed in the section below is to anticipate which biases are likely to occur given the planned elicitation. This step, *Anticipate the biases,* is followed by others in this and later chapters:

- *Redesign the Planned Elicitation To Make It Less Prone To the Anticipated Biases—Step 2* (chapter 8).
- *Make the Experts Aware of the Potential Intrusion of Particular Biases and Familiarize Them with the Elicitation Procedures—Step 3* (chapter 10).
- *Monitor the Elicitation for the Occurrence of Bias—Step 4* (chapter 10).
- *Adjust, in Real Time, to Counter the Occurrence of Bias—Step 5* (chapter 10).
- How to analyze the data for the occurrence of particular biases (Step 6) is discussed generally in chapter 14.

In chapter 3, *Determining Which Steps to Apply,* we describe how the reader can determine which of the above-mentioned steps to use. A description of how to anticipate bias follows.

Anticipate Which Biases Are Likely to Occur in the Planned Elicitation—Step 1. Applying this step presupposes the selection of a motivational or cognitive view of bias, as described under the subheading *Determining Which Steps to Apply* found in chapter 3. While we consider the cognitive and motivational views to be equally valid definitions of bias, we believe that one way of construing bias may be more useful than another for a particular project. We suggest that the reader select and use only one view of bias at a time to avoid creating contradictions.

To anticipate some of the biases that can occur in a particular elicitation method, determine which components of elicitation will be used and find them in the table *Index of Selected Biases* (Table 1). (The elicitation techniques are not listed in the index because some, like general interviewing, can be performed in such varying ways that they would be subject to different biases.) The index lists nine frequently encountered biases from motivational and cognitive sources. To obtain more information on the motivational and cognitive biases, locate the item following the index in *Definitions of Selected Biases* (Table 2).

Another option is to skim Table 2, *Definitions of Selected Biases,* to learn if the planned elicitation is susceptible to one of the selected biases. The elicitation situations and components prone to these biases have been underlined in this segment for easy access. This segment also provides information on *why* the elicitation is prone to a particular bias.

By consulting the *Index* and *Definitions,* the project personnel will become aware of the biases' existence and of their own tendencies to introduce or acerbate them. We hope that their awareness will alleviate their tendencies toward bias.

TABLE 1
Index of selected biases

Elicitation Component	View of Bias	Source
Elicitation Situations		
Individual Interview	Motivational	Social pressure, data gatherer
	Motivational	Wishful thinking
	Cognitive	Inconsistency
Delphi	Motivational	Wishful thinking
	Cognitive	Inconsistency
	Cognitive	Anchoring
Interactive Group	Motivational	Social pressure, group think
	Motivational	Wishful thinking
	Cognitive	Inconsistency
Modes of Communication		
Face to Face	Motivational	Social pressure, data gatherer
	Motivational	Wishful thinking
	Cognitive	Underestimation of uncertainty
Telephone	Motivational	Social pressure, data gatherer
	Motivational	Wishful thinking
	Cognitive	Availability
	Cognitive	Anchoring
	Cognitive	Underestimation of uncertainty
Mail	Motivational	Social pressure, impression management
	Motivational	Wishful thinking
	Motivational	Misinterpretation by analyst
	Cognitive	Inconsistency
	Cognitive	Availability
	Cognitive	Anchoring
	Cognitive	Underestimation of uncertainty
Response Modes		
Complex ones such as probabilities, Bayesian updating, and uncertainty measures	Motivational	Misinterpretation by expert
	Cognitive	Inconsistency
	Cognitive	Underestimation of uncertainty
Aggregation		
Behavioral Aggregation	Motivational	Social pressure, group think

TABLE 2

Definitions of selected biases

Motivational Bias. We define motivational bias as the altering of the experts' responses through social pressure (from the data gatherer, group think, or impression management) and misinterpretation. In addition, the experts' motivation (e.g., wishful thinking) can influence their responses. Wishful thinking bias differs from the other selected motivational biases in that the experts' expressions are not necessarily changed by the experts themselves or by the knowledge engineer. However, we have included wishful thinking in the motivational biases because it originates from the expert's motivation.

Social pressure. Experts may alter the descriptions of their thoughts because of social pressure, the desire to gain acceptance and to be seen in the most positive light possible. This altering can take place consciously or unconsciously. The social pressure can come from those physically present, such as the interviewer, knowledge engineer, or the other experts, or it can come from the expert's internal evaluation of others' reactions.

Social pressure from the data gatherer. This form of social pressure is most likely to occur in those methods in which the interviewers or knowledge engineers meet face to face with the experts, such as in the individual interview and the interactive group situations. In face-to-face situations, data gatherers can intentionally or unintentionally influence the expert through their body language, facial expression, intonation, and word choice. We expect this source of bias to be weaker in telephone conversations and weaker still in communications by mail. These last two communication modes do not allow some of the above-mentioned means of expression that the face-to-face mode does. In addition, social pressure bias is more pronounced when the data gatherer asks leading questions. Thus, the bias' influence is weaker when the data gatherer uses one of these elicitation techniques—the verbal report, the verbal probe, or ethnographic phrasing of questions (Meyer, Mniszewski & Peaslee, 1989). The concurrent verbal report requires the experts to think aloud as they solve a problem—this process almost eliminates the chances of leading the experts because it does not involve questioning them. The verbal probe uses general, nonleading phrases and the ethnographic technique uses the expert's own words to formulate the questions. We would also expect other knowledge acquisition techniques, such as KADS (Wielinga, Akkermans, Schrieber & Balder, 1989) and SORTAL (Regoczei & Hirst, 1989), which are driven by the expert's terms and meaning, to be less prone to social pressure from the knowledge engineer.

Group think. Group think is another form of social pressure. Social pressure from others in the group induces individuals to slant their responses or to silently acquiesce to what they believe will be acceptable to the group (Meyer, 1986, p. 89). Zimbardo, a psychologist, explains that it is due to the basic needs of people to be loved, respected, and recognized that they can choose or be induced to behave in a manner which will bring them affirmation (1983). There is abundant sociological evidence of conformity within groups (Weissenberg, 1971). Generally, individuals in groups conform to a greater degree if they strongly desire to remain a member, if they feel satisfied with the group, if the group is cohesive, and if they are not a natural leader in the group. Furthermore, individuals are generally unaware that they modify their judgments to conform with the group's views. (Note that we make a distinction between unthinking conformity and experts changing their minds as a result of new information received. The former is group think; the latter an illustration of knowledge in action.)

The theory of cognitive dissonance helps explain this unconscious modification of opinion. ***Cognitive dissonance*** occurs when individuals find a discrepancy between thoughts that they hold or between their beliefs and actions (Festinger, 1957). For example, cognitive dissonance results when an individual holds an opinion that conflicts with that of the other group members and he or she has a high opinion of the others' intelligence. Individuals often resolve the discrepancy by unconsciously changing their judgment or the reporting of that judgment to be in agreement with that of the group (Baron & Byrne, 1981). For example, Janis' study (1972) of fiascoes in American foreign policy illustrates how presidential advisers often silently acquiesce rather than critically examine what they believe to be the group judgment. This phenomenon has also been called *group think bias* and the *follow-the-leader* or *bandwagon* effect.

Group think is only likely to be a concern in an interactive group situation. Group think will also tend to occur in situations in which behavioral aggregation is used because this type of aggregation requires the encouragement of pressures toward conformity.

Impression management. Impression management, another type of social pressure, occurs when the expert imagines the reactions of those not physically present. This effect can occur in any elicitation situation. However, impression management is more noticeable when it is not covered by other effects, such as social pressure caused by the knowledge engineer. For this reason, its occurrence is most noted in mail surveys or Delphi situations. Individuals may try to answer in a manner that brings the most approbation, such as from the person who wrote the questions. Similarly, they may try to respond in such a way as would be acceptable to their employers or to society in the abstract. For this reason, this type of social pressure has been termed impression management (Goffman, 1959). Payne (1951) found evidence of individuals giving the responses that they perceived to be the most socially acceptable rather than those which accurately portrayed their thoughts or actions. For example, people answering surveys claim that their education, salary, and job title are better than they are. Often there is a 10% difference between what is claimed for reasons of prestige and what objectively *is* (Meyer, 1986, p. 90).

Misinterpretation. Misinterpretation occurs when the methods of elicitation and documentation alter the expert's responses. This effect, while prevalent, has not received much attention. Misinterpretation frequently occurs as a result of the response mode. If the experts cannot adequately translate their judgments into the response mode, misinterpretation will result. We have noticed that experts seem to have more difficulty with the following response modes: probability distributions, ranks, and percentiles.

Misinterpretation is also more likely with elicitation and documentation methods written from the knowledge engineer's, rather than the expert's, viewpoint. For example, we have all experienced the frustration of trying to force our views into the limited response options of a mail survey.

Wishful thinking. Wishful thinking occurs when individuals' hopes influence their judgment (Hogarth, 1980). What the subjects think should happen will influence what they think will happen. To illustrate, presidential election surveys show that people predict the winner to be the candidate that they expect to vote for (Armstrong, 1985, p. 79). The above instance is one in which the subjects stand to gain very little personally from their answer. The wishful thinking effect is stronger where the subjects are personally involved or would gain from their answers. Hence, this bias is also called *conflict of interest.* In general, people exhibit wishful thinking about what they can accomplish in a given amount of time—they overestimate their productivity (Hayes-Roth, 1980).

Wishful thinking is not particular to any elicitation method. Instead, its occurrence relates to the selection of experts and their assignment to specific questions or problems. If experts have a special interest in an answer, wishful thinking will likely occur regardless of the elicitation method (individual interview, interactive group, or Delphi) or the mode of communication (face to face, telephone, or mail). For this reason, we suggest selecting those experts who have the least to gain from their response. However, often those most qualified will also be those with the most at stake. For this reason, Boose and Shaw (1989, p. 73) advise obtaining data from experts holding divergent views. One approach to obtaining diverse expertise is to select experts that represent different organizations (e.g., government, academia, and industry) and the various theoretical stances on the subject.

In general, the effects of wishful thinking will be most pronounced when the experts do not have to explain their reasoning. The experts' highly optimistic responses are checked by having them disaggregate the problem and explain their problem solving. For example, Hayes-Roth (1980) found that having people disaggregate the tasks that they had earlier thought they could accomplish in a given time led to more accurate estimates.

Cognitive bias. This book defines cognitive bias as occurring when data fail to follow mathematical and logical standards because of inconsistency, anchoring, availability, or underestimation of uncertainty.

Inconsistency. Inconsistency bias occurs when people are inconsistent in their solving of problems, especially through time. Of all of the biases mentioned in this section, this is the most common. As the experts' thinking evolves, their current thoughts or solutions may contradict those that they expressed earlier. In addition, individuals often unintentionally change definitions, assumptions, or algorithms that they meant to hold constant throughout the problem-solving process. Inconsistency in individuals' judgments can stem from their remembering or forgetting information during the elicitation session. For example, individuals may remember some of the less spectacular pieces of information and consider these in making judgments later in the session, or they may forget that particular ratings were only to be given in extreme cases and begin to assign them more freely toward the end of the session.

As Dawes, Faust, and Meehl (1989, p. 1671) have noted, such factors as fatigue, recent experience, or seemingly minor changes in the ordering of the information or in the conceptualization of the ". . . task can produce random fluctuations in judgment. Random fluctuation decreases judgmental reliability and hence accuracy." These inconsistencies may result in answers that do not make logical or Bayesian sense. For instance, a series of answers proposing that factor A was more critical than B, B more than C, and C more than A would not make logical sense. Similarly, if an expert gave the same probability of A for two situations, one of which involved an influential factor C and one which did not, his or her answers would not be coherent from a Bayesian viewpoint.

Several conditions acerbate the natural tendency toward inconsistency, including memory problems, fatigue, and confusion. During elicitation sessions of more than 30 minutes, people often forget the instructions, definitions, or assumptions that they initially received. For example, the experts may forget that a rating of nine meant a near certainty and therefore assign it more easily than the definition specified. Thus, unstructured elicitations, which do not have periodic reviews of the problem information, are more likely to have a high inconsistency. Inconsistency can occur between experts' answers (e.g., the experts meant different things by the same numerical answer) or within an expert's answer (e.g., sometimes the expert gave a specific rating more easily than at other times). Also, inconsistencies tend to occur in situations in which the expert's understanding through time cannot be easily monitored. These situations include the Delphi or mail survey.

Confusion can also lead to inconsistency. Thus, any of the more complicated response modes, such as probability distributions and percentiles, are more prone to this problem. To prevent such confusion, experts should receive training in the use of these modes (step 3). In addition, if the experts must mentally juggle more than five to seven things, such as in rating them, they are likely to become confused and inconsistent. The Saaty pairwise comparison mode (see Saaty, 1980) is used to prevent such confusion even though it is more time consuming than some of the other response modes.

Anchoring. Anchoring occurs when an individual fails to adjust sufficiently from his or her first impression in solving a problem. We would rate it next to inconsistency in terms of frequency of occurrence. Sometimes this tendency is explained as the failure to adjust a judgment in light of new information as much as it would be in terms of Bayes' mathematical formula (Meyer, 1986, p. 88). Spetzler and Stael von Holstein (1975) and Armstrong (1985) describe how people tend to anchor to their initial response, using it as the basis for later responses. Ascher (1978) has found that this problem exists in forecasting where panel members tend to anchor to past or present trends in their projection of future trends. Ascher determined that one of the major sources of inaccuracy in forecasting future possibilities, such as markets for utilities, was the extrapolation from old patterns that no longer represented the emerging or future patterns. Another example of anchoring occurs when group members' last estimates are closer to their initial impressions than these would have been had they fully taken earlier group discussions into account.

Anchoring is most prevalent in situations where the expert is not likely to experience the opposite bias of being influenced by the knowledge engineer or the group, such as in the Delphi method. In addition, those modes, such as mail or telephone communications, in which the expert's thoughts cannot be easily monitored by having the expert think aloud, are prone to this bias. Finally, we have noticed that experts tend to remain loyal to their anchor if they have described it orally or in writing and fear losing face for changing their position.

Availability. Availability bias arises from the differing ease with which events can be retrieved from long-term memory. Data involving catastrophic, familiar, concrete, or recent events tend to be easier to recall. Contributing to this bias is the human tendency to think that "we know more than we do and also that what we don't know must be unimportant" (Boose & Shaw, 1989, p. 70). Availability bias affects people's ability to accurately estimate frequencies and recall other aspects of the event. For example, the incidence of severe accidents in reactors tends to be overestimated in part because of their catastrophic and newsworthy nature.

Availability bias is more common when the expert does not receive any information from others and, thus, does not have a chance of triggering other, less accessible memory associations. As a result, the individual interview is the most prone to availability bias, whereas the interactive group is the least prone to the bias. With individual interviews, a series of different scenarios is often used to help the experts enlarge on the sample of thoughts contributing to their final answers.

Availability bias is also more common with telephone and mail modes of communication because the expert is usually not given much background before being asked for the answer. A structured hierarchical presentation of the information, such as from the general to the specific, can alleviate this weakness.

Underestimation of uncertainty. Underestimation occurs when people underestimate the amount of uncertainty in the answers that they provide. For example, when people are asked to put a range around an answer such that they are 90% sure that the range encompasses the correct answer, their ranges only cover about 60% of the probable range of values (Capen, 1975). Even when they are given quizzes and feedback on their performance, they cannot break the barrier of covering only 70% (Capen, 1975, p. 846). A popular explanation for this effect is that we are uncomfortable with the amount of uncertainty in our lives and, thus, minimize it. In particular, humans may avoid confronting the large uncertainties in their judgments. In addition, in planning activities, they may underestimate uncertainty because of the chain-like nature of planning. They may ignore factors which could cause a delay because each factor seems unlikely by itself (Boose & Shaw, 1989, p. 72).

Although underestimation of uncertainty is very widespread, Martz, Bryson, and Waller (1985, p. 72) have noted that it is more pronounced with probability and chance estimates than with some of the other response modes. Chance estimates, also called *odds*, are given as 1 chance in a total (such as 1 in 1000).

Redesign the Planned Elicitation to Make It Less Prone to the Anticipated Biases—Step 2. The definition section (Table 2) can be used to redesign the elicitation, if the project personnel as a result of anticipating bias wish to do so. For instance, suppose that the data gatherers became concerned with the potential for wishful thinking bias as a result of reading the *Index* and *Definitions* in step 1. They could consult the definition of wishful thinking and replan their strategy for sampling the experts. Data gatherers who consult the definition section would receive two suggestions: either select experts who have the least to gain from their responses or sample experts from different organizations and schools of thought.

In addition, some suggestions for countering selected biases (step 5 in chapter 10) could be incorporated into redesigning the elicitation. For example, one of the ways to counter wishful thinking is to ask the experts to explain their thinking in detail. The project personnel could incorporate this requirement into their elicitation design (e.g., plan to have the problem disaggregated into its component parts and to have the experts explain their reasoning on each part).

Documentation During and After the Elicitation Sessions

How to record the expert data is just one of the considerations that guides the tailoring of the elicitation. The other considerations mentioned so far have been the logistics and cost of different elicitation situations and modes of communication, the structuring of the elicitation, and the treatment of bias. Before we can present the options for who performs the documentation and when, we need to describe what can be documented.

What Documentation Can Include

1. **The statement of the question in its final version.** This statement would include (1) any background information that clarifies the question, such as scenarios, and (2) definitions or assumptions that the experts are to use.

   Stating the question often proves difficult because the statement of the question constantly changes until the actual beginning of the elicitation, or in some cases until the expert responds. For example, in the NUREG-1150 reactor risk study, the experts continuously developed their own disaggregations of the questions (essentially their statements of the question) until the moment that they gave their judgments.
2. **The identity of the experts.** The experts' names may need to be recorded to answer questions about the experts' responses or to update their responses after the elicitation. For example, if the experts are to review the project personnel's documentation of their responses, records must be kept of each expert's response, name, and date of elicitation. In addition, records of the

experts' identities and responses will be needed for report writing. However, even when all of this information is recorded, it need not be divulged in a report. There are three ways of presenting the experts' identities depending on the level of confidentiality that has been selected. (See chapter 6, *Will the Judgments Be Anonymous,* under *Motivating the Experts Through Communication of the Intrinsic Aspects of the Project.)* These three methods are to either list (1) the organizations or offices that contribute experts, (2) the experts' names and affiliations, or (3) the experts' names and affiliations in connection with their responses. The last of these ways is the most demanding in terms of the records that will have to be kept.

The identity of the experts can also refer to aspects of their professional background, such as how long they worked in an area of expertise and where they attended school. We recommend recording and testing this information to learn if it correlates with the expert's answers or other factors. In our experience, people intuitively expect the experts' answers to correlate to some aspect of their background. Although we have found no such evidence of correlation (Booker & Meyer, 1988*a*; Meyer & Booker, 1987*b*), we suggest documenting this information for testing.

3. **The methods used to obtain expert data.** The elicitation methods need to be documented if any of the project personnel intend to use them again, if the project is likely to be reviewed by outsiders, or if a written report is a required product of the study. In general, we suggest recording two types of information when documenting methods in a report: a summary of what methods were used and how they were applied, and an explanation of why each method or combination of these was selected. Support for the use of the methods can include references from the literature and other considerations such as the project's data-gathering objectives and the need to reduce costs, time, or the occurrence of particular biases. This support can also include descriptions of how the methods were pilot tested or rehearsed and revised. In addition, the supporting information can include statements on what the documentation of the expert judgment represents. Readers of the report may not know that the documentation represents the expert's state of knowledge at the time of the elicitation. For this reason, it may be necessary to state that the expert might, with time and new knowledge, give a different answer to the same question.
4. **The experts' responses.** Several schemes can be used to document the experts' responses. The documentation schemes, described in chapter 7, basically consist of two types: only the experts' answers are recorded or both the experts' answers and thoughts used to arrive at these answers are recorded. The second type can be documented in at least three ways. The first is a summary documentation in which the experts' answers and a few sentences or paragraphs on their thinking are recorded. The next method, detailed verbatim documentation, involves a written record of everything that the expert says or does. Often this scheme involves mechanically recording the elicitation session and then transcribing it. The last and most involved means of

documentation is the detailed structured scheme. This method usually includes a format to guide the person doing the documentation. The format includes blanks labeled for recording those aspects of the elicitation that are considered most important, such as the sources of information (e.g., code simulations, references, experiments, and personal communications), and the assumptions, algorithms, and equations that the expert used in solving the problem.

The documentation can be used in a report in its original form or it can be rewritten to be more general; that is, the report can include fewer or more general data than were gathered but not the reverse. For example, if the experts' names were not initially recorded with their responses, this data will not be available for inclusion in a report.

Logistics of Who Will Record and When

Project personnel and external experts can either work together or independently to document expert judgment. For example, a data gatherer can write the judgments and an expert can then review the records for accuracy and completeness. In addition, there is sometimes an option for timing when particular types of information are documented—before, during, or after the elicitation session.

Step 1. Determine Which Persons Are Best Qualified and Motivated To Do The Documentation. The following information is listed according to what is being documented.

1. **The final statement of the question.** Project personnel, particularly the data gatherers, are usually the most qualified and motivated persons to document this information. The one exception would be if the external experts had been totally responsible for the refining of the question. In such cases, the experts would be the most qualified, although not necessarily the most motivated, to record this information. (The experts are generally not as concerned with documenting the elicitation as they are with solving the problem.) For this reason, even when the experts are the most qualified to document the statement of the question, the project personnel often do it. The project staff may request the latest version of the question from the experts and then record it.
2. **The experts' identities.** Writing the experts' names by their responses does not demand as much qualification and motivation as recording the types of information mentioned, described in items (1), (3), and (4). Anyone recording the expert judgment can note the experts' identities. If the experts use a special format to record their own judgments, they can simply enter their names on the predesignated blank. If the data gatherers use formats (e.g., write the expert's responses in an organized manner on the board or use interview guides), they can label the responses by the expert's name. If the data gatherers mechanically record individual interview sessions, they can

label the tapes. In general, labeling should be double checked because the documenter can forget to record a name or record the wrong one. Occasionally the data gatherer records characteristics of the expert, in addition to the experts' names and responses. These characteristics are gathered when later analysis is planned, such as for learning if the characteristics correlate to the experts' data. Some commonly recorded attributes include the experts' education, years of experience, job title or position, area of specialization, and psychological test results.

3. **The elicitation method.** We consider the project personnel, specifically those who conduct the elicitation, to be the most qualified and willing persons to record information on the elicitation methods. Information on the elicitation methods is the only one of the four types of information under *What Documentation Can Include* that project personnel can document both before and after the elicitation. The elicitation methodology can be initially written up after its design is last revised (chapter 9). After the elicitations, the description of the methods can then be updated because they may have been conducted slightly differently than planned.
4. **Experts' responses.** Who is best qualified to record the experts' responses depends on how detailed or technical this data are. Almost anyone can write down a short answer (e.g., a probability estimate, rating, or ranking). Frequently, the project personnel record this information if it is given verbally in an interactive setting (e.g., interactive group or individual interviews). If the elicitation is conducted over the telephone, the interviewer records the answer. On simple mail surveys, the expert records the answer, using the desired response mode.

   However, if the expert data include detailed problem-solving data, the expert is the most qualified to document it. Unfortunately, experts are not usually as well motivated as the project personnel to document this information. Generally, they are more willing to deliver the data verbally than they are to take the time to record it in written form. Thus, documenting problem-solving data must be made as easy as possible to motivate the experts to do it. Documentation formats (also called interview instruments or forms) are often used for this purpose. (See sample forms included at the end of this chapter.) These formats can help guide the data gatherer or the expert to document the information. If the experts complete the formats, have the project personnel check each format before the experts leave. (We assume that an individual interview or interactive group situation is being used to gather detailed information on the experts' problem-solving processes.) Experts frequently fail to document their thinking as they were instructed to do. They may have misunderstood the instructions or simply become tired of writing their data. Similarly, if the project personnel did the documentation, the expert should check the notes for accuracy and completeness. For example, in the NUREG-1150 reactor risk study, an extra precaution was taken—the experts signed the final documentation to show that they had approved it. On the other hand, if the elicitation is tape recorded, someone should check that the expert's voice is audible and understandable, especially if this will be the only record of the session.

If the data gatherer is unfamiliar with the field and if complex questions are being asked of the experts, the data gatherer may no longer be the most qualified to document. Such cases may require arrangements beyond those already discussed. For example, in the NUREG-1150 reactor risk study, the elicitations were performed by decision analysts who possessed experience in the elicitation but not in the technical areas. For this reason, project personnel who were knowledgeable in the question areas attended the sessions and, like the decision analysts, provided written records. In addition, all sessions were tape recorded and the expert was encouraged to document the reasoning for his or her judgments immediately after the elicitation (Wheeler, Hora, Cramond & Unwin, 1989, pp. 3-7).

Step 2. Determine Whether The Persons Tentatively Selected To Perform the Documentation Can Actually Do It. The selected person may be too busy eliciting the expert data or solving the question to do the documentation. A list of tasks that people can do simultaneously appears in Table 3 to help make this assessment.

TABLE 3

Elicitation and documentation tasks that a person can simultaneously perform

---

Experts

- Experts can solve the problem and verbally deliver or write down the answer
  - *If* only a short answer is required.
  - *If* the answer is given in a manner (response mode) that the expert has grown accustomed to.
- Most experts can solve the problem and verbally describe their problem-solving processes. However, a few individuals have difficulty doing these two tasks simultaneously, as noted in chapter 7's description of the verbal report.
- Most experts <u>cannot</u> solve the problem and write down their thought processes in solving it
  - *If* their thought processes are to be recorded in any detail.

Interviewers or knowledge engineers:

Working with a single expert

- The interviewer/knowledge engineer can request the expert's answer, record it on a documentation format, and run the tape recorder.

- In addition to the above tasks, the interviewer/knowledge engineer can usually also document the expert's problem-solving process

  If the interviewer/knowledge engineer is familiar with the subject matter.

Working with multiple experts face to face

- The interviewer/knowledge engineer can request the expert's answer and record it on a documentation format

  If obtaining the expert's answer does not require in-depth elicitations.

  If the experts will give their answers in an orderly fashion, as in a structured elicitation situation.

  If the interviewer/knowledge engineer is not also running mechanical recording devices.

- In addition to the above tasks, the interviewer/knowledge engineer can usually document the experts' problem-solving processes

  If only one or two sentences are needed (e.g., on the references that the experts used or the group's rationale for their answer).

  If the interviewer/knowledge engineer is familiar with the subject matter.

Working in any elicitation situation

- Interviewers/knowledge engineers cannot write all of an expert's problem solving as the expert is verbalizing it

  If the problem-solving processes are being elicited in great detail (because people usually verbalize their thoughts much faster than the average person can write. People's rate of speech is one of the reasons that tape recorders and stenographers are used to supplement taking notes).

Note: Running a tape recorder may prove more difficult than expected. In all sessions, someone needs to turn the tape recorder on, check that the tape head turns properly, and change the tape cassette as needed. In group situations, the omnidirectional microphone needs to be turned on and have its batteries checked. In individual interviews, someone needs to remember to attach and detach the microphone from the expert's lapel.

Step 3. If The Documenter Must Do More Than What Was Listed As Feasible In Step 2, Consider The Three Following Aspects of Documentation. Manipulating One Or More of These May Lighten The Documenter's Load.

1. **Who does the documentation and who else can assist?** For example, if the data gatherer is to do the documentation, could the expert help? Could additional project personnel (e.g., a stenographer, secretary, or editor familiar

with the technical area) assist the data gatherer by taking notes or running the tape recorder during the session?

2. **The timing of when the documentation is done.** Some documentation can proceed or follow the elicitations. Frequently, the statement of the question, the elicitation methods, and the experts' identities can be documented in preliminary form before the elicitation. During the elicitation, brief notes can then be taken on any subsequent changes. For example, the statement of the question often evolves or changes during the process or elicitation. Similarly, the list of experts assigned to the questions may change if some of the experts decide not to participate and substitutes are used.

- ***The leeway in timing the documentation of the experts' problem-solving processes.*** The problem-solving data cannot be recorded in total before the elicitation because it often changes significantly during the session. However, it is possible to document some of the basics of the expert's problem solving if this information has been written before the elicitation. For example, in the NUREG-1150 reactor risk study the experts disaggregated their questions before the elicitation sessions took place. However, even in this project the experts changed their disaggregations and thoughts during the elicitation. Waiting to document the problem-solving data until after the elicitation does not work either because portions of it will be forgotten or lost.

  Tape recording the problem-solving data, although helpful, is not a panacea because much of what people say is unclear without their expressions or gestures. (We recommend that tape or video recording be used only as a backup to taking notes because the mechanical recordings often malfunction and they consume much time during playback, such as to clarify particular points or to transcribe.) Thus, problem solving must be documented, at least in part, during the elicitation if these data are to be recovered and reported later.

- ***The amount of time needed to verify the experts' data.*** Generally, verifying expert data takes more time after the elicitation than it does during the elicitation. For example, if the experts' problem-solving processes were documented after they had left, they would have to be contacted to verify what had been written. Therefore, verifying documentation long after the session is more time consuming than verifying it during or shortly thereafter by having each of the experts read and initial the record while they are still present.

3. **The level of documentation.** If none of the suggestions mentioned above resolve the documentation problem, simplifying the documentation demands may be the last resort. Would a less detailed or structured documentation scheme still provide the necessary data? The general documentation schemes listed in chapter 7 are listed in order of the least demanding to the most demanding.

Sample Documentation Formats

Formats serve as guides for the project person or the expert who records the information. For instance, formats have been developed for questioning the expert on the subject area (Spradley, 1979), for obtaining background on the expert (McGraw & Harbison-Briggs, 1989; Meyer, 1987; Meyer & Booker, 1987*b*), and for obtaining short explanations of the expert's problem-solving process (Meyer, 1987; Meyer, Cullingford & Peaslee, 1981). Formats that have been designed specifically for knowledge acquisition projects are given in McGraw and Harbison-Briggs (1989).

Two generic formats for recording the experts' answers and their problem-solving processes are illustrated at the end of this chapter. The first format (example 8.1) helps document experts' ratings and only gathers a few notes on why the expert gave these ratings. The second format (example 8.2) focuses more on the experts' problem-solving processes. It is used to document the experts' disaggregation of the problem, their answers, and their reasons for giving both of these.

Presentation of the Question—A Quick Check

Although chapter 5 covers this topic, we readdress it in this section to ensure that the presentation of the question fits the revised elicitation. In chapter 5, the question was refined by considering the information that the experts needed to answer it (e.g., the background, assumptions, and definitions), the order in which the information needed to be presented (e.g., general to specific or specific to inclusive), the decomposition of the question, and the phrasing of it. Now these same aspects of presenting the question need to be reconsidered in terms of the revised elicitation.

In general, project personnel who follow this chapter's suggestions should note the changes that they make on the elicitation and decide whether the presentation of the question still fits. A few considerations follow:

1. **Are both the components of elicitation and the presentation of the questions structured to a comparable degree?** For example, if the elicitation components were structured to provide a clear field for observation, an unstructured presentation of the question could compromise this aim. For example, in an unstructured presentation the interviewer would ask a question without having determined which wording was clearest and what information would be needed to answer it.
2. **If the basic elicitation situation was modified to include a combination of the other situations, will the planned presentation need changing?** For example, imagine that the data gatherers decide to break an interactive group situation into individual interviews to avoid the possibility of group think during the elicitation. If the data gatherers had planned to present the question (i.e., its background, definitions, and assumptions) only once to the group, they would have to rethink this presentation for the individual. The question could be announced to the group before individual interviews began, but if too

much of a delay occurred before interviewing the last expert, the question would have to be presented again to refresh the expert's memory. In such cases, the data gatherer could present the question on paper and then verbally review it at the onset of each elicitation.

3. **If the telephone or mail was selected as an alternate mode of communication, will the method for presenting the question need to change?** For example, if the plan was to save on costs and time by eliciting the expert's solutions over the telephone, could a simple and brief presentation of the question be given over the telephone? If not, perhaps the experts could receive the questions by mail, giving them time to develop their answers before the subsequent telephone sessions.
4. **Does the documentation scheme support the presentation of the problem and vice versa?** For example, if a detailed structured documentation scheme was selected, the presentation of the question should exhibit comparable care.

Pilot testing can determine any remaining conflicts between the presentation of the question and the elicitation methods. Pilot testing, which chapter 9 describes in detail, is the last stage in refining the elicitation process before the elicitation sessions occur.

Common Difficulties—Their Signs and Solutions

Difficulty: Realizing that the Time Allotted to Planning the Elicitation Is Not Enough. This is one of the most common problems that we have observed. People often view the planning stage as an unnecessary delay to starting the elicitation. Project personnel usually notice signs of insufficient time for planning when looking at the project schedule. The schedule may show that the elicitation, not its planning, should begin *now*.
Solution: Determine if the planning can be dropped without creating severe consequences later. The need for planning diminishes if only a few experts will be interviewed, especially face to face. In face-to-face interviews, the interviewers do not have to coordinate groups of interacting experts nor polish elicitation materials, such as to send them to experts in a Delphi situation. Also, in face-to-face interviews, the data gatherers have greater license to plan as they go because there are fewer things requiring advance preparation. Sessions that use unstructured techniques also require little if any planning. As mentioned earlier in the chapter, the more highly structured techniques require more planning. However, planning may prove important even in those applications that are not structured. If sufficient time for planning has not been allowed, consider modifying the schedule.

Difficulty: Ignoring the Possibility of Bias. The possible occurrence of bias is disregarded frequently because of the project staff's ignorance. They may have been totally unaware of its existence, or they may have chosen to ignore it. Both of these are understandable responses. The topic of bias in expert judgment has not commonly been addressed, and the physical scientist in particular is not accustomed to dealing with it.

However, ignorance is definitely not bliss. We have seen an increase in external reviewers' criticism of expert judgment studies. Generally, these reviewers have been sensitive to the following biases: social pressure from data gatherers, group think, and wishful thinking.

Solutions: One approach would be to use this chapter's section on anticipating bias to design the elicitation before proceeding with the elicitation. Another narrower approach is to selectively focus on handling the three biases (social pressure from the data gatherer, group think, and wishful thinking). Yet another approach is to proceed with the idea of gathering enough data to test for the presence of selected biases after the elicitation. Chapter 14 describes the analysis techniques for testing bias.

For example, on a magnetic fusion project (Meyer *et al.*, 1982), two tests were run to examine the problem of wishful thinking. The purpose of the study was to determine if those individuals who worked on the fusion project could meet scheduled milestones. After the elicitation, the sample of experts was divided into those with more to gain from their answer (the managers) and those with less to gain (the scientists working on the hardware). Their data were analyzed. These two groups could not be found to give significantly different answers to the same questions. The experts' answers were also compared to the wished-for outcome. That is, the experts' predictions of when they would meet various scheduled milestones were compared to the planned schedule and found to differ significantly.

The last approach, not considering bias until during the analysis, has one advantage over doing absolutely nothing about bias. If reviewers of the work raise questions about a particular source of bias, the results of the analysis can be presented. If the results show no evidence of the bias and the reviewers find no fault with the methods of analysis, the reviewers will focus their attention on other matters.

EXAMPLE 8.1: *Sample format for documenting answers*

Expert's name:____________________

Date:____________________

Time:__________ to __________

Interviewer:____________________

Definition of question:____

____

	Variable 1	*Variable 2*	*Variable 3*	*Variable 4*
Variable w				
Variable x				
Variable y				
Variable z				

Expert's comments on reasons for giving estimates:

Variable w and *1*:____

Variable x and *2*:____

Variable y and *3*:____

Variable z and *4*:____

■

EXAMPLE 8.2: *Sample format for documenting answers and problem-solving processes*

Expert's name:____________________

Date:____________________

Time:__________ to __________

Interviewer:____________________

Question name:____

Information defining the question: ________________________________

Sketch and label each question, subquestion, and branch, as shown in the example.
Provide a few sentences on the expert's reasoning for each point, as shown in the example.

Footnote any references that the expert uses.

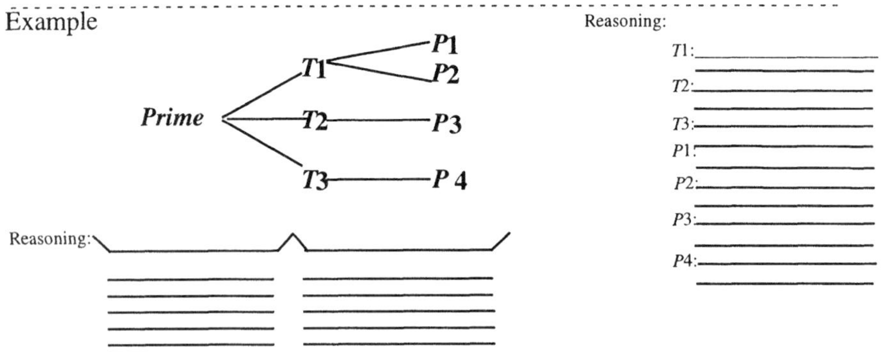

9

Practicing the Elicitation and Training the Project Personnel

This chapter provides the last check on all aspects of the elicitation design before the elicitation is conducted (chapter 10). The logistics, ease of use, interface, and timing of the parts of the elicitation process are examined to find any remaining glitches and resolve them. The project personnel identify any remaining problems by practicing and pilot testing the different parts of the elicitation.

Practicing the Elicitation

Project personnel frequently practice the following aspects of the elicitation: the briefings given to the experts, the elicitation procedures, the documentation, aggregation, and entry of the data into a model or data base. ***Practice*** is defined loosely in this book to mean rehearsing some act to become more proficient in it. Practicing serves another purpose—the training of project personnel. One subset of practice is pilot testing. ***Pilot testing*** involves taking a very small sample of the expert population, presenting them with the aspect of the elicitation to be tested, obtaining these experts' feedback, and revising the elicitation accordingly.

What Needs to Be Practiced?

The following items should be rehearsed to resolve any difficulties and to learn how long they will take.

1. **Introducing the experts to the elicitation process.** This process includes familiarizing the experts with the general steps and schedule to be followed, the questions or problems that they will address, the response mode that they will use, and the biases to which they may be prone. If the experts were not heavily involved in developing the elicitation, they will need clear and concise briefings on their particular roles during the session. Conducting such briefings does not occur naturally—it requires practice. If the experts are confused about any of the above-mentioned topics, their confusion could make them more prone to bias (e.g., inconsistency). Their confusion could also make them reluctant to participate. If the project personnel practice these

topics, they will communicate them more clearly to the experts and feel greater confidence.

The following suggestions will help introduce the experts to the elicitation process. This step is critical in establishing the project's credibility and the experts' understanding of their tasks. (The most appropriate time to introduce the experts to the elicitation process is *after* they have been briefed on the general elicitation procedures.)

- We suggest that the experts receive sample questions to work so that they can practice using the response mode. If there are any techniques for properly using the response mode, they can be introduced and practiced here. For example, if the response mode is probability distributions, consult Hogarth's (1980, p. 149) eight keys to assigning probabilities.
- The experts should be briefed on the biases identified in chapter 8 as being likely to occur. This briefing should include an explanation of why the selected biases occur and of how the experts can reduce their tendency to commit them. (Table 2 in chapter 8 provides examples of this type of information.)

  The briefing on bias should include exercises designed to evoke the selected biases. The interviewer or knowledge engineer can read the correct answers and allow the experts to score their own exercises. These exercises can convince the experts that they, like everyone else, are prone to these biases. If the briefing is given without exercises, we have noticed that the experts are not as effective in countering their tendencies toward bias, perhaps because they were never convinced that they too would be biased.

2. **The elicitation procedures.** Even carefully planned elicitations can have flaws, so it is best to test them on sample experts. For example, these tests often show that the elicitations last longer than expected and that the sample experts become fatigued during the sessions.
3. **The documentation procedures.** Pilot testing the documentation provides information on the logistics and the format, such as whether or not these should be revised. For example, the pilot tests may show that the interviewer or knowledge engineer cannot conduct the elicitation and simultaneously complete the documentation format.
4. **The mathematical aggregation of the expert's answers (if this is planned).** Although mathematically combining the expert's responses may seem straightforward, many methods are available and some may not be appropriate to the elicited data (chapter 16). Therefore, project personnel should practice the chosen method. At the very least, the project personnel can learn whether the procedure takes too long—this in turn can lead the personnel to automate it. The data acquired from limited pilot tests can be used to practice the aggregation. This step should also be practiced when the experts will disaggregate the problem and the project personnel will reaggregate the experts' responses (e.g., the NUREG-1150 reactor risk study).

5. **The entry of the data into a model for analysis or end use.** The data gathered from the limited pilot tests, as described below, are used for rehearsing the entry of the data into a model. Frequently, this rehearsal reveals a major discrepancy between the form in which the expert's data was obtained and that necessary for the model.

What Requires Pilot Testing?

The procedures that require pilot testing are those in which the experts participate (e.g., giving their judgment). The experts' understanding is critical to their participation. Therefore, anything that the expert must understand makes for a good pilot testing candidate. However, a smaller set of all those things that the expert must understand is usually pilot tested because pilot testing is time and expert intensive. As a general rule, pilot testing is done on the experts' understanding of instructions, whether these are given orally or in writing (e.g., how to fill out a survey). The following parts of the elicitation are suggested for pilot testing.

1. **The experts' understanding of the problem or question.** In some cases, a final statement of the question or problem exists long before the experts provide their solutions. In other projects, such as the NUREG-1150 reactor risk study, the experts develop their own statement of the problem from an initial version. In the first case, the pilot test provides information on the experts' interpretation of the problem. If the project personnel did most of the question development, their interpretation is likely to differ noticeably from that of the experts. In the second case, the pilot test checks that the experts understand that they should take the basic problem area and refine it.
2. **The experts' use of the response mode.** The pilot test allows the interviewer or knowledge engineer to make sure that the expert understands how to use the mode. The expert's response can be checked against both the instructions and logical or mathematical standards. For example, probabilities for all outcomes, previously defined as mutually exclusive events, should sum to one.
3. **The experts' understanding of the elicitation procedures that they must follow.** For example, if verbal report is chosen, it would be important to make sure the experts knew that they must *think aloud* and that they had the capability to do so.
4. **The experts' ability to use any documentation format.** The pilot test provides feedback on the experts' understanding of the directions on how to complete the format. (e.g., on a mail survey). We have found that experts typically do not provide as thorough a documentation of their judgment as they are requested to do.

How to Pilot Test

Sample Sizes and Selection

Ideally, pilot testing uses experts who represent the range of experts who will later participate in the actual elicitation process. For instance, if the experts have been drawn from positions in government, private industry, and academia, the pilot sample should contain members of each of these positions as well. Consider the factors used to select the expert populations and use these to choose the sample for pilot testing. In selecting an expert pilot sample, consider that this selection decreases the pool of experts whose judgments can be elicited later. In general, experts who participate in the pilot test or who have otherwise assisted in the method's development should not participate in the actual elicitation. Similarly, we would advise against using the advisory experts in the pilot sample. The advisory experts will not approach the test materials from a fresh perspective if they helped develop them.

The size of the test sample depends on the size of the expert population available for the elicitations. Test samples for expert elicitations rarely, if ever, have the large sample sizes (10% of the total) associated with traditional mail surveys. Typically, expert samples for pilot tests are five persons or less because the largest expert population does not usually exceed 50. Because of these small sample sizes, the strategy for pilot testing in expert judgment studies has been to test intensively. In other words, try to obtain as much benefit or feedback as possible from these few experts.

Two types of pilot test are mentioned below. We recommend conducting the intensive pilot test first. The ***intensive pilot test*** was developed (Meyer, 1986, p. 92) to trace the expert's understanding. It consists of structured, in-depth interviews and observations. The other type of pilot test is called ***limited*** to distinguish it from the traditional tests of high sample size and from the intensive pilot tests mentioned above. Limited pilot tests work best after intensive pilot tests have been performed and the elicitation revised. The limited pilot test allows the project personnel to practice the elicitation procedures, time their duration, and check how these procedures fit together.

Sequence for Pilot Testing

The sequence for performing these pilot tests follows:

1. Apply part one and part two of the intensive pilot test (as described on the next page) to the elicitation.
2. Use the pilot test results to revise the tested parts.
3. Practice the entire elicitation process and do limited pilot tests on those parts that will eventually involve the experts. For example, the project personnel would rehearse their introduction to the elicitation in front of the experts for the limited pilot tests. Project personnel would then practice the elicitation procedures on these experts. The expert data from the practice elicitations would be used to check the other steps of the elicitation. Specifically, the data would be documented, aggregated, and used as input just as it would be when the elicitation is actually performed. Each of the elicitation procedures is timed during the practice.

How to Conduct Intensive Pilot Tests

To our knowledge, this type of pilot test is the only one that allows people's thinking to be tracked through information presented in written form. The intensive pilot test provides two kinds of information: how the experts progress through the information, their general impressions, and when and why they decide to respond to particular questions; and how the experts specifically interpret each direction, statement of the question, or response mode option.

First, the materials to be pilot tested are selected using the list provided under *What Requires Pilot Testing?* A typical selection would include the set of problems, any written directions on assumptions or definitions that the expert must use, and directions on what information the expert should document on the form and in what manner, such as on a continuous linear scale.

Intensive pilot testing takes place with one expert at a time in a face-to-face situation. The interviewer sits on the opposite side of the expert's handedness to allow easy viewing of the expert's writing; that is, the interviewer sits to the expert's left, if the expert is right handed.

Intensive Pilot Test—Part 1. For the first part of the intensive pilot test, the expert is instructed to complete the written materials as he or she would if no observer were present. The expert is also asked to *think aloud* (verbal report) as he or she reads through the written materials. The interviewer has a copy of the same materials given to the expert for recording the data.

While the expert pages through the written materials, the interviewer records the order in which the expert looks at the materials, his or her pauses, gestures, facial expressions, and words. For example, the expert may skim through the introduction, cover letter, or directions and then flip through the rest of the materials before returning to read each of these more thoroughly. The expert may also make a variety of comments, including *I have problems with this page and will probably let it sit on my desk for several days.*

In addition, if limited pilot tests will not follow intensive pilot tests, the interviewer can record the expert's starting and ending time on the first part of the intensive test. The limited pilot tests described in the next section provide better time estimates because the expert does not have to think aloud throughout the interview. However, gathering some data on time is better than gathering none, so this data should be gathered during the intensive pilot tests if they will not be gathered by other methods. If the intent is to obtain time data, save the questions of the expert until the second part of the intensive test. Otherwise, these questions will inflate the time estimates of how long the expert takes to respond to the written materials.

The intensive pilot test provides better time estimates than might be expected. The time estimates obtained from the first part of the intensive pilot test would be high because the expert continuously verbalized his or her thoughts (and may not do so for the actual elicitation). However, we have found that this measurement provides an adequate estimate of the upper limit of time needed. This occurs because the experts selected for the pilot test may not represent the range of experts, some of whom could take a significantly longer time to finish. For instance, we have seen pilot tests that last two hours while the actual elicitations range from 1 hour to 2 hours and 15 minutes.

The expert is frequently allowed a brief break between the first and second part of the intensive pilot test.

Intensive Pilot Test—Part 2. During the second part of the intensive pilot test, the expert is asked to paraphrase in his or her own words the meaning of each direction, question, or response option. This information allows the interviewer to track the expert's interpretation in detail. It has always amazed us that something could be interpreted in so many different ways, ways that we had not thought of because we *knew* what we meant. The interviewer can also question the expert about any reactions, such as a look of puzzlement, noted at one of the questions during the first part of the pilot test. This questioning can jog the experts' memories so that they can give a more thorough description of their impressions.

Even though the interviewer records all information on a copy of the written materials, he or she still collects the expert's hard copy. The expert is thanked and asked not to discuss the details of the pilot test with anyone because it could influence their responses. The data from the intensive pilot test is examined and used in revising the elicitation procedures or the logistics of who performs the different tasks. If limited pilot tests are planned, they are the next step.

How to Conduct Limited Pilot Tests

In limited pilot tests, the elicitation procedures are conducted as closely as possible to the way that they will be done for the actual elicitation. There would be little point in examining the expert's responses to a form or a procedure that did not resemble the one to be administered. Thus, in the limited pilot test, the experts are given the briefings or the introductions that have been planned for the larger population of experts. Similarly, the sample experts receive whatever forms and instructions on providing their responses that the other experts will subsequently use during the elicitation sessions.

The interviewer elicits the data in the planned manner and obtains the experts' responses. In addition, data are gathered on how long each elicitation lasts so that these can be added to other time estimates to produce a total. If the elicitations are being done in person, the interviewer may also record the amount of time it takes an expert to respond to a particular problem.

The following illustrates how timing data are used. The practice briefing of the group of sample experts lasted 2 hours and one elicitation lasted 2 1/2 hours. This time included the experts' review of the documentation of their responses. Assuming that only one elicitation team performs 10 individual interviews, about 27 hours would be needed just to perform the elicitation.

As with the other pilot test, experts receive thanks for their cooperation after completing the session. The data from the pilot test are used in practicing the other stages of elicitation, including mathematical aggregation and modeling, as mentioned in the *Sequence for Pilot Testing*. In addition, the limited pilot test may prove useful in training project personnel, as will be discussed in the next section.

Training Project Personnel

Project personnel are frequently lay persons in the area of expert judgment and require training in elicitation methods. Even those experienced in expert judgment may need training if they are unfamiliar with the specific methods selected. Training provides the personnel with instructions (directions) and the opportunity to practice until they become proficient.

Training is done in the same areas that are practiced.

1. Introducing the experts to the elicitation process.
2. The elicitation procedures.
3. The documentation procedures
4. The mathematical aggregation of the experts' answers (if this is planned).
5. The entry of the data into a model or data/knowledge base if the data is to be used in this way.

The training can be done to differing levels of proficiency, depending on what the project demands and the personnel desire. For instance, if the experts felt reluctant to give their judgments, we would recommend that the data gatherers be trained to a higher level of proficiency to obtain the judgments. In addition, particular situations make training necessary.

When Training Is Needed

Training the project participants is especially necessary under the following conditions.

1. **When different people plan the elicitation than will conduct it.** In many projects, experienced persons or those selected by a committee design the elicitation methods. However, the task of implementing the methods is often given to those who are less experienced and, perhaps, those who did not participate in the selection of methods.
2. **When the persons who will perform the elicitation and documentation feel uncomfortable with the assignment.** This reaction may stem from the situation mentioned in item 1. From these persons' viewpoint, it may seem that they are being forced to implement methods that they did not help select. In addition, those assigned to perform the elicitations may believe that they lack the experience to do a competent job. Training can make them feel more confident in their ability to perform these tasks.
3. **When more than one person or team will gather and document the expert data.** The more people, the greater the chance that each member will perform the tasks differently, which in turn will result in data inconsistencies. Training the data gatherers promotes consistency in these tasks. In addition, the process of providing instruction often allows the instructors to identify any vagueness in the procedures and to provide stronger guidelines.

4. **When more than one project person participates in an experts' elicitation.** More than one person may perform the elicitation if one cannot do all the tasks simultaneously. In addition, the elicitation can require very different skills, such as interviewing skills and knowledge on the technical subject area. In many cases, several people participate in the elicitation because training one person in both skills (e.g., training the technical person to interview) simply takes too much time. Thus, one person may elicit one type of information from the expert and a second person, another. For example, in the NUREG-1150 reactor risk study, a decision analysis method of elicitation was selected, and three decision analysts performed the elicitations. However, the information being elicited on reactor phenomenology was so technical that a separate project person needed to participate in the elicitations to answer the expert's technical questions about the problem, to record, and to check the expert's technical data for gaps. In another project examining tank tactics (Meyer, 1987), one person obtained the expert's reasoning as he or she responded to a computer-simulated battleground, and the other ran the computer. The computer person, who was familiar with the computer program, ensured that the computer ran properly and answered all the expert's questions on it. If a two-person interviewing situation exists, the project persons must rehearse their roles together. Otherwise, they can appear like a slapstick comedy routine—bumping into one another, interrupting and confusing each other and the expert.

How to Train

Different methods can be used to help train project personnel to execute the five items mentioned in *What Needs to be Practiced.* The paraphrased items are listed below and followed by training suggestions.

1. **Introducing the experts to the elicitation.** The project personnel can simply rehearse this talk in front of an audience, preferably one unfamiliar with the planned elicitation.

2 and 3. How to run the elicitation and document the data. Training personnel in these two areas can be difficult because the areas are more complex than those mentioned in 1, 4, and 5. Elicitation and documentation involve interactions with the expert and perhaps simultaneously with another data gatherer. The instructions given may not encompass all that could happen in the elicitation. Therefore, these tasks must be rehearsed with someone playing each role as it will be done in the actual procedure.

Three training options and their advantages and disadvantages.

- **The first option** is to instruct the trainees in the procedures and then have them conduct the limited pilot tests as practice. This option's advantage is that it does not require a large pilot test sample or much time.
- **The second option** is to have the experienced data gatherers perform the first limited pilot tests while the trainees observe quietly and then have the trainees conduct the latter pilot tests. This option's advantage is that

trainees learn by observing before actually doing the procedures. The disadvantage of this option is that it requires a large enough sample to do more than a few tests.

- **The third option** is to have the trainees observe the limited pilot tests or videos of the tests and to practice their skills by taking turns role playing the interviewer and the expert. This option's advantages include benefits derived from role playing the experts and no special requirements on the number of pilot tests. The trainees who play the experts gain insight into the elicitation from the expert's perspective and become better interviewers. When they play the experts, they learn first hand how the interviewees wish to be treated. For example, the trainees as interviewees may view some of the elicitation procedures as condescending and dictatorial.

4 and 5. How to perform mathematical aggregation and enter the data. The trainees are given the data from the limited pilot tests, instructed on what to do with it, allowed to perform and time these operations, and asked to report any problems.

Common Difficulties—Their Signs and Solutions

Difficulty: Pilot Tests Show that the Sample Experts Experience Significant Difficulty with the Response Mode. If experts experience difficulty in using the response mode, the intensive pilot tests usually indicate this. The sample experts may have paused on the section that describes the response modes—they may have appeared confused, remarking that they did not understand. In addition, the answers may give further evidence of the experts' difficulties. For example, it may be that the response mode is to follow particular logical or statistical standards. If the sample experts' responses consistently violate these, a problem is indicated.
Solutions: One approach is to rethink the briefing on the response mode that was given to the experts. Perhaps this briefing or the written instructions on the response mode were not clear. If this is the case, a new sample of experts should receive a revised briefing and the response mode should undergo intensive pilot testing again.

If this sample of experts experiences the same amount of difficulties, reconsider using this response mode. In selecting another response mode, consider what would fit with the other phases of the elicitation, such as the aggregation and modeling of the responses.

Difficulty: Pilot Tests Indicate that the Elicitation Will Likely Need More Time Than Is Available. It is a common occurrence to find that the elicitation requires more time than has been allotted. This discovery is evidence that all of us are prone to wishful thinking (to overestimating how much we can do in a given amount of time), and to underestimating uncertainty (forgetting those things that can take more time). Signs from the limited pilot test may indicate that the interviews by themselves or perhaps in combination with the other stages, such as aggregation, sum to more time than was scheduled.

Solutions: A first option is to review the information from the limited pilot tests on how long the different stages take and decide which of these stages could be done differently to cut time; based on this information, the project personnel can repeat the stage, pilot test the elicitation again, and record its new duration. A second option is to change the schedule. In our experience, people have often chosen the second option because it does not require the careful consideration that the first option does.

Difficulty: Project Personnel Resist the Training. Personnel can resist elicitation training for several reasons: they do not view elicitation as part of their job, they do not feel qualified to perform elicitation, they fear being blamed if the elicitation fails, and they resent having to do something that they did not help plan. If those receiving the training appear uncooperative or object to the training, the first step is to find out *why*.
Solutions: Any solution rests on learning the person's reasons for resistance and addressing those reasons. If only a few persons seem to inspire this reaction, addressing the concerns of the minority may resolve the problem.

Difficulty: The Rehearsal Shows that the Documentation Scheme Does Not Meet the Project's Other Requirements. The documentation can fail to mesh with the rest of the elicitation process if it is at a different level of detail (granularity) or form than needed for the later reporting, aggregating, inputting, or analyzing of the data. If the documentation scheme was not pilot tested, this difficulty is often not discovered until after the elicitations have been conducted.

Signs of the above situation may occur during the pilot testing as processes that do not fit or flow. For example, the project personnel may try to aggregate the experts' solutions mathematically while taking into account their differing assumptions so as not to *mix apples and oranges*. Imagine, however, that these data are missing or that the data are in a fuzzy form and cannot be used for comparing the experts' solutions. Another example is when the documented data cannot be entered into the model—the data are in the wrong form. For instance, the experts gave solutions on a linear scale of 1 to 5 and the computer model requires probability distributions with confidence ranges. Similarly, the project personnel may have tried to analyze the effect of some variable on the expert's problem solving only to find that they did not have the necessary data. Perhaps they used a summary documentation scheme and could not find any record of this variable.
Solutions: If this problem is caught before or as a result of pilot testing, the solution is simple. Select and develop another documentation scheme (chapters 7 and 8) that will address these needs. Perform limited pilot tests again and use the data gathered in them to test the aggregation and the entry of the data into the analysis model.

If this problem is not detected until after the elicitations have been completed, the options are limited. Basically, the aggregation, report, or analysis cannot be done as planned. If the documentation was in the wrong form, perhaps it can be translated into the desired one. However, in translating the data from one form to another, there is always the risk of misrepresenting it or of making an assumption that is not valid, such as concerning its distribution.

Difficulty: The Documentation of the Experts' Problem Solving Has Been Done Differently or at Different Levels. We have frequently encountered this difficulty and have observed many other situations in which it has occurred. Generally, inconsistency in the documentation either occurs when multiple project people or experts have done the documentation or when the documentation includes questions of differing complexity.

For example, in the NUREG-1150 reactor risk study, some of the final documentation was done by the decision analysts who performed the actual elicitation and by the project staff who were familiar with the technical area of the experts' problem solving. Given the differences in these documenters' backgrounds and the information that each was to record, they could not (in all likelihood) have produced the same documentation. In another case, our documentation proved inconsistent even though we used the same person to gather the data and the same documentation format. The experts' means of solving problems were gathered in detail—first on classical statistical questions and then on judging the adequacy of a computer model in simulating reactor phenomena (Meyer & Booker, 1987*b*). While the statistical questions were not simple, they were simpler than the evaluations of the computer model. The experts seemed to think and verbalize very differently about solving the simpler versus the more complex questions. Thus, documenting the same level of detail on these two questions was a most difficult task.

Solutions: If this problem has been detected during the practice elicitations, the rehearsals have served one of their purposes. However, several actions can help promote consistency in the documentation. First, the guidelines on the documentation can be tightened. One of the reasons that people have often done their documentations differently is that their instructions on the format have been open to differing interpretations. Second, additional training on documentation could be offered. If project personnel will perform the documentation, they can rehearse it by doing limited pilot tests and taking turns at playing the experts. They could deliver their practice documentation for review and feedback on which aspects of their elicitation still required changes. If the experts do the documentation, the instructions and format given to them should be carefully pilot tested for clarity.

As a final check on documentation, a qualified person can be appointed to review each record as it becomes available after the actual elicitations. The project person who will aggregate, model, or program the expert data or write the report would be an ideal choice for this task. The designated individual could be asked to check each record, preferably before the experts have left. The appointed person could quickly check that all the required information was present and that it was legible and comprehensible. This type of check is particularly necessary if the experts document their own solutions or problem solving.

If the documentation problem was not discovered until after the elicitation (as is the case most often), the options for resolving it are more limited.

The first option is to try to reconstruct the elicitations to address the gaps. This entails using all the records, both written and mechanically recorded, to find the missing information. If the information cannot be found, recontacting the experts may be necessary, perhaps even to have them rework some parts of the problem. This option is so tedious and embarrassing that it is rarely done.

The second option is to adjust what and how the data are reported so there will be no gaps. Usually this means reporting or modeling the data at a more general level (a coarser granularity). For example, if some of the notes on the expert's problem-solving processes are more detailed than others, the less detailed set will have to be used as the common denominator.

10

Conducting the Elicitation

This chapter provides details on how to schedule, prepare for, and conduct the elicitation. The different elicitation situations and the three techniques for eliciting problem-solving data are covered. In addition, a section provides guidance on how to make the experts aware of the potential for bias and on how to monitor and adjust for bias during the elicitations. These topics are respectively steps 3, 4, and 5 of the program for handling bias.

Part III supplies information on how to analyze the data collected from the elicitations.

Scheduling the Elicitations

There are several fine points in scheduling and confirming meetings that will make this phase go more smoothly. In general, these points are simple courtesies that set the stage for good relations with the experts. The next two sections—*Scheduling the Meetings* and *Confirming the Scheduled Meetings*—may be skipped if meetings with the experts will not take place (e.g., Delphi situation or telephone or mail modes of communication).

Scheduling the Meetings

To schedule the meetings (e.g., individual interviews or interactive group situations), the data gatherer calls the expert and follows the steps given below:

- **The interviewer introduces him or herself and the affiliation for this project.** For example, *Hello, Dr. Jones, this is Mary Smith calling for the Division of Risk Analysis, Nuclear Regulatory Commission* (the name of the organization that is funding the study).
- **Ask if the expert can conveniently talk for a short, specific amount of time.** An example of how to ask this is, *Is it convenient for you to speak to me for about five minutes at this time?* We have all experienced the frustration of answering the telephone when we are busy, having the caller speak nonstop, and not having the opportunity to interrupt and explain that we cannot talk at this time. Similarly, it can be irritating to the experts to be called when they are leaving for a meeting or holding one in their offices, and their irritation may emerge later (e.g., in their reluctance to be scheduled for an interview). For this reason, we recommend checking that the call is at a convenient time. The

following illustrates such a check: *Have I called at a convenient time, or am I interrupting something?* If the expert answers in the negative, quickly ask when it would be a good time to call back.

- **State the purpose of the call.** Briefly explain to the expert that the call's purpose is to schedule a future interview. In general, we have found that it is best to say *interviewing* rather than *obtaining answers.* The latter seems to lead many experts to think that they will be asked abruptly for their answers. This imagined scenario appears to make them uncomfortable or leads them to doubt the validity of the caller's intentions (e.g., *anyone who believes that these answers can be given so easily must not possess much knowledge*).
- **Tell the expert how long the interview will last.** This time estimate will allow the experts to decide when they can block out an appointment of the necessary length. If pilot tests were performed (as described in chapter 9), estimates of the length of an elicitation will be available. We frequently say something like *the session will likely last from one to two hours, depending on the individual.* (Later, if the experts complain about the amount of time that their sessions took, emphasize again that the duration depends on the expert, and then thank the experts for their *thoroughness.*)
- **Select a time and date convenient with the expert's schedule.** We have found it effective to emphasize our wish to schedule the meeting at the expert's convenience. This courtesy initially promotes the expert's good will. If the meeting will include other experts, mention the times that the other experts have suggested and ask this expert which are the most convenient. (Continue this process until one preferred time and one alternate time are found that are acceptable to all the experts.) If the meetings will involve only one expert, ask the expert to select a convenient time. For example, *Could you pick a convenient time for a two-hour meeting within the next two weeks?* Request to schedule the appointment in the next few days, few weeks, or months, depending on the deadline for completing the interviews.
- **After the expert selects a time and date, verify this information by repeating it.** Repeating this information provides both the caller and the expert time to record the appointment and to notice any misunderstandings. In addition, repeat the location of the meeting, especially if it will be held outside the expert's office. If this is the case, either give or request directions to the meeting place.
- **State that the appointment will be confirmed.** For example: *I will call on Thursday, the day before, to check that the meeting time remains convenient.*
- **Thank the expert.**

Confirming the Scheduled Meetings

- **The interviewer introduces him or herself and states the intention of confirming the meeting.** For example: *Hello, this is Mary Smith calling about our meeting from 9:00 A.M. to 11:00 A.M., tomorrow, the 31st, in your office.*

Is this meeting time still convenient? Confirm the schedule one working day before the meeting takes place. (If the experts will be traveling to the meeting, confirm the meeting a few days before their trip.) This simple reminder tends to eliminate many frustrations and wasted trips.

- **If the meeting involved only one expert who then needs to cancel, reschedule the meeting. If the meeting involved more than one expert, offer to call the expert back after talking with the other participants.** Consider when rescheduling a meeting: whether other experts or only this one will be unable to attend, whether the meeting could be video taped, whether a cancellation will affect costs (e.g., travel, lodging, or meeting room), and whether it is possible at this late date to contact the other experts to cancel the meeting. Offer to call the first expert back after talking with the other participants. If only one expert cannot attend, he or she could be shown a video tape of the session and interviewed later. If several experts cannot attend, consider using the alternate (backup) date.

Preparing for and Conducting the Elicitations

Tips on Preparing for the Elicitations

Preparing for the elicitation means bringing the necessary papers and supplies and physically arranging the meeting room. Many of these preparations do not critically affect the success of the elicitations; however, they do make the elicitations easier and more pleasant. The preparations are listed below and can be used as a checklist or memory aid.

How to Prepare for an Individual Interview

We recommend assembling the following information or materials for an individual interview:

1. **The expert's name,** because forgetting it in the midst of the elicitation can be embarrassing. In addition, determine the address and telephone number of the meeting place to avoid becoming lost or arriving late. Maps of the area can also prove useful.
2. **A short letter on the project,** including who sponsors it, who conducts it, and what its product will be. (A longer version of this letter is described in *Motivating the Experts Through Communication of Intrinsic Aspects of the Project,* chapter 6). This information on the proper letterhead helps establish the interviewer's and the study's credentials and refreshes the expert's memory on the project. In addition, we recommend stating this information to the expert, rather than just handing it to him or her for reading. Conveying this information verbally will get the interview started and ensure that the expert has heard the necessary information at the beginning of the interview.
3. **The documentation format, list of question topics, or any form** that will help guide the questioning. These can be labeled with the interview's date and the expert's name or identification number before the interview.

4. **Copies of the questions and/or background materials** (references, charts, tables, etc.) that the expert will use. The expert will need one copy and the interviewer will need another to follow what the expert views or comments on. Again, the expert's identity can be recorded in advance on these papers.
5. **Extra pencils or paper** for taking notes.
6. **A mechanical recording device (tape recorder or video cassette recorder), cassettes, extra batteries and/or an extension cord.** The tape cassettes can be labeled in advance with the expert's name and the time and date of the elicitation. This labeling is particularly important if an expert is interviewed more than once on the same question because his or her thinking is likely to change with time.

How to Prepare for an Interactive Group Situation

A visitor or meeting coordinator can help with the support materials that follow. We recommend having the following papers and supplies ready for the experts:

1. **Materials that the group moderator/interviewer will need**, such as lists of participants, the introduction to the elicitation, copies of the experts' background materials and statements of the problem, the moderator/interviewer's question topics, and documentation formats.
2. **The mechanical recording devices (tape recorder or video cassette recorder), cassettes, extra batteries and/or extension cords.** When video recording a session, set up the equipment and the experts' seating to test the equipment's functioning. Make sure the seating arrangement does not hamper the equipment's reception of sound or picture. Label the cassettes with the session's date, time, and any other relevant information before the meeting begins.
3. **Name tags for the experts,** especially if they are meeting the session moderator or each other for the first time. For large groups, have the experts print their names across each side of 8 1/2- by 11-inch papers that have been folded lengthwise. Have the nameplates placed in front of the experts for easy viewing. If the experts are supposed to sit in particular seats, their name tags can be set out in advance to show them where to sit. Otherwise, it is very effective to present each expert with a packet of the materials listed below and labeled by his or her name tag.
4. **The program schedule.** A general schedule includes the goals and deadlines. More detailed schedules include the names of speakers, topics of discussion, and their times. If a project will have presentations, we recommend listing the speaker, title, times of the talks, and the person in charge of each session. This procedure has been effective in keeping elicitation meetings on schedule. Please note that time overruns are sometimes necessary, such as when the experts are clarifying a question. The person listed as being in charge of the session can be asked to decide if the discussion should continue or if it is being unproductive and should end.

5. **The experts' copies of background materials, statements of the questions, and documentation formats,** if the experts will be recording their own data. Often the experts will wish to provide current references to be distributed at the meetings. It is much easier to receive and copy these in advance of the meeting than during it.
6. **Refreshments.** If the experts will be meeting for more than a few hours *or* for more than one day, the use of caffeinic beverages and snacks can prolong their productivity. Sweet, fatty foods, such as doughnuts, should be avoided because of their sedating effects.
7. **A list of restaurants and things to do** (sights, tours, and activities), if the meetings will last more than a day and are being held outside of the experts' work places. We have found that the extra effort of treating the experts like VIPs is more than rewarded by their good will and favorable impression of the project. As a professional visitor coordinator told us, "People may not remember the technical content of the meetings but they will remember how they were treated."
8. **Extra notepaper and pens** for the experts' use.

How to Prepare for a Delphi Situation

Preparing for a Delphi situation differs from the other two situations because its communications will be by mail and/or by telephone. As mentioned in chapter 7, one of the greatest problems with mail communications is that it has a low response rate. We therefore advise putting as much effort as possible into these communications with the experts to increase response rates.

If the experts will receive and return the questions by mail, the following need to be prepared.

1. **A cover letter for the set of questions.** The cover letter contains an abbreviated version of the letter first sent to motivate the experts to participate. (See chapter 6, *Motivating the Experts Through Communication of the Intrinsic Aspects of the Project.*) The cover letter should be carefully composed because it is one of the most important tools for encouraging the experts to answer.
2. **Copies of the set of questions.** These sets include instructions on how to respond to the questions, explanations of the response mode, and directions on how to return the expert judgments. If the expert needs to return data by postal mail, provide a self-addressed stamped envelope. Self-addressed envelopes tend to increase response rates on mail surveys.
3. **The experts' names, addresses (electronic or postal mail), and telephone numbers.** If the address list is likely to be outdated, call the experts to verify current mailing addresses.
4. **Advance publicity to increase the response rate.** If the experts work for the same organization, design a brief article for the organization's news bulletin or have the management prepare a memo. If the experts are not all in the same place, call them before sending the set of questions. In general, we recom-

mend calling the experts in addition to publicizing the study through the use of articles and memos. Frequently, so much time will have elapsed between the time when the experts were selected and when they receive the set of questions that they will have forgotten about the study.

If the experts will give their answers to the interviewer over the telephone, follow the directions given above, with only slight modifications. With a telephone interview of this sort, the experts are usually mailed the information and then called to obtain their judgments. The interviewer discusses each question over the telephone with the expert. Please note that telephone communication between the expert and interviewer should not last more than about fifteen minutes per call. As with scheduling interviews, the interviewer should ask the expert if it is a convenient time to talk before proceeding. The interviewer can use the written cover letter as a guide in introducing the expert to the elicitation. To avoid sounding repetitive and flat, do not read the cover letter over the telephone.

Tips on Conducting the Elicitations

For convenience we have divided the elicitation sessions into four parts: introducing the experts to the elicitation process, monitoring the session for bias, gathering and recording the data, and countering bias. The following suggestions may help interviewers perform each part of the elicitation. However, we recommend relying on the lessons learned from practicing or pilot testing the elicitation methods.

Introducing the Experts to the Elicitation Process

This section includes step 3 in the program for handling bias, as discussed in chapter 3 (*Program for Handling Bias).* If the project will not use a program for handling bias, we recommend the following procedures because they generally prove beneficial to the expert's performance. If the program for handling bias will not be used, omit the procedure of briefing the experts on the biases to which they may be prone.

Make the Experts Aware of the Potential Intrusion of Particular Biases and Familiarize Them with the Elicitation Procedures—Step 3. The experts should understand the potential for bias and be acquainted with the elicitation procedures. Regarding bias, they should know what it is and why it happens. Without this information, the experts will be unable to combat their own tendencies toward bias. The interviewers or knowledge engineers can use the *Index* and *Definitions* (chapter 8) to inform the experts about bias.

Also note that making the experts aware of bias helps but does not completely alleviate the problem. For example, if experts understand how their responses may be misinterpreted or influenced by the data gatherers, they can help monitor for these biases. However, in some cases the cause of the bias, such as the experts' underestimation of uncertainty, is too ingrained to be completely overcome. In other cases, the experts will not effectively attempt to counter the natural tendency toward bias. People typically believe that others, not themselves, will suffer from the biases described. With some biases, such as anchoring and underestimation of uncertainty, the experts can participate in tests designed to evoke the bias. Frequently, almanacs are used to construct test questions, such as: *How much rain fell in St. Paul, Minnesota in 1987?* While the experts will not know the answers to such questions, the data gatherers can research the correct answers. They can read the answers and allow the experts to score their own. Such a demonstration often convinces the experts that each of them is prone to the bias.

The experts also need to understand the elicitation procedures. If experts are confused about how and when they should respond, the data gathered as well as the expert's cooperativeness may be negatively affected. One confusing aspect of elicitation is the response mode, especially if the expert is not accustomed to using it. Rehearse the use of unfamiliar response modes with the experts during the training session. [If the response mode is probability distributions, Hogarth (1980, p. 149) provides eight keys to its use.]

The following information describes how to conduct step 3 of the bias program based on the elicitation situation. Items 3 through 5 provide information for the individual interview; items 3 through 5 on the next page provide information for the interactive group; and the two paragraphs following these sections apply to the Delphi method.

How to Introduce the Experts to an Individual Interview Situation

1. **The interviewer introduces him or herself,** if unacquainted with the expert. The interviewer then gives the expert a cover letter (see *How to Prepare for an Individual Interview*) to establish his or her credentials. If the interviewer intends to question the expert on classified matters, he or she must be prepared to show further identification and provide names of personal references. After the expert has glanced at the cover letter, the interviewer should quickly deliver this same information verbally. (Giving the information verbally ensures that all the experts will receive the same information, something that cannot be said of their reading it.) For example, we usually mention the sponsor of the study, how the expert was selected, what the expert will be doing, how long it is likely to take, how the expert's data will be protected, and the anticipated product of the project and the expert's access to it. At the end of this description, the interviewer should ask the expert if he or she has any questions. Start any recording devices at this time.
2. **Start with questions on the expert's professional background**. These simple questions will allow the experts to get accustomed to the interviewing process and will reassure them that they are capable of answering later questions. Examples of these questions include: *How many years have you worked in your present position? What educational degrees do you hold? In what fields are your degrees?*
3. **Give the expert some sample questions** to familiarize him or her with the use of the response mode.
4. **Brief the expert on any biases** identified as being likely to occur (chapter 8). Give the expert ideas on how to try to counter the tendency toward these

biases. (The section *Definitions of Selected Biases* in chapter 8 provides examples of this type of information.)

5. **Give the expert the set of questions and verbally go over any instructions.** Ask the expert if he or she has any questions.
6. **Tell the expert that he or she can begin.** If the elicitation process requires time estimates, record the beginning and ending times for each session.

How to Introduce the Experts to an Interactive Group Situation

1. **Distribute the materials described in** *How to Prepare for an Interactive Group Situation.* Turn on any recording devices at this time.
2. **Introduce the meeting moderator/interviewer, the project staff, and the experts.**
3. **Review the project's purpose, its schedule, and, in general, the elicitation procedures for the benefit of the experts.** Some descriptions of the elicitation procedures follow: *You will meet for one week to develop detailed statements or representations of the problems. On the last day, Friday, you will vote on what you think the answers should be.* A more detailed overview of elicitation procedures follows: *You will meet here three times: first, to become familiar with the project and the elicitation procedures; second, to present up-to-date technical information and refine the rough-drafted questions; and third, to give your expert judgment in private meetings with an interviewer.*
4. **Give the experts sample questions to work** so that they can practice using the response mode. Any techniques on properly using the response mode can be introduced and practiced at this time.
5. **Brief the experts on those biases identified in chapter 8 as likely to occur.** This briefing should include an explanation of why the selected biases occur and of how the experts can help prevent introducing them. In addition, the briefing on bias should include exercises designed to evoke the selected biases. After the experts have completed the exercises, the moderator/interviewer can read the answers and allow the experts to correct their own. If the briefing is given without these exercises, we have noticed that the experts are not as effective in countering their tendencies toward bias, perhaps because they were never convinced that they, too, would be vulnerable.
6. **Ask if the experts have any questions.** Afterwards, state that the introduction is concluded and that the sessions will now begin.

How to Introduce the Experts to a Delphi Situation. If the Delphi is conducted entirely by mail, the experts will not be introduced to the elicitation by the moderator/interviewer in person. Instead, the experts will receive the cover letter and set of questions as described above.

If part of the Delphi is conducted by telephone, call the experts to help them understand the set of questions or to directly obtain their answers. Use the items listed for individual interviews as a basis for introducing the experts to the elicitation process.

Monitor the Elicitation for the Occurrence of Bias—Step 4. Many of the selected biases have signs that indicate their occurrence. Before the elicitation sessions, the data gatherers can read about the signs that signal the occurrence of particular biases. (See Table 1, *Signs of Selected Biases.*) The interviewer, knowledge engineer, or a trained observer can watch for these signs during the elicitation. In general, monitoring biases, as described in this book, requires that the experts verbalize their thoughts and answers. Without this feedback, we have found the monitoring to be much more difficult.

TABLE 1
Signs of selected biases

Group think. There are several signs that predict the development of a group-think situation. In general, no one voices a difference of opinion; the experts appear to defer to one or more members of the group (Meyer, 1986, p. 95).

Wishful thinking. Wishful thinking is indicated if the experts were previously judged to have something to gain from their answers and if the experts appear to answer quickly and with very little thought.

Inconsistency. A number of signs can indicate inconsistency. The interviewers can detect many inconsistencies when the experts verbalize their thoughts and answers. In particular, the interviewers can discern when a response mode or rating is being applied more easily through time (Meyer, 1986, p. 94). Experts tend to apply the extremes of a rating scale more easily as they become fatigued. The interviewers can also hear when the experts contradict assumptions that they made earlier. For example, a tank expert chose two very different routes through the mapped terrain because the second time he unconsciously assumed that his company was the main effort and therefore *had to push hard.*

Inconsistency can also be monitored by the use of Bayesian-based scoring and ranking techniques. During the elicitation, the experts' judgments can be entered into a scoring and ranking program, such as that of Saaty's Analytical Hierarchical Process (1980), to obtain a rating of their consistency. Then, if the inconsistency index from this method remains too high, indicating significant inconsistency, the experts can redo their judgments.

Availability. A potential problem with availability bias is indicated if the experts do not mention more than one or two considerations in giving their answers. If the experts only consider a few things, these were probably the most easily remembered and their answers are likely to be skewed to reflect these few.

Anchoring. Suspect anchoring bias when an expert receives additional information from experts or other sources during the elicitation but never waivers from his or her first impression. For example, reactor code experts were asked to compare the performance of their computer models to plots of experimentally generated data. They often commented on their first impression. When they examined the plots more closely, they typically found places where the computer model did not capture the experimental phenomena. However, the experts usually adjusted upward or downward from their initial assessment rather than revising it completely (Meyer & Booker, 1987*b*).

Gathering and Recording Expert Data

Using the Individual Interview, Interactive Group, and Delphi Situations. As a general rule, allow the results from the practice runs or pilot tests (as described in chapter 9) to serve as a guide for this phase. If the elicitations were not pilot tested, we recommend reading *Common Difficulties—Their Signs and Solutions* at end of this chapter before proceeding. Reading the section on common difficulties may provide ideas on how to avoid some of these problems.

Using the Three Techniques for Eliciting Problem-Solving Data. The three techniques for eliciting problem-solving data—verbal report, verbal probe, and the ethnographic technique—are frequently used in combination with individual interviews. The verbal probe and ethnographic technique are occasionally used with the interactive group situation to gather a few sentences on how the experts solved the problem. Details are provided below on how to administer these techniques.

Verbal report. To review, verbal report involves instructing the expert to think aloud as he or she progresses through the problem. The expert usually receives a written copy of the problem. The expert's verbal report resembles someone talking to him or herself. This technique comes from psychology. The following suggestions will help conduct the verbal report.

- **The interviewer should sit slightly behind the expert and opposite the expert's writing hand** (i.e., the interviewer would sit by the expert's left side if the expert is right handed). This positioning offers several advantages. First, the expert will not be as able to see when the interviewer is taking notes. If the expert becomes aware of the interviewer's particular interests, this awareness may bias the response. Second, the interviewer can observe the expert's actions and reactions, including what he or she looks at, marks up, or writes down.
- **Initially design the questions so that they fit the expert's expertise.** Otherwise, there is little point in using the expert. Organize the elicitation sessions to resemble (as much as possible) the environment in which the expert usually solves such problems. Otherwise, factors may be introduced into a particular elicitation session that change the expert's usual way of thinking and make him or her inconsistent. The sessions can take place in the expert's

customary work place, if interruptions can be controlled. For instance, telephone calls can be handled by call forwarding.

- **Obtain copies of whatever visual aids (e.g., tables, graphs, and equations) or references the expert will use.** This practice allows the interviewer to follow what the expert views even when he or she cannot see over the expert's shoulder because of the distance and the print size. The interviewer also uses the copy for recording what the expert looks at and marks. Hard copies should also be obtained if the expert is solving problems on a computer. In the latter case, the hard copies are copies of the computer screens.
- **Take notes rather than rely solely on the expert or recording devices.** Experts should not be asked to provide detailed notes because this activity is likely to be done at the expense of their thinking. If an expert is required to take notes, his or her resultant information may be sparse and unreliable. Recording devices do not provide complete records. For example, they do not show the marks that experts make on their papers as they think. In addition, recording devices malfunction, so a backup copy, that of the interviewer, is likely to be needed.
- **Emphasize that the expert should work through the question rather than talk hypothetically about how it could be solved.** Experts are rarely aware of or accurate on how they actually solve a problem. Therefore, they provide more reliable (less biased) information if they verbalize while they solve the problem. The message on actually working the problem need only be delivered the first couple of times that the interviewer works with the expert.
- **Stress the importance of thinking aloud when instructing the expert to solve the question.** The first time an expert solves a question may remind him or her of a test situation in school. As a result, the expert's tendency may be to rush through the question to obtain the solution. The interviewer must convince the expert to think aloud while being careful not to emphasize the importance of verbalizing at the expense of problem solving. The latter alters the way the expert would normally work the problem and is therefore undesirable.

EXAMPLE 10.1: ***Illustration of the verbal report***

The interviewer in this example wishes to learn, in general, how the expert plans a feed program, what information on the horse and on the types of feed are needed, and what concepts (e.g., the horse's metabolism) are considered. The expert receives a written copy of the problem:

> What feed program would you start this colt on? The colt is 6 months old, 550 lbs., has an average metabolism, and will receive light exercise through ponying. Assume that the colt has normal good health.

Interviewer: Solve this problem in the same manner as others that you receive in this field. Please try to think aloud as you work your way through the problem. Your thinking aloud is as important to me as the answer that you reach.

Expert: *(The expert slowly reads the question aloud.)* The first thing that I'll need to find out is what this colt would weigh full grown. *(The expert scans some xeroxes that were previously copied from reference books.)* Let's see, at six months and 550 lbs., he will be about 1100 to 1300 lbs. full grown.

Next, I need to find the balance between hay and grain that I'd feed a horse of this age. *(The expert looks at another table.)* Four lbs. of hay and 8 lbs. of grain. However, I like to feed more hay, so I will aim for 6 lbs. of hay and 7 lbs. of grain.

Other constraints that I have are the amount of protein in pounds needed for a colt that will mature to a 1100- to 1300-lb. horse. *(The expert refers to another table.)* I'll want to balance between 1.74 and 1.89 lbs. of protein per day.

(The expert examines another chart.) I also need to balance protein in the overall diet to about . . . 14.5%, calcium to . . . 6%, and phosphorus to . . . 0.45%.

Foods that I like to feed in the diet are oats, corn, and soybean meal. I also like a feed supplement for extra protein, Horsecharge, and alfalfa and timothy hay. I'm getting their percentages of protein, calcium, and phosphorus from the charts. *(The expert ceases thinking aloud, records the percentages of these feeds, and begins using a calculator.)*

The expert writes a list of feeds and weights then relaxes in the chair, signaling that the exercise has finished.

■

- **Experts frequently stop thinking aloud and need prompting to resume. Several types of prompts follow.** One reason for using different prompts is to avoid repetitiveness that can irritate both the expert and the interviewer. Another reason is to use the most subtle prompt possible to remind, rather than to distract, the expert. Distraction can cause experts to lose their place and then solve the problem differently than they would have otherwise. The following prompts are arranged from most-to-least subtle.

  1. Look intently at the expert and lean forward slightly.
  2. Look at the expert, lean forward, and clear your throat.
  3. State *please try to think aloud.*

  Use the first prompt with an expert who only needs an occasional reminder to think aloud. The expert perceives the interviewer's forward movement almost subliminally and then responds. The second prompt creates a slightly stronger impression because it triggers associations concerning the expert's throat (e.g., *my throat should also be making sounds*). Use the third prompt only on experts who fail to notice or respond to the first two prompts.

  - **Avoid prompting an expert who is counting or otherwise performing calculations.** The expert could lose his or her place. If the interviewer believes that prompting is necessary, he or she should start with the first prompt. The nonverbal aspect of the first prompt allows the expert to note it without being distracted.

Verbal probe. The verbal probe is questioning done at a particular time and in a specific manner. The type of verbal probe discussed in this book is used immediately after the expert has reached a solution, focuses on only one problem, and is indirectly phrased so as to minimize influencing the expert's thinking. For example, immediately after the expert solves the problem and gives an answer, the verbal probe helps determine why the expert gave that particular answer.

Interviewer: Why did you give that answer—that feed program?
Expert: Well, it provides the right amount of protein, calcium, and phosphorus for the horse to grow at this age.

- **If the verbal probe is used in a face-to-face individual interview, set up as described earlier for the verbal report.** As a general rule, individual interviews can obtain more detailed information than group interviews.
- **If there will be multiple experts as in an interactive group setting, the group moderator should tell the experts in advance that they will be asked to provide their reasons for giving their answers.** The experts should provide this information verbally because people, in general, do not usually provide good written records of their reasoning. The experts' written notes are usually sketchy and, for this reason, more likely to be misinterpreted by the interviewer than their more complete verbal counterparts.

EXAMPLE 10.2: ***Illustration of the verbal probe***

Assume that the expert has just finished solving the problem. After the expert gives an answer, the interviewer will administer a verbal probe to learn why the expert gave that answer.

Interviewer: What was your answer?
Expert: The diet, per day, that I would recommend for this horse is 4 lbs. of oats, 1.5 lbs. of corn, 0.5 lbs. of soybean meal, 1 lb. of Horsecharge, 4 lbs. of alfalfa, and 2 lbs. of timothy. This is, of course, only the starting point.
Interviewer: Why did you give that answer—that feed program?
Expert: Well, this feed program provides the right amounts of protein, calcium, and phosphorus for a horse to grow at this age.

■

- Word the verbal probe in the past tense (e.g., *Why did you give this answer?*) to emphasize wishing to know how the expert actually solved the problem. Some experts will begin, even after solving a problem, to describe how they or a colleague *could* have solved it. The interviewer's response to such a beginning should be something like: *I am interested in your thinking and how you solved this problem.*

- Check for tautological responses. Tautologies are reasons that do not truly provide the *why* information being sought. For example, *I gave that feed program because it seemed right* is a tautological response. Although, this particular tautology is obvious, tautologies can be difficult to discern in an unfamiliar domain, vocabulary, or when fatigue sets in. Frequently, the desired information can be obtained by using the ethnographic technique to probe on the tautology.

Ethnographic technique. The ethnographic technique involves restating the expert's responses into questions. For example, the interviewer could use the ethnographic method to probe on one of the expert's responses to obtain an operational definition that could then be entered into the expert program. Assume that the expert has just said that *the colt's feed program may need adjustment if the colt is not keeping his weight on.*

Interviewer: Not keeping his weight on?
Expert: Yes, not gaining as he should at this time.
Interviewer: At this time?
Expert: At his age, 6 months, he should be gaining between 1.5 and 2 lbs. per month.

- The physical setup of the ethnographic technique is usually determined by the setup for the other techniques used or by the situation in which it is used alone. For example, if the ethnographic technique were used with the verbal report, the verbal report's setup would be used. If the ethnographic technique was used by itself, no special preparation would be needed. For example, if the interviewer intended to use the ethnographic technique on one expert at a time, this technique would be physically arranged as a conversation (e.g., with the interviewer sitting adjacent to or across from the expert).

EXAMPLE 10.3: ***Illustration of the ethnographic technique***

The expert has just answered the verbal probe explaining his or her answer. The ethnographic technique could be used either to investigate the response to the verbal probe or to information given earlier. Earlier, the expert had given the feed program and added, *Of course, this is only the starting point.* For the purpose of this illustration, the ethnographic technique will be applied to the expert's reply to the previous verbal probe and then to the subject as it develops.

Interviewer: A starting point?
Expert: Yes, because the rations may need to be adjusted, if the colt is getting fat or if he's not keeping his weight on.
Interviewer: Not keeping his weight on?
Expert: Yes, not gaining as he should at this time.

Interviewer: What do you mean by "at this time?"
Expert: At his age, 6 months, given his expected total weight, he should be gaining between 1.5 to 2 lbs. per day.

The ethnographic technique was used to follow a line of the expert's thoughts to a detailed level where the expert's operational definition of *not keeping his weight on* was obtained. The interviewer stepped down three levels in questioning, as shown in example 10.4. The ethnographic technique is effective at taking the questioning to the desired level of granularity, in this case to a definition that could be mathematically represented and entered into a computer program. In another case, the ethnographic technique could be used to branch across several topics at a more general level.

EXAMPLE 10.4: ***Illustration of a series of ethnographic questions***

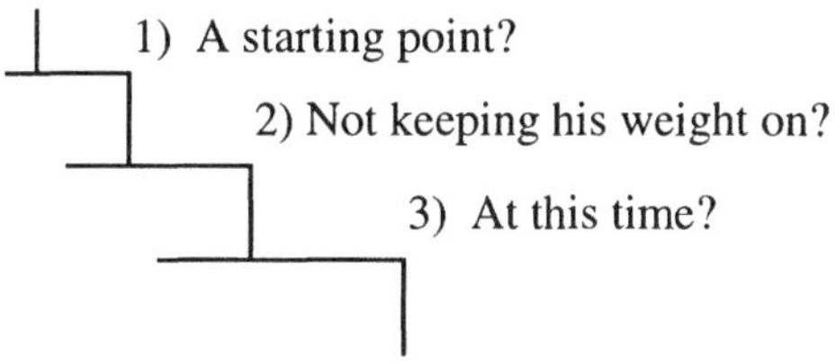

- **Vary the length of the ethnographic questions so that they do not become tedious.** For example, if the expert said, *The books provide samples and I start with these,* a short ethnographic query would be, *samples?* For variety, a longer query would be, *What samples do the books provide?*
- **If the problem-solving session involves the ethnographic technique, use it only after the expert reaches the point in the problem-solving process that satisfies the interviewing goals.** Otherwise, the expert will likely lose what he or she has stored in short-term memory while responding to the questioning. If the interviewer then resumes questioning where the interruption occurred, the expert will have to guess his or her previous thoughts. Bias can result. For this reason, the ethnographic technique is best used after other techniques or by itself. The ethnographic technique can be occasionally inserted during the use of another technique but only if it is asked quickly and not pursued to fine granularity.

Adjust, in Real Time, to Counter the Occurrence of Biases—Step 5. Bias can intrude during the gathering and recording of data mentioned above. This section includes step 5 of the program that we propose for handling bias.

Table 2 mentions some of the techniques that we have used to adjust for bias. In general, our approach has been to either impede those factors contributing to a particular bias or to employ the opposite bias. For example, the data gatherer's tendency to lead or influence the expert contributes to social pressure bias. When following the first approach, we advocate using elicitation methods to curb this tendency (Meyer *et al.*, 1989). An example of this approach in countering inconsistency would be to focus on the fatigue and confusion that can contribute to this natural tendency. The elicitation sessions can be scheduled to stop before the point when most of the experts become fatigued. The basis of the second approach, fighting bias with bias, comes from the grandfather of survey design, Stanley Payne (1951). Payne believed that all interviewing was biased and that interviewers should therefore aim for equal but opposite biases. An example of this approach is to have experts anchor to their own judgments in an attempt to counter a group-think situation. One means of having the experts anchor to their own judgments would be for the interviewer to write each expert's viewpoint by name on a viewgraph.

Step 5 is perhaps the most delicate one in the program for handling bias because if done incorrectly it could confuse the results, affecting later analyses. The interviewer or knowledge engineer must decide in advance on the timing of the adjustment because these adjustments change the conditions under which the data are gathered. To avoid any problems, correct conditions that may lead to bias before the actual data collection takes place. Mixed data result, however, when analyzable data are gathered under two conditions: when bias occurs and when bias is corrected. Testing for the presence of a particular bias is not possible unless the situations can be clearly separated, such as before and after the correction of this bias.

TABLE 2

Suggestions for countering selected biases

Group think. Two approaches can help counter the social pressure from group think (Meyer, 1986, pp. 95-96). Using the first approach, the interviewer can attempt to prevent those factors that contribute to group think. For instance, the interviewer can stop the elicitation and warn the group members about group think. If the group contains an official or even a natural *ex officio* leader, the interviewer can ask for his or her responses either last or in private. In addition, if someone other than the interviewer leads the group meeting, this person can be encouraged to be nondirective during the meetings. An explanation of the group-think phenomenon usually convinces such leaders that better discussions and data will result from their avoiding leading.

The other approach is to try to counter the effects of group think with an opposite bias—anchoring. One technique for fostering anchoring is to require the group members to write down their judgments and reasoning. In this way, they are more likely to anchor to their own judgments rather than silently acquiesce to someone else's. If the experts discuss their judgments, each person can record and report his or her response before the floor is opened to discussion. Once individuals have publicly announced their view, they are unlikely to spontaneously modify it. (However, they will still modify their view if someone raises a valid point that they had not previously considered.)

Wishful thinking. The tendency toward wishful thinking can be countered by making it more difficult for the expert to indulge in it. If the experts must explain their answers in detail, it will become apparent whether there was any objective basis for their responses.

Inconsistency. Two techniques can help reduce inconsistency. The first technique addresses the aspects of the elicitation that contribute to the inconsistency.

As mentioned earlier, fatigue contributes to inconsistency. If the interviewer notes that the experts have become more inconsistent with time, he or she can quickly end the meeting or schedule a break. In general, experts can participate in discussion or problem solving for a maximum of two hours before becoming tired. (Experts often signal their fatigue either by providing briefer responses or by leaning forward or back in their chairs.)

Faulty memory also contributes to inconsistency. If at the beginning of every session the statement of the question, definitions, assumptions, and response mode are reviewed, the experts will be more consistent in their judgments (Meyer, 1986, p. 96). They will be more consistent between and within themselves. In addition, if much time elapses between this first review and when the experts' judgments are requested, the question can be worded to include some of the above information. For example, *What rating would you give to the importance of element X over Y to the reaching of goal Z?* If they use a response mode (in this case a Saaty pairwise comparison) they will need to have the definitions of the scale available.

Another technique that helps reduce inconsistency is to have the group members monitor their own consistency (Meyer, 1986, p. 96). This technique was successfully employed in a simple interactive group elicitation where the experts were able to watch the interviewer's monitoring of inconsistency and then to mimic the interviewer's performance (Meyer, Peaslee & Booker, 1982). The experts received copies of a matrix of the elements being judged, the criteria on which these elements were being judged, and their past judgments. When experts monitor their own consistency they may wish to change an earlier judgment to reflect their current thinking. If their reasoning does not violate the logic of the model or the definitions, they can be allowed to make the change. This process often helps the experts discover that they had forgotten to include some pertinent information. After this addition, some of the judgments may require redoing.

If Saaty's Analytic Hierarchy Process (1980) had been applied and its results indicated high inconsistency, the experts could review and redo the affected judgments.

Availability. Stimulating the expert's memory associations can counter availability bias. In general, group discussion will cause the expert to think of more than just the first readily accessible information. In addition, free association can be introduced to single experts or those in groups. [Free association has the expert or experts quickly generate any elements that might have bearing on the question (Meyer, 1986, p. 94).] Free association is similar to brainstorming or the Crawford Slip method (Boose & Gaines, 1988, p. 38). The experts are asked to refrain from being critical to generate the widest possible pool of ideas. (The number of ideas is later narrowed to those judged to be most pertinent to the question.) A related technique is to hierarchically structure the presentation of question information so that it flows from the general to the specific. In this way, the expert can consider the pertinent information before reaching a solution. This strategy fires as many memory associations as possible so that the maximum number of relevant associations enter into the expert's final judgment.

Anchoring. Techniques similar to those used to counter availability bias are used to counter anchoring. In particular, giving the expert input from other experts as in a Delphi situation or an interactive group makes it more difficult for the expert to anchor to his or her first impression. Another technique involves asking the experts for extreme judgments before obtaining their likely ones (Cleaves, 1986, pp. 9-10).

Three additional ideas for countering availability bias from Boose and Shaw (1989, p. 70) follow: (1) ask the experts to describe how other experts might disagree with their responses, (2) ask the experts to temporarily forget recent events, and (3) aggregate outcomes with small probabilities into a single, larger class to reduce the perceived joint impact in cases in which probability estimates are elicited.

Underestimation of uncertainty may be reduced by asking the expert to further disaggregate the parts of the question and give estimates of the quantities of interest for each small part. In this way, the experts are less likely to overlook something; the knowledge engineers can check at this point whether the details of each of the experts' thinking correctly reflect their answers. If the underestimation occurs in a planning problem, a comprehensive list of events that could upset the plans can be elicited from the expert (Boose & Shaw, 1989, p. 72). Creating this list may make the experts realize that they should take some of these possibilities into account in their problem solving.

Step 5 is the program's last step for handling bias that can be applied during the elicitation. For this reason, step 6, *Analyze the Data for the Occurrence of Bias,* is discussed generally in the Analysis section (Part II, chapter 14).

Common Difficulties—Their Signs and Solutions

Difficulty: The Experts Resist the Elicitation Process or Resist Giving Judgments Under Uncertainty. We have seen this situation develop in a few interactive group situations, especially among engineers who were unaccustomed to thinking in terms of uncertainty. One of its first signs is the experts' reluctance to give their judgments. They may mutter amongst themselves or they may openly criticize the elicitation procedures and refuse to give their data.

Experts who are reluctant to give their judgments should not be forced for two reasons: they may not be experts in this area if they do not feel qualified to give their judgments; and they may react negatively to such action, detrimentally affecting their views of the entire project. This reaction is illustrated by a reviewer's statements about a study: "The participants were forced to provide unsubstantiated guesses as input" (Benjamin *et al.*, 1987, p. F-5, 6).

Solutions: In general, the solution to this problem must rest on addressing the experts' reasons for resistance. We recommend that the objecting experts be individually taken aside and asked in a pleasant manner why they are reluctant to give their judgments. Ask the experts separately because asking them as a group may not reveal the individual reasons and it may reinforce their resistance. We provide the following reasons for experts' resistance: (1) they have misgivings about an elicitation process that they helped develop; (2) the experts do not trust the applications that will incorporate their final data ; (3) that they thought there would be new data related to the question and that they would they would not be able to use it in making their judgments; (4) they feared that their judgments would be misinterpreted, taken out of context, and unfairly criticized; and (5) they did not think they could give reliable estimates, given the high uncertainty of the subject area.

The means for resolving this difficulty lie in addressing its underlying causes. For instance, if the experts implicitly feel that they did not have sufficient input into developing the questions, as described in (1), allow them to refine the questions at this point. In addition, let them know of the design considerations and constraints that led to the creation of the current methods. This information will help convince them that the elicitation was not ill conceived and will save the project personnel from having to accept everything that the experts suggest in refining the questions. In other words, the project personnel will be able to point out which of the experts' proposed modifications would not meet the project's constraints.

The other reasons for resistance mentioned above can be similarly resolved.

If the experts do not trust the use to which their judgments will be put (2), show them an outline of the project report or arrange for the eventual users of their judgments to brief them.

If the experts were concerned about using the most current information in making their judgment (3), help them circulate this information before the elicitation sessions. Explain that their judgment is a snapshot of their knowledge at the time, that it is likely to remain in a state of change, and that there must be some cutoff point for writing up the results of the project or for first entering the expert data into a model or data base.

If the experts were worried about misinterpretation of their judgments (4), explain that they will review the documentation of their judgments after the elicitation situations. If possible, show them an outline on how their judgments will appear in the report or model.

Address the experts' doubts about giving judgments in areas of high uncertainty (5) by acknowledging the validity of this concern. Explain that the high uncertainty in the field is one of the reasons why their judgment must be elicited—that other data are lacking. Also summarize how the elicitation practices will help them give more reliable estimates.

Questioning the experts on their reasons for resistance also helps identify those experts who are the most vocal in their reluctance. In many cases, only one or a few experts cause the rest to question the elicitation. We have found it effective to first address the concerns of these few natural leaders. Again, we suggest talking to these experts separately and privately. If it is not possible to resolve the leaders' concerns, it may be feasible to offset their concerns by examining the positive repercussions of the project. For example, one expert expressed a concern that the funder of the project would use the expert data inappropriately, something which we could not absolutely prevent (i.e., people can always pull data out of context and misuse it). Aside from emphasizing the good that would come from the study (the opportunity for the experts to meet together, to pool their most current data, to identify gaps, to conduct further research, and to cause the field to progress), we could not completely assuage this expert's worries. However, we managed to convince this expert to proceed, and thus the others, by privately appealing to him for his assistance. We explained that he was a natural leader in this group, that the others were being influenced by his views, and that we needed him to agree to an elicitation session.

If all else fails, consider doing as much of the elicitation process as possible with one expert at a time. For example, perhaps individual interviews can be used instead of interactive group situations to obtain the experts' data. In the former, the interviewer has only to deal with one uncooperative expert at a time and can be more flexible in responding to their individual needs.

Difficulty: The Experts Appear Confused About What They Should Do or How They Should Do It. The experts may progress slowly through the elicitation procedures and look puzzled. Interviewers may find it difficult to discern if the experts are confused about the elicitation or simply stalling because in both cases they will take much time to provide their judgments. If the experts were introduced to the process by working sample questions and using the response mode, these exercises will provide the first evidence of their confusion. If they have not been given any explanation of what they are to do, then the existence and source of their confusion can almost be assumed. Intensive pilot tests can also provide warning as to which parts of the elicitation are likely to cause confusion.

Solution: The first step, as in the above-mentioned difficulty, is to identify the cause of the expert's reaction. The cause may stem from the general procedures for the elicitation, such as when the experts meet together and when they are separately interviewed; it may stem from the response mode that they are to use; on the other hand, it may be confusion over some instructions. One of the special problems with confusion is that people often cannot express the source of their confusion. To help identify a source of confusion, the interviewer could quickly go through the information already presented and ask the expert to signal the points that were unclear. The interviewer should then convey that the problem has resulted from a lack of clarity in the presentation rather than from a lack of understanding on the part of the expert.

We urge working with the experts until their confusion is resolved because not doing so can have severe ramifications. In the one situation where the experts were pushed to provide data when they were confused, the experts later criticized the elicitation methods and raised doubts as to the data's validity.

Difficulty: The Final Statement/Representation of the Question or the Expert's Latest Data Were Not Documented. The main cause of this common difficulty is the evolutionary quality of elicitations. The phrasing of the question often evolves gradually during a group session, and no special sign signals its entry into its final form. In addition, during an individual elicitation, the expert may try to solve the question in different ways, backtracking a few times, before settling on a process and arriving at a final judgment. Thus, the failure to record the final form of the question or its solution often goes undetected—unless those who worked on it take notice. To detect this difficulty, the involved persons must review the documentation while their memory of this information remains sharp. We recommend that the interviewer request the group or the expert to review the final form while it can still be easily corrected.
Solution: If this difficulty is detected immediately following an elicitation session, the group or the expert and the interviewer can update their copy by using their memories. If not, the interviewer will have to replay the mechanical recordings to update the documentation. The latter is usually very time consuming and difficult because the communications are not as clear in retrospect as when they occurred.

Difficulty: Bias May Have Occurred but Its Presence Was Not Monitored During the Elicitations. Bias has not usually been considered in designing or conducting elicitations. For the majority of studies, the possibility of bias is not considered until the end of the elicitation and only then because someone has raised this question. Raising the question of bias late in the process reduces the options for its detection (e.g., anticipating which biases were likely to occur and then monitoring the session for their intrusion). However, if sufficient data were gathered, it may be possible to test for the presence of a particular bias.

Solution: Chapter 14 provides a detailed description of how to analyze the expert data for bias. If this analysis is being performed at the request of persons external to the project, they often identify which biases to test for. Our experience shows that people worry about three biases: those that occur when the interviewer leads the expert's thinking processes; those that occur when the experts lead fellow experts (e.g., group-think situation); and those that arise from a conflict of interest. These three can be considered motivational biases because they arise from human needs and desires.

Analyzing the data for the presence of particular cognitive biases tends to be an easier procedure (Meyer & Booker, 1989, p. 13). For example, we tested for the underestimation of uncertainty on one project. The experts had estimated the likelihood of achieving national magnetic fusion milestones within particular time periods. They gave probability estimates, such as 0.90, and ranges, such as $\pm$ 0.10. The experts' ranges were analyzed and found to be within one standard deviation of the set of probability estimates. This result indicated that the experts thought they were adequately accounting for uncertainty when they were only accounting for about 60% of uncertainty (Meyer, Peaslee & Booker, 1982).

As a general rule, we recommend that the data be analyzed for the presence of whatever biases are possible and that these results be included in the project report.

Difficulty: There Is Wide Disagreement Between the Experts' Data. Although differences in the experts' data may not constitute a problem, they are frequently perceived as being such. Some view interexpert disagreement as an indication that the elicitation process failed—that it did not produce the one *correct* answer or means of solving the question. However, as mentioned in chapter 1, experts can legitimately solve the question in different ways, and the ways that they solve the question affect the answer that they reach (Ascher, 1978; Booker & Meyer, 1988*a*; Meyer & Booker, 1987*b*). Handling differing expert data can pose problems if this data later require mathematical aggregation or organization, such as for inclusion in an expert or knowledge-based system.

Solutions: Before considering solutions, determine if a problem truly exists—in other words, determine if the disagreement resulted from a weakness in the elicitation or from the natural differences between the experts. To answer this question, data are needed on the experts' problem-solving processes, in particular their definitions and assumptions. If the experts' questions were only loosely defined, the experts will often make their own definitions and assumptions in clarifying the question. In so doing, they can create different questions, use different problem-solving processes, and give correspondingly different answers. If this occurs, there is indeed a problem. However, if records show that the experts used the same definitions and assumptions (i.e., answered the same questions), any differences can be assumed to result from their use of acceptable but different means of solving the question. In this case, if there is a problem, it is one of perception.

If the experts gave different responses because they essentially answered different questions, there is little that can be done. One option is to separate the experts' answers according to the questions that they answered. This operation is sometimes used to declare the answers of one or more experts as being inapplicable to the question. That is, their answers differed from the rest because they answered a different question, one that is outside the study's selected questions. Another option is to recontact those experts who used the discrepant definitions and assumptions and ask them to use the agreed-upon definitions or assumptions and solve the question again. Recontacting the experts to have them solve the problem again can be very time consuming and can, unless carefully done, cause others to question the competence of the project staff.

If the experts legitimately differed, there still may be problems—those of other's perceptions. Those funding the project or external reviewers frequently interpret expert differences as a negative sign. We recommend that they be made aware that expert disagreement is valid and that differences can arise from variation and uncertainties in the experts' answers. Also, they should be provided with any evidence that the elicitation played no role in inducing the differences (e.g., the experts used the same definitions of the questions).

For detailed information on how to aggregate experts' answers, see chapter 16. Chapter 16 provides mathematical techniques for combining the experts' answers by commonly used aggregation estimators: the mean, the geometric mean, and the weighted mean. Comparisons and recommended uses of these estimators are also provided. Because aggregation also involves the combining of uncertainties and ranges of answers, probability distribution approaches (including data based, Monte Carlo simulations, and Bayesian approaches) are also discussed.

Suggestions on how to organize differing experts' knowledge for use in expert systems is given in chapter 9 of *Knowledge Acquisition* (McGraw & Harbison-Briggs, 1989). For example, one of McGraw and Harbison-Briggs' suggestions is to document the experts' names and circumstances of their data as a means of determining where to include each expert's input in the knowledge base. They also describe two approaches to accommodating differences in the experts' data: (1) have the experts reach consensus (behavioral aggregation) and enter only this one piece of expert data in the expert system; and (2) use manual or computer-based techniques for overlaying the multiple experts' input and investigating areas of agreement and disagreement. In some cases, the expert system itself can select from among the differing expert data encoded within it and present the most appropriate expert advice to a user.

Part III

Analysis Procedures

11

Introducing the Techniques for Analysis of Expert Judgment Data

Some analysts view the analysis of expert judgment data as ad hoc. There are no standard forms of modeling and analysis that are applicable to all types of problems in the world, and there are no standard forms for analyzing all expert judgment data. Therefore, Part III contains a compendium of techniques whose applications vary depending upon the design of the elicitation and upon the goals of the study. This chapter provides a detailed explanation of these techniques for those unfamiliar with them. Chapters 12 through 18 discuss the various statistical and computational techniques and provide suggestions for their use.

Some statistical concepts and some notation conventions are needed to understand the suggested analyses and techniques. The first section of this chapter presents a list of letters and symbols establishing the notation convention for Part III. The glossary provides basic definitions of the statistical concepts. The second section of this chapter presents some discussions that cover some basic statistical concepts and topics on ***random variables***, ***probability distributions***, ***estimators***, and ***significance level.***

The remaining sections of this chapter discuss specific analysis techniques used in analyzing expert judgment data. These techniques cover three basic areas of statistical methods: ***simulation*** techniques, data analysis techniques, and ***Bayesian methods/approaches***. The simulation section describes the uses, advantages, and disadvantages of ***Monte Carlo*** and ***bootstrap simulation*** methods. The data analysis section describes ***multivariate analysis*** techniques such as ***correlation, general linear models, cluster analysis, factor analysis, discriminant analysis, analysis of variance,*** and a ***decision analytical*** tool called Saaty's method. The third section describes the methods developed from Bayes theorem and its applications. In all sections the descriptions are brief and limited to the applications for expert judgment data analysis.

Notation for Part III

The following list describes the letters and symbols used throughout the analysis section. Some letters and symbols have more than one meaning (e.g., the symbol

x). The appropriate description for x is given in the text each time that it is used. Some descriptions have more than one representation (e.g., ***variance***). All descriptions are based upon common usage in the literature.

The following list provides the symbol and a corresponding definition:

a	minimum value for a distribution such as the uniform or for a range
$B(,)$	mathematical beta function
b	coefficient in a model usually a linear regression
	maximum value for a distribution such as the uniform or for a range
c	values for the random variable representing a component's reliability
df	degrees of freedom (parameter) of an F distribution, which is used in the analysis of variance technique
$E( )$	expected value of the quantity in parentheses
e	unit vector of 1s
F	value from an F probability distribution, which is used in analysis of variance
f	abscissa values or frequencies on a histogram
$F(,)$	value from an F distribution with parameters df_1 and df_2 degrees of freedom
$f(x)$	probability distribution function (pdf)
$f(x\|\theta)$	the likelihood or data distribution with the data conditioned on the parameter θ
$F(X)$	cumulative probability distribution function (cdf)
$h( )$	general notation for a functional relationship or model
l	coefficient in a linear discriminant analysis model
MS	mean square of the values of a random variable, which is used in analysis of variance
m	estimate of the mean or vector of estimates of the means
n	total number of trials, observations, or samples taken
	one of the parameters of the beta distribution (representing the number of trials)
n_0	one of the parameters of a prior beta distribution (representing the number of trials)
N	total number of sample runs specified for a computer simulation
$n(\sigma^2,\rho)$	number of equivalent experts with common variance (σ^2) and correlation (ρ)
$P( )$	probability of the quantity in parentheses
$P(\|)$	probability of the quantity before the |, which is conditioned on the quantity after the |

r	Pearson correlation coefficient
S	matrix of variance and covariance estimates
s	estimate of the standard deviation
s_{ij}	estimate of the covariance between two variables, i and j
s^2	estimate of the variance
SS	sum of squares of random variable; values used in the analysis variance technique
t	time interval
$var(\,)$	variance estimate of the quantity in parentheses
w	weighting value or factor
X	conventional symbol for a random variable
x	unspecified value of the random variable (X)
	number of failures in a binomial process
	one of the parameters of the beta distribution (representing the number of failures)
	independent variables in GLM
$\bar{x}$	mean of the variable X
x_0	a parameter of a prior beta distribution (representing the number of successes)
y	dependent variables in GLM
	unspecified value of the random variable (Y)
z	unspecified value of the random variable (Z)

The following list provides the symbol and corresponding definition:

α	level of significance or type-I error
ε	the random error (residual) term in a model
λ	failure rate, usually the number of failures per total observations or per time
θ	general parameter for a probability distribution
$\hat{\theta}$	estimate for the general parameter for a probability distribution
$\hat{\theta}(\alpha)$	estimate of the α–th percentile of the distribution of θ
μ	value or vector of mean parameters
ρ	correlation coefficient parameter
Σ	matrix of known variances and covariances (parameters)
Σx	sum of all the x values
Σx^2	sum of all the squared x values
σ^2	known value of a variance, or the variance parameter

Description of Some Basic Statistical Concepts

Random Variables and Probability Distributions

Random variable means that for each quantity of interest measured or observed, there is a set of possible values that the quantity of interest can assume, *take on*, or be equal to. That quantity of interest is called a random variable, or simply ***variable***. When a measurement is obtained, an observation is made, or a datum is gathered, the random variable is assigned one of those possible values. The symbol for a random variable is usually a capital letter such as X. The values that x can have are generically listed as X. There are ***probabilities*** associated with these values of the random variable. The ***probability distribution function*** is a real-valued function for assigning the probabilities to ranges of values of a random variable. The symbol $f(x)$ denotes the distribution functions and designates the value of that function at $X=x$.

Probability distribution functions (pdfs) or, more simply, probability distributions, are nonnegative functions that allocate mass to the real line. Pdfs have two characteristic properties: each pdf value is greater than or equal to zero and the area under the entire curve of the pdf is equal to 1.0. In terms of probability, $f(x)$ does not represent the frequency or probability of x; instead $f(x)$ represents the probability per small increment of x (e.g., dx).

The following represents an easier probability interpretation: the probability of a range of values for the random variable is equal to the area under the pdf curve for that range. The function resulting from such an accumulation or integration is the ***cumulative distribution function*** or cdf. The notation for the cdf is $F(x)$, and the interpretation for $F(x)$ is the probability that $X \leq x$. In other words, the cdf gives values for the probability that the random variable is less than or equal to a value of the random variable.

The pdf and cdf have two relationships. The cdf is the integral of the pdf. The pdf is the derivative of the cdf. Therefore, one can be calculated from the other by using calculus.

Examples of random variables include (1) an event probability, (2) a component's failure rate, (3) the amount of time required to repair a component, (4) the time to failure of a component, (5) the number of failures for a component, (6) an expert's age, (7) the colleagues working with the expert, (8) the ***rank*** of one alternative relative to another, and (9) the odds of winning a race. Random variables can be anything else that is determined in an experiment, an observation, an elicitation, or other data-gathering process.

The following five examples represent probability distribution functions, $f(x)$, for a random variable, x. The first, called the ***normal***, Gaussian, or bell-shaped curve, has one ***mode*** (hump) at the center, with half of the curve cut at the mode forming an exact mirror image of the other half (called symmetry), with pdf

$$f(x) = \frac{e^{[-(x-\mu)^2/2\sigma^2]}}{\sqrt{2\pi\sigma^2}},$$

where x and μ are values between $-\infty$ and $+\infty$ and σ^2 is positive. The second example is called the ***lognormal*** distribution with the $\log_e(x)$ having a normal distribution and a single mode shifted to the left (skewed right) such that the right tail is long and drawn out, with pdf

$$f(x) = \frac{e^{[-(\log(x) - \mu)^2/2\sigma^2]}}{x\sqrt{2\pi\sigma^2}},$$

where $\sigma > 0$. The third example is the ***beta*** distribution, whose possible values of X are restricted to the range of 0 to 1.0 and whose many possible shapes include U-shaped, horizontal line, decreasing curve, skewed right curve, and bell-shaped curve, with pdf

$$f(x) = \frac{1}{B(x_0, n_0 - x_0)} x^{(x_0 - 1)}(1-x)^{(n_0 - x_0 - 1)},$$

where

$$B(x_0, n_0 - x_0) = \int_0^1 y^{x_0}(1-y)^{(n_0 - x_0 - 1)} dy$$

is the mathematical beta function. The fourth example is the ***uniform*** distribution which assigns equal probability to each value of X such that the distribution is a straight horizontal line, with pdf

$$f(x) = \frac{1}{(b-a)},$$

where $a \leq x \leq b$. The fifth example is the exponential distribution, where the values decrease in a decay function shape (reversed J), with pdf

$$f(x) = \lambda e^{-\lambda x},$$

where $x > 0$ and $\lambda > 0$.

The pdfs and random variables discussed to this point are continuous pdfs and continuous random variables. These two concepts mean that x and $f(x)$ can be any value along the real number line. For random variables that can only *take on* a finite number or countable number of possible values, the corresponding pdf is a discrete distribution function. For discrete distributions, the value of $f(x)$ is the probability that $X = x$, $P(X=x)$. Cumulative distribution functions for discrete distributions are found by summing all the probability values rather than by integration.

Examples of discrete distribution functions include the binomial and the Poisson distributions. The binomial characterizes the number of successes as x in a given set of n trials, where each trial has the probability of success, p. The pdf for a binomial random variable, X, is

$$f(x) = \frac{n!x!}{(n-x)!} p^x (1-p)^{(n-x)},$$

where x is either 0 or 1 and $0 \leq p \leq 1$. The Poisson process characterizes the number of occurrences (failures), x, in given, fixed t units of time where the failures are independent and at a constant rate, λ. The pdf for a Poisson random variable, X, is

$$f(x) = \lambda^x e^{-\lambda}/x!,$$

where $\lambda > 0$ and x is an integer (0, 1, 2, etc.).

Estimators and Significance Tests

Relationship of Estimators and Distributions

According to the definition of random variables given above, the ***estimates*** or ***answers*** given by experts are specific values of random variables which have underlying probability distributions. These underlying and unknown distributions (often called ***populations***) consist of a universal set of values from which the observed values of the random variable can come. These distributions have certain features or characteristics called ***parameters***. Values for these parameters are unknown; however, parameters can be estimated from the ***sample*** (i.e., the data gathered). Estimates calculated or determined from the sample data are called ***statistics***. Estimators refer to the formulas or calculations used to obtain the estimates of the unknown parameters.

Classical statistical philosophy defines the parameters as fixed but usually unknown characteristics of the underlying (but also usually unknown) distribution of the observed values of the random variable. In this philosophy, estimators are random variables and have pdfs. Because parameters are considered fixed quantities (not random), they have no pdfs. However, there is a Bayesian philosophy where parameters are considered quantities measured with uncertainty and therefore have pdfs. Therefore, parameters remain characteristics of the underlying distribution from which the data came; however, they are not considered fixed values. Both philosophies prove useful when analyzing expert judgment.

Both philosophies use estimators to determine the center of the distribution, the ***dispersion*** or spread of the distribution, and the fractional areas under the distribution. Estimators commonly used for determining the center of a distribution are the ***mean, median, geometric mean,*** and mode. Because some distributions have more than one hump or mode, the mode is not a very good estimator of the distribution's center. The median is the 50th ***percentile***, estimating the point of the distribution such that half the area is to the right and half the area is to the left of the median. Other percentile estimators estimate how to cut the distribution into other areas such as the middle 90% by using the 5th and 95th percentiles. Some commonly used dispersion estimators are the variance, ***standard deviation***, and the ***range***. Example 11.1 illustrates how to calculate these estimators from a simple data set.

EXAMPLE 11.1: ***Estimator calculations***

Ten experts provided answers to a probability question:

(0.0001, 0.00001, 0.0001, 0.000001, 0.0000001, 0.0001, 0.0001, 0.00000001, 0.000001, 0.0001) .

The following procedures use the data to estimate the central measures of the underlying distribution. To estimate the median and the mode, the answers are first placed in ascending order:

	0.00000001
	0.0000001
	0.000001
	0.000001
	0.00001
	0.0001
	0.0001
	0.0001
	0.0001
	0.0001
total	0.00051211

To estimate the mode, determine the most frequent value, which in this case is 0.0001. To estimate the median, determine the value exactly in the middle. For an even sample size, the median is the halfway point between the two middle values. In this case, the median is between the 5th and 6th values,

$$(0.00001 + 0.0001) / 2 = 0.000055.$$

To estimate the mean, average the ten values:

$$\overline{x} = \frac{\sum_{i=1}^{n} x_i}{n} = \frac{\sum_{i=1}^{10} x_i}{10} = \frac{0.00051211}{10} = 0.000051.$$

To estimate the geometric mean, determine the product of the ten numbers raised to the 1/10th power:

$$\left(\prod_{i=1}^{n} x_i\right)^{1/n} = \left(\prod_{i=1}^{10} x_i\right)^{0.1} = \left(1.0\text{x}10^{-52}\right)^{0.1} = 0.0000063.$$

The mode is a very large number and serves as a poor indicator of the center; the mean and median are nearly identical; the geometric mean is an order of magnitude smaller than the mean. For normally distributed data the mean, the median, and the mode are nearly identical. Because this data does not have that feature, it is most likely not normally distributed.

The following procedures estimate dispersion measures. To determine the range, subtract the minimum value from the maximum value:

$$(0.0001 - 0.00000001) = 0.0001 \quad \text{(after rounding)}.$$

To estimate the variance, calculate the ***sum of squares*** of deviations of each value from the estimated mean and divide by the sample size minus one:

$$s^2 = \frac{\sum_{i=1}^{n} (x_i - \overline{x})^2}{(n-1)} = \frac{\sum_{i=1}^{10} (x_i - 0.000051)^2}{9} = \frac{0.0000000024}{9} = 0.0000000027.$$

To estimate the standard deviation, take the square root of the variance:

$$s = 0.000051.$$

For normally distributed data, the standard deviation can be approximated by dividing the range by 4 or by 6. However for nonnormal data, such as this data set, that approximation gives too small a value for the standard deviation.

Percentile values represent proportions of the area under the distribution function. Because the sample data represent the underlying distribution, percentiles can be estimated from the proportions of the sample data. For example, the value that cuts the data set into a 10%/90% split is the 10th percentile. For a sample of only 10 numbers, that cut occurs at the first value (when the values are listed in ascending order), 0.00000001.

Multivariate Estimators

The estimators calculated in example 11.1 characterize the underlying distribution of a single random variable. In expert judgment applications, data is gathered on many random variables. For example, each expert is asked several different questions. The answers to each question are different samples; estimators are calculated for each answer random variable.

With multiple variables (multivariate), the means of each of the variables are placed together in a vector. The variances are placed on the diagonal elements in a matrix. The off-diagonal elements of this matrix are the ***covariances*** between pairs of variables. The resulting matrix is called the ***variance/covariance matrix***. The covariance for two variables, x and y, is estimated by

$$s_{xy} = \frac{\sum_{i=1}^{n} (x_i - \bar{x})(y_i - \bar{y})}{(n-1)} .$$

This formula resembles the estimator of the variance with one exception: the numerator is not a squared quantity for one variable; rather, it is a cross-product quantity of two variables. Example 14.7 of chapter 14 explains how to use a mean vector and variance/covariance matrix for three variables.

Correlation is another type of estimator related to the covariance used to characterize the relationship between two variables. Correlation is the covariance divided by the product of the standard deviations of each variable. The correlation is a standardized covariance and has values between -1 and 1.

Correlation also refers to the linear relationship between two variables. If the two variables have values that are completely identical, their correlation is 1.0. From a graphical perspective, if the values of two variables are plotted, and they fall exactly on a line with a positive slope, then the correlation is 1.0. If the values fall exactly on a line with a negative slope, then the correlation is -1.0. In most applications, all values do not fall exactly on a line. Instead, the values of the two variables form a general scatter with either a positive trend, negative trend, or no trend. In these cases the correlation is near 1.0, near -1.0, or near 0.0, respectively. The closer that the correlation values are to 1.0 or -1.0, the stronger the linear relationship between them.

The most common measure of correlation is the Pearson product-moment correlation coefficient (r). For two variables, x and y, that coefficient, r, is calculated by the following equation:

$$r = \frac{\Sigma(x-\bar{x})(y-\bar{y})}{\left[\sqrt{\Sigma(x-\bar{x})^2 \Sigma(y-\bar{y})^2}\right]},$$

where $\bar{x}$ and $\bar{y}$ are the mean values of x and y, and the summations are over all values of x, y, and xy pairs. Example 11.2 illustrates how to calculate the correlation coefficient.

EXAMPLE 11.2: ***Correlation analysis***

Ten experts provided the following answers to two different questions. Determine if the two answer variables are correlated.

Expert	Answer 1 (X)	Answer 2 (Y)
1	0.10	0.15
2	0.05	0.00
3	0.15	0.20
4	0.11	0.10
5	0.10	0.12
6	0.14	0.10
7	0.00	0.05
8	0.00	0.10
9	0.15	0.20
10	0.09	0.09

$$\Sigma x = 0.89 \qquad \Sigma y = 1.11$$
$$\Sigma x^2 = 0.11 \qquad \Sigma y^2 = 0.16$$
$$\Sigma xy = 0.12$$

The following is a convenient calculator formula for the correlation coefficient:

$$r = \frac{\Sigma xy - \Sigma x \Sigma y / n}{\sqrt{\left[\Sigma x^2 - (\Sigma x)^2/n\right]} \bullet \sqrt{\left[\Sigma y^2 - (\Sigma y)^2/n\right]}} = \frac{0.12 - 0.89 \bullet 1.11/10}{\sqrt{(0.11 - 0.89^2/10)} \bullet \sqrt{(0.16 - 1.11^2/10)}}.$$

Using the above values and the calculator formula, $r = 0.64$. For 10 experts, an r value greater than or equal to 0.63 is considered greater than $r = 0$, meaning no correlation. Therefore, the relationship between X and Y is positive in nature (as X increases, Y increases). This conclusion is based upon a significance test discussed in the remainder of this section.

Although the calculator formula is convenient for hand and pocket calculator applications, its use can result in a significant error in the value of r as a result of rounding problems. This error can easily occur with very small or very large numbers. To avoid this error, we recommend using the original formula for the correlation.

■

Other correlation measures exist, such as ***nonparametric*** measures based on ranks and agreement between X, Y pairs. If two random variables, X and Y, are normally distributed and their Pearson correlation is zero, then X and Y are statistically, conditionally independent. Chapter 14 provides additional discussion on the correlation and dependence of variables.

Significance Testing of Estimators

Estimators conveniently summarize the characteristics of the data and estimate the characteristics of the underlying population. In addition, theoretically-based statistical tests can compare the values for estimators against specified values.

The statistical theory behind these tests requires the user to know or assume the form of the underlying distribution function of the estimator (e.g., a normal distribution). The test involves comparing the value of the data-obtained estimator to some hypothesized or standard value. The test answers the following question: how much different can the estimated value be from the hypothesized value before we can actually claim a difference? The test results indicate whether there is a statistically based difference or whether there is not enough evidence to claim a difference exists.

There is a danger in claiming that a difference exists when it really does not. However, the analyst can control the chance of making such an error in results. This chance is a preset value chosen before the test and has several names: the ***type-I error***, the significance level, or the ***α level***. Example 11.3 illustrates how the significance level is used to determine whether an estimate for the mean differs from a hypothesized value of 0.0.

EXAMPLE 11.3: ***Test for the mean estimator***

The mean estimated from a set of data (a sample) is theoretically distributed as a normal random variable. The ***central limit theorem*** (CLT) forms the basis of this theory. The CLT states that the sample follows a normal distribution that has two parameter values: a mean and a variance. These parameters of the normal are the mean of population and the variance of the population divided by the sample size. They can be estimated from the sample. The normal parameter for the mean is estimated by the sample mean; the normal parameter for the variance is estimated by the sample variance divided by the sample size (n).

To test whether this estimated mean differs from some specified or hypothesized value of the population mean, specify the value of the population mean. In most cases the specification is based upon some standard value for the mean. However, determining whether the mean differs from zero can prove interesting. The following example will perform such a test.

The following steps discuss the idea behind the theory for the test of the mean. Testing for the mean is only one type of significance test. Other tests compare other estimators from the sample to hypothesized values. In all such tests, the ideas outlined in the steps below are similar; however, the distributions and calculations will differ.

Step 1. Choose which estimator to test (compare) with a parameter. In this case the estimator is the sample mean, $\bar{x}$, to compare to the population mean, μ.

Step 2. Establish a hypothesized or standard value for the parameter being tested. In this case, $\mu = 0$.

Step 3. Determine the nature of the comparison between the parameter and the estimator. The choices for comparisons are called one-sided and two-sided comparisons. The one-sided comparison tests whether the parameter is greater than the estimator or whether the parameter is less than the estimator. The two-sided comparisons cover both one-sided comparisons simultaneously by determining if the estimator is not equal to the parameter. In this case, the one-sided comparison is whether the parameter is less than the estimator.

Step 4. Choose the significance level—the chance of claiming that the estimator is not the same as the parameter value when, in truth, they are the same. In this case, the α-level is chosen to be 0.05 or 5%.

Step 5. Use statistical theory to determine a test statistic or calculation that has a known distribution and that compares the estimator value to the hypothesized parameter value. In this case, the distribution is normal, with a mean of 0 and a variance of 1; the comparison is the scaled difference between the estimated value and μ. To perform the scaling, divide the difference by the estimated variance of the mean. The following calculation is for this case:

$$\textit{normal variable} = \frac{(\bar{x}-\mu)}{\sqrt{\frac{s^2}{n}}} = \frac{(0.000051-0)}{\sqrt{\frac{0.0000000027}{10}}} = 3.10\,.$$

Step 6. Determine the value on the distribution such that the area to the right corresponds to the percentage specified by the significance level. This value is called the ***critical value***. In this case, that area is 5%, which corresponds to a critical value of 1.96 on the normal distribution. That area is marked on the normal distribution in the figure below.

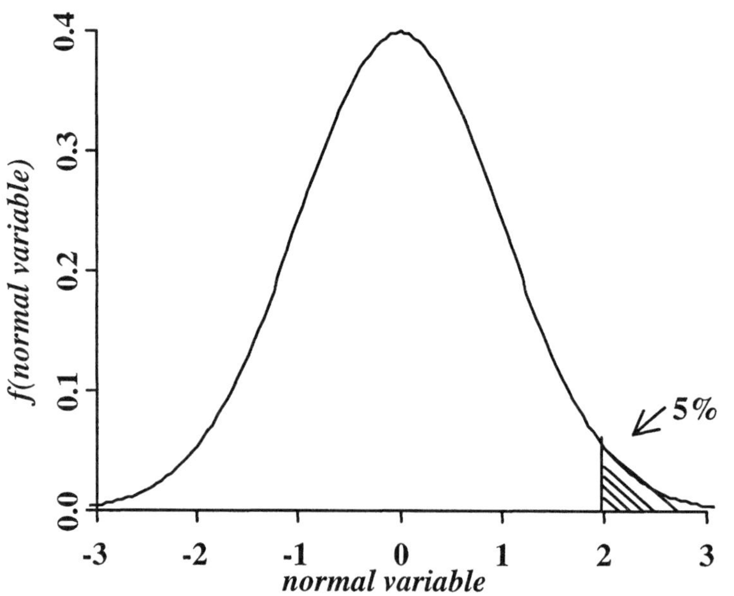

Step 7. Compare the value of the calculation in step 5 to the critical value on the distribution marking the significance level. If the calculated value appears in the shaded area, then the estimator is considered statistically or significantly different from the parameter value. If the calculated value does not appear in the shaded area, then there is not enough evidence to say that it differs from the parameter value. In this case, 3.10 falls in the shaded area. Therefore, we can conclude that the estimated mean value significantly differs from 0, and there is a 5% chance that this conclusion is incorrect.

■

The statistical test in the above example represents one of many different types of such tests. Some tests are used to determine the equality of several parameters, such as the equality of different means in an analysis of variance, while others test for different parameters, such as the variance, the correlation, or the slope of a ***regression*** line. Theory dictates the use of different types of distributions such as the ***F distribution*** for variances and the t distribution (which is similar to the normal) for regression slopes. For example, a significance test was performed on the 10 experts in example 11.2. The conclusion drawn is based on a comparison between the estimated r value of 0.64 and the critical r value of 0.63 determined from the appropriate probability distribution for a 5% significance level. Because the estimated r exceeded the critical value, we conclude that this bivariate correlation is a significantly positive relationship.

Example 11.3 does not provide enough instruction on how to perform statistical tests. The example merely illustrates some of the ideas involved in testing. Consult a statistician to establish and perform the appropriate statistical test.

Part III suggests and describes other statistical techniques in enough detail for use by nonstatisticians. The following sections of this chapter present these techniques for use by analysts.

Descriptions and Uses of Simulation Techniques

With the availability of fast, personal computers, the use of simulation techniques has recently grown. New data sets can be formed from the original data set (***bootstrap*** simulation) or from specified distributions representing the data (standard ***Monte Carlo simulation***). Both can help gain a new understanding about the data set and help form estimates that might not be available otherwise.

Monte Carlo Techniques

Monte Carlo simulation (also known as traditional simulation) forms new samples or distributions of data either from existing samples of data or from specified distributions. To initiate the formation process, randomly select values from the existing samples or specified distributions, make some calculations, perform this several hundred or thousand times, and collect the hundreds or thousands of calculated values into a table or distribution for ***inference*** (statistical) purposes.

The following steps illustrate how to perform a simple Monte Carlo simulation to solve a problem that has no tractable mathematical solution. The following summarizes these steps:

Step 1. Specify the desired quantity. For example, suppose that the desired quantity is the product of two random variables. Each variable has a specified distribution; however, the product of these distributions does not have a closed or known form.

Step 2. Determine the distributions from which sampling is done. For example, each of two random variables is normally distributed, one with mean 0, variance 1 and the other with mean 1, variance 0.5.

Step 3. Find or code a computer program that randomly selects a value from each of the specified distributions from step 2. The random selection of values from the distributions in step 2 requires a random number generator, mapping each value onto the specified distributions. Many such codes are available. Appendix B provides a code that selects values from beta distributions. Appendix C a provides a code that forms empirical (data-based) distributions for simulation. For a more complete guide on simulation techniques consult Ripley (1987) or Johnson (1987).

Step 4. Determine the number of samples (N) to be taken. With most modern computers, 1000 samples is not too expensive or time consuming—it also gives accuracies of one or two decimal places.

Step 5. Construct a program for taking the N samples. For each of the *N* samples, choose the random values and map them onto the function or distribution, then calculate the desired quantity (e.g., the product of the two normals from step 1). Completion of the N samples results in N values of the desired quantity.

Step 6. Collect the values of the desired quantity for making inferences. The *N* values of the desired quantity form a distribution of possible values for that quantity. Estimations are possible using this distribution. For example, estimates for the mean and the variance of the products of the two normals are available. First, collect the 1000 (N=1000) values for the 1000 products of two normals, then order the 1000 values. The mean of these 1000 values is the estimate of the desired mean quantity, and the variance of these 1000 is the estimate for the desired variance quantity. This method can also establish estimates of percentiles. Sometimes the analyst's interest is in the resulti distribution rather than in an estimator (such as a mean, variance, or percentile). Quantities of interest in the resulting distribution include the shape, center, tails, spread or range, and modes (humps).

Example 11.4 applies Monte Carlo simulation to a ***reliability analysis*** problem.

EXAMPLE 11.4: ***Monte Carlo simulation***

The reliability of a component (c) is often characterized by the beta probability distribution function:

$$f(c) = \frac{1}{B(x, n\text{-}x)} c^{(x-1)}(1-c)^{(n-x-1)},$$

where the parameters x and n-x are interpreted as the number of successes and the number of failures, respectively, in n trials. B is the mathematical beta function given earlier in this chapter (Martz & Waller, 1982).

For a system composed of two components (c_1 and c_2) in series, the system reliability is the product of the reliability functions. If the reliability function for each component is distributed as a beta distribution, then the system reliability is the product of two beta distributions. Simulation is used to find the distribution of this product that represents the system reliability.

Previous tests indicated that the first component failed in four out of seven trials; therefore, $x_1 = 3$ and $n_1 = 7$. No test data is available on the second component, but an expert estimates that its average reliability is 0.90 and that the 95th percentile value for the reliability is 0.99 (i.e., there is a 5% chance that the reliability is greater than 0.99). The parameters of the beta with that mean and 95th percentile are $x_2 = 11.7$ and $n_2 = 13.0$. The notation for this beta is beta (x, n-x) or, in this case, beta (11.7, 1.3). The code in Appendix B determined the parameters.

A comparison between the two components yields the following observation: the reliability of the first is much worse than the second. Their beta distributions illustrated reflect this.

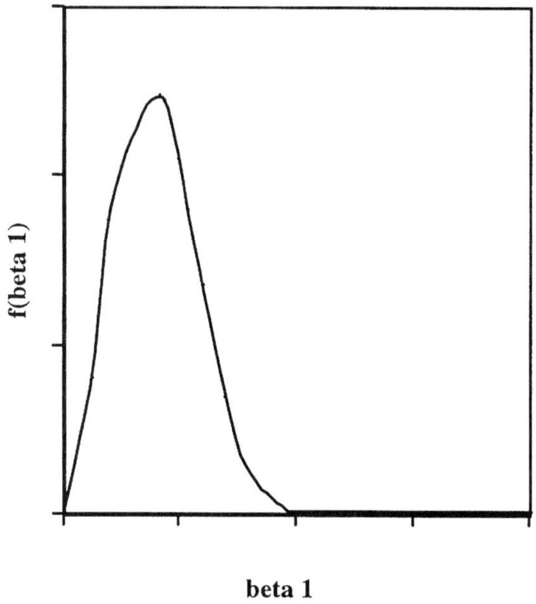
f(beta 1)
beta 1

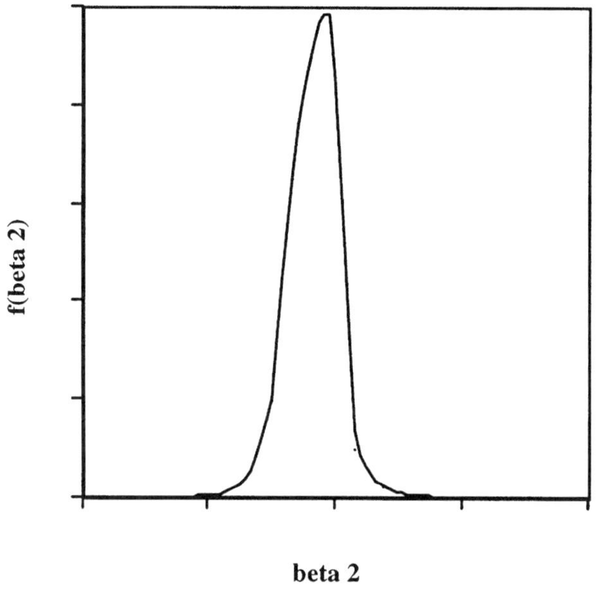
f(beta 2)
beta 2

To determine the system reliability, use the following simulation steps:

Step 1. Specify the desired quantity; the product of the two reliabilities ($c_1 \bullet c_2$).

Step 2. Specify the distributions for the reliabilities of the two components: the reliability of component 1 is distributed as a beta (3, 4), and the reliability of component 2 is distributed as a beta (11.7, 1.3).

Step 3. Find or write a computer code with a uniform random number generator for selecting values of c_1 and c_2 from the two betas.

Step 4. Choose the value of N (the number of samples) as 1000.

Step 5. Write a computer code to determine the random values of c_1 and c_2 from their respective beta distributions. A uniform random number generator chooses a random value for the probability of the reliabilities of c_1 and c_2. The code determines the values of c_1 or c_2 by mapping the chosen probabilities through each beta distribution. This mapping is an inverse process. In other words, random numbers are used to choose the probability values for the beta distribution function; then they are mapped onto the reliability axis to determine the reliability value, c. This process is done for c_1 and again for c_2 in each of the N samples. In each sample, the chosen value of c_1 is multiplied by the chosen value of c_2. Appendix B provides the code for the beta sampling and simulation.

Step 6. The 1000 product values are stored and sorted. The values are plotted to indicate the shape of the system reliability distribution. The following graph is a plot of the results.

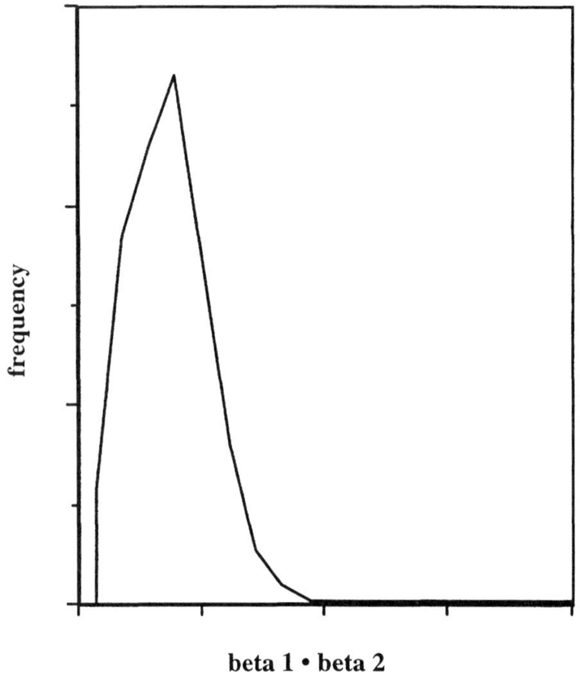

Mean of this simulated distribution:	0.36
Standard deviation:	0.16
5th percentile:	0.12
95th percentile:	0.65

Therefore, the system reliability is not very good because it is dominated by the poor reliability of the first component.

The product of two beta distributions does not usually produce a well-known distribution function. However, research (Bruckner & Martz, 1987) indicates that this final distribution is almost in the beta distribution family.

■

Advantages and Disadvantages

Monte Carlo simulation has one major advantage: it allows the analyst to determine estimates of desired quantities that would be either impossible or extremely difficult to determine by theory or mathematical computation. (Some analysts follow the philosophy, "when in trouble, simulate!") This advantage is important in ***risk analysis*** and reliability applications in which many system components interact, yet the performance of the entire system is under study. Monte Carlo simulation is also advantageous for calculations in which uncertainties are combined to determine the uncertainty of a larger system.

Therefore, simulation can help increase the understanding of the data. In many cases, simulation can identify new features in the data. Applying simulation to sensitivity studies can determine which characteristics or components are most important to the system.

The disadvantages of simulation are related to the advantages listed above. For example, programming can be difficult if it requires complex distributions or if packaged distribution programs are not available. Bad random number generation or improper use of sampling techniques can wildly distort results. Exercise care when programming and using packaged routines.

Monte Carlo simulation, especially the sampling and collecting steps, is relatively easy to program. Many such programs already exist, and most computers have random number generators available. Simulation can be expensive and consumes much computer time; however, current technology has significantly reduced these two problems.

Uses for Monte Carlo Simulation

Risk and reliability studies widely use this technique. A large number of components (random variables) comprise a system in these studies, and system behavior is required for making ***general inferences***. Each component can affect the system or other components. The result is a complicated expression describing the system behavior in terms of the various components.

Each component can be characterized by a probability distribution function. The system is usually a complex combination of these component distribution functions. Simulation provides a way of obtaining information about the system. This situation also describes how simulation can be used in an environment characterized by uncertainties (chapter 17 provides more details on characterizing uncertainty).

Monte Carlo simulation represents the backbone of other data-based analysis techniques, such as the bootstrap simulation method. These data-based analysis techniques also have advantages in performing exploratory data analysis and handling small sample-size problems.

Bootstrap Sampling and Estimation

One of the recent developments in the simulation community is the increased use of empirical distributions. Empirical distributions are distributions resulting from the original data set. Constructing distributions from the data without making assumptions about the form of the distributions (e.g., that the data are normally distributed) falls under the general heading of nonparametric techniques. Because the original data are used over and over again in a simulation method, construction of empirical distributions is often referred to as a resampling technique. The bootstrap method (Efron, 1979) is a commonly used nonparametric and resampling technique.

Advantages and Disadvantages

The bootstrap relies solely on the original data without assuming that the data follow a particular probability distribution. Bootstrap simulation allows the formation of distributions of any desired quantity (e.g., the sample median). The resulting simulated distribution of that quantity provides estimates of the variance and percentiles of that quantity. The bootstrap estimation process can produce statistical biases with estimable effects. (Statistical bias means that the ***expected value*** of the estimated quantity does not equal the parametric or theoretical value.) The extremely small samples commonly encountered in expert judgment can prove troublesome with both parametric or nonparametric techniques. However, the bootstrap does reasonably well for smaller samples (Efron & Gong, 1983).

The major disadvantage of the bootstrap technique is that the formation of samples in the simulation is limited to the range of the original data. This limitation tends to form empirical distributions with truncated tails (at both ends). In addition, the bootstrap may introduce statistical bias in the estimation process. In other words, the bootstrapped value for a chosen estimator may not be unbiased. However, analysts should monitor the statistical bias as part of using the technique.

Uses for the Bootstrap

In general, the bootstrap technique provides estimates of parameters that would usually be obtained by assuming a distributional form for the data or for a parameter of a distribution (in a Bayesian context). The bootstrap avoids the necessity of these assumptions and therefore can be used in any application or problem setting. In expert judgment problems, using the bootstrap method proves ideal because information regarding distributions of expert judgment data is lacking. The bootstrap is also useful for the small sample sizes that are usually found in expert judgment applications.

This book uses the bootstrap method in three different ways. First, it will provide methods of investigating the correlation structures and biases (motivational and cognitive) in the data. Second, it will characterize and analyze uncertainties in the experts' estimates. Third it will provide a convenient way for aggregating the expert's multiple estimates into a single estimator and corresponding distribution.

How to Implement the Bootstrap

Any programmable computer or calculator can easily implement the bootstrap method. The original sample data (the responses to a single question) are denoted as $\{q_1, q_2, \ldots, q_n\}$ for the n experts. The parameter to be estimated is denoted by θ. The estimate of θ from the bootstrap results will be denoted as $\hat{\theta}$. To perform a simulation, randomly form many samples, N (usually of the original size, n), from the original data. The samples are formed by placing sampled data points back into the original data set so that they can be sampled again. In other words, q_3 can be chosen randomly more than once, appearing more than once in the same sample. This method is called sampling with replacement. The basic idea of the simulation is to calculate the estimate of θ for each sample formed. The various values of $\hat{\theta}$, denoted as $\hat{\theta}_j$ ($j = 1, 2, \ldots, N$), then form an empirical distribution for θ in the same manner that the resulting distribution of values is formed in Monte Carlo simulation. The following steps summarize the bootstrap technique.

Step 1. Specify the desired quantity to be estimated. For example, the estimator or quantity of interest is the sample median of a sample of size n.

Step 2. Decide how many bootstrap samples to take (N). One thousand will usually yield a reasonable standard deviation.

Step 3. Form N random samples of size n from the original data by sampling with replacement (replace each sampled value back into the data set so that it is available to be chosen again). The random samples are formed using a uniform random number generator to select which of the data points goes into each sample.

Step 4. Do N times: Calculate the desired estimator for each sample (e.g., the median). These individual estimators are $\hat{\theta}$.

Step 5. Calculate the overall bootstrap estimator of θ, $\hat{\theta}$, using

$$\hat{\theta} = \frac{\sum_{j=1}^{N} \hat{\theta}_j}{N} .$$

Step 6. Calculate the variance of $\hat{\theta}$ using

$$\widehat{var(\theta)} = \left[\frac{\sum_{j=1}^{N} (\hat{\theta}_j - \hat{\theta})^2}{N-1} \right] .$$

Step 7. Order the N estimates of $\hat{\theta}_j$ to find (1-2α)% putative central coverage intervals for θ. ***Putative intervals*** represent the central (1-2α)% area of the bootstrap distribution. In the case of the median, for $\alpha = 0.05$, then this area corresponds to the central 90% of the median values generated in N bootstrap samples.

Step 8. *Find the two values of the* $\hat{\theta}$ *that correspond to the* α% *and* $(1-\alpha)$% *in the ordered list.* These two values, $\hat{\theta}(\alpha)$ and $\hat{\theta}(1-\alpha)$, are the α-th and $(1-\alpha)$-th percentiles of the bootstrap distribution for θ, and they form the $(1-2\alpha)$% putative central confidence interval.

Step 9. *Calculate the statistical bias induced by the bootstrap technique by calculating the value of* $\hat{\theta}$ *from the original sample,* $\hat{\theta}_0$. The statistical bias in the method is the difference between the value of the estimator calculated from the original sample, $\hat{\theta}_0$, and $\hat{\theta}$ calculated from step 5. If the bias is large (more than 10% of either of the two values), then the use of the bootstrap technique might not be appropriate.

These steps are easily programmed into a code. Appendix D contains a FORTRAN code of these steps, and that code is illustrated in example 11.5.

EXAMPLE 11.5: ***Bootstrap simulation***

Fourteen experts provided value estimates for a quantity on a continuous ***linear scale*** from 0 to 1:

(0.07, 0.50, 0.28, 0.63, 0.95, 0.70, 0.62, 0.70, 0.58, 0.78, 0.40, 0.68, 0.43, 0.60).

A plot of the data reveals that the distribution of these values is not symmetric. The median is a commonly used ***mathematical aggregation*** estimator to represent such asymmetric data. The median of this particular data equals 0.61. However, because the exact form of the distribution of this data cannot be assumed, there is no available estimator for the variance of the median. The bootstrap sampling method provides such a variance estimate, and more.

Following are the steps outlined in the bootstrap method:

Step 1. The specified parameter of interest is the median.
Step 2. The specified number of bootstrap samples is 1000.
Step 3. Applying the bootstrap code in Appendix D, 1000 random samples of size 14 are taken from the original set of data .
Step 4. To form each sample, 14 data points are selected from the original set, with replacement. For each sample, the sample median is calculated; this results in a set of 1000 medians.
Step 5. The average of these 1000 medians equals 0.597.

Step 6. The *variance* of these 1000 medians equals 0.0034, with a standard deviation (the square root of the variance) equal to 0.058.

Step 7. The 1000 medians were sorted in order.

Step 8. To form the 90% putative interval for the median, the 5th and 95th percentiles are found by the 50th and 950th values from the ordered median set. These values equal 0.465 and 0.680, respectively.

Step 9. The bias is calculated by the difference between the sample median and the median from step 5, 0.610 - 0.597 = 0.013. This bias is judged acceptable because it is less than 10% of either value.

As a result, the estimate for the median is 0.597, with a variance of 0.003, and the 90% putative interval for the median is 0.465 to 0.680. This forms a complete description of the chosen aggregation estimator for this data set.

■

Descriptions and Uses of Data Analysis Techniques

Multivariate Techniques

Multivariate analysis techniques refer to statistical methods designed for the analysis of data sets with many random variables (multivariate). Most data sets have a multivariate structure; however, possible relationships among the variables are not often done, or the relationships among the variables are assumed without analysis. In such cases, drawing conclusions may not include important characteristics of the data.

Two types of variables exist: ***answer***, response, or ***dependent variables,*** and ***ancillary, conditional***, or ***independent variables***. The names differ, but the relationship between these two types is the same. The independent variables are measured or fixed variables that influence or functionally relate to the dependent variables. The independent variables are variables that can be controlled by the analyst in the data-gathering process; however, the dependent variables are variables that are unknown and uncontrolled. As a result, the dependent variables rely on the independent variables.

Regression Analysis

Chapter 15 describes general linear models. Regression is a commonly used subset of this class of models. Regression usually refers to the cases in which one continuous, dependent variable (Y) is modeled as a function of one or more continuous, independent variables (X_i). Therefore, regression is a multivariate procedure that models a dependent variable with multiple independent variables. There is also a multivariate version with multiple dependent variables. The following represents a regression model:

$$Y = b_0 + b_1X_1 + b_2X_2 + \ldots + b_mX_m + e.$$

The model includes a random error term (e) that contains the information in Y that was not modeled by the X_is. Each X_i contributes to Y weighted by the coefficient (b_i) that is determined by a technique called least squares. The least squares approach finds the coefficients in a manner that minimizes the sum of squares of errors. A single independent variable example is given below. Example 11.6 also illustrates the relationship between regression and correlation.

EXAMPLE 11.6: ***Regression***

The data from example 11.2 consisted of two random variables. If one variable represented an answer variable, then this is the dependent variable, Y. If the other variable represented the proportion of time spent on a type of problem, then this is the independent variable, X. The regression model for this data would be

$$Y = 0.04 + 0.76\,X.$$

A significance test can determine whether the coefficient for X (the slope of the regression line) is different from zero. In other words, the test will determine if X contributes to Y, and if there is a significant relationship between X and Y. In this case, X and Y have a positive relationship. The following graph illustrates this point:

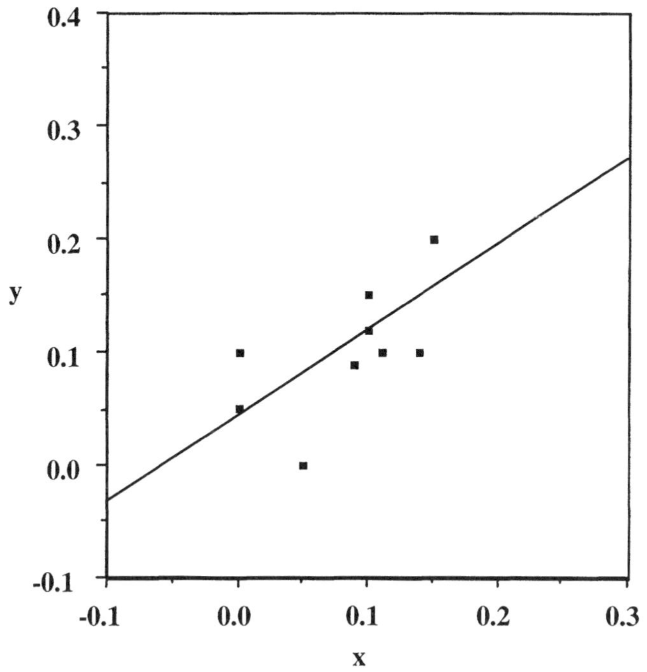

This result is consistent with the result from the correlation analysis in example 11.2. In general, if two variables have a significant correlation, that relationship will again be validated from a regression analysis.

■

Using regression analysis requires two assumptions about the data. The first assumes that the random errors are distributed with a normal distribution that has a mean of zero and some common but unspecified variance. The second assumes that the independent variables are measured without error.

Tests for normality of the errors are readily available to verify that the first assumption is not severely violated. Among these tests is a simple graphical technique. On special normal probability graph paper, the errors (the difference between the observed values of *Y* and the values of *Y* predicted from the model) will plot as a straight line. Most statistical computer packages have the W test for normality which is easy to use and interpret. In most applications, the errors are normal or near normal in distribution. In addition, violation of this assumption does not have as severe consequences as violation of the second assumption.

The second assumption requires independent variables to be either fixed quantities or to be measured without error. Including the answer variables in the model as Xs violates this assumption because the answers are considered variables measured with error. Sometimes it is difficult to determine whether a variable is measured without error. Therefore, variables that can be used in regression modeling are restricted by compliance with this assumption. Including variables with errors in the model affects the estimation of the variance of the dependent variable. This variance, which is used in making predictions by the model, will tend to be underestimated. Chapter 15 discusses this topic further and covers models for expert judgment data as well.

Cluster Analysis

Clustering refers to the data's grouping structure. Clusters form according to how the values from a single variable (quantity of interest) are grouped, or how the different variables are grouped. Groups form based upon how close the values are to each other. Cluster analysis traditionally refers to methods of determining how the values of one or more variables can be grouped together. Most methods form hierarchical clusters—beginning with all the data in one cluster, splitting it into two, then three, etc., until each datum forms its own cluster.

Many different methods determine cluster formation (Duran & Odell, 1974). Basically, the methods compute and then compare some measure of the distance between clusters. For example, some methods form clusters by maximizing the distance between cluster means. If the squared Euclidean distance is used, then the method is the centroid method. Other methods, called linkage methods, form clusters by maximizing distances between individual observations in the clusters. Still other methods minimize variances to determine clusters.

Of particular interest are clusters formed by grouping similar variables in multivariate data sets. Variable cluster analysis uses the correlation or covariance matrix of the variables to determine the variables' clusters. Again, many techniques for variate cluster determination are available.

Both the data and variable hierarchical clustering processes begin with the formation of one large cluster and end with each observation or variable in its own cluster. The analyst must decide which set of clusterings between these extremes to use for interpretations and conclusions. Many packaged programs provide graphical trees to aid in this decision. The distance measurements are plotted against the cluster groupings to help determine which grouping to use. Example 11.7 illustrates how this is done using the SAS® software (a registered U.S. trademark SAS Institiute, Inc., Box 8000, Cary, NC 27512) for data clustering (using the CLUSTER procedure) and for variable clustering (using the VARCLUS procedure).

EXAMPLE 11.7: ***Cluster analysis of variables and of data***

Eleven experts provided the following data. This example illustrates the uses of cluster analysis in three different clustering situations. Three different cluster analyses are given in this example illustrating the uses of cluster analysis.

YRSEXP	--	total years of professional experience
YRSED	--	total years of college education
YRSJOB	--	number of years on the current job/project
DEGAREA	--	code (1-3) for discipline of highest degree/education
EXPAREA	--	code (1-5) describing major area of experience
ANSWER	--	answers to the question on a continuous number scale [0,1]

Expert	*ANSWER*	*YRSEXP*	*YRSED*	*DEGAREA*	*YRSJOB*	*EXPAREA*
1	0.11	6.00	4	1	0.75	5
2	0.12	4.50	6	2	1.00	5
3	0.90	8.50	8	3	1.25	5
4	0.78	3.00	6	3	2.25	4
5	1.00	5.00	6	2	3.00	5
6	0.17	4.75	5	2	1.00	4
7	0.14	4.25	6	2	3.25	3
8	0.83	5.00	6	2	3.00	3
9	1.00	7.00	6	2	2.50	2
10	0.88	8.25	7	1	3.00	2
11	0.20	4.00	5	1	1.00	1

I: Cluster Analysis of All Six Variables

Cluster analysis of the six variables indicates the following cluster formations. Each formation is based on the proportion of variance explained by the clustering—the higher the proportion, the tighter the individual clusters and the larger the separation between clusters.

Proportion of Variation	Cluster Formation	
0.37	All 6 variables in one cluster	---
0.63	*YRSEXP, YRSED, YRSJOB, ANSWER*	Cluster 1
	DEGAREA, EXPAREA	Cluster 2
0.76	*YRSEXP, YRSED*	Cluster 1
	DEGAREA, EXPAREA	Cluster 2
	YRSJOB, ANSWER	Cluster 3
0.85	*YRSEXP, YRSED*	Cluster 1
	EXPAREA	Cluster 2
	YRSJOB, ANSWER	Cluster 3
	DEGAREA	Cluster 4

The proportion of variance values can be used to determine which cluster formation to use. The proportion is doubled from the all-in-one cluster formation to the two-cluster formation. This increase makes the two-cluster formation an attractive choice. The remaining formations do not indicate as great a change in the proportion.

Cluster interpretation is just as important as deciding which cluster formation to use. For example, if the clusterings in a particular formation do not make sense, then using that formation makes no sense even if the clustering looks good with a large proportion of variance. In the example, the two-cluster formation does make sense. Cluster 1 contains the years-related variables and the answer. Cluster 2 contains the variables related to the experts' areas. Based on interpretation and proportion of variance, the two-cluster formation is the logical choice.

II: Cluster Analysis of the Experts Using the Answer Variable

Note that the experts provide answers that nearly cover the entire available range on the offered number line from 0 to 1. Some experts give high values while others give low values (experts do not give middle values). A cluster analysis on the answer variable can be done in a number of different ways. One way is to use only the answer variable itself, disregarding the other variables. Another way is to use the other variables to help determine how the answers cluster. Yet a third way is to use a subset of the other variables. A natural choice for such a subset would be the set of variables that provide similar information about the answers. From the above variable cluster analysis, this subset would include the three variables describing the number of years for various items.

Again, a number of viable cluster formations result from the cluster analysis of the answer variable. Criteria, such as distance measures between clusters, can determine which formation is a logical choice. Having a logical or reasonable interpretation of the results is as important as the results themselves. Only the answers help determine the following cluster formations from an analysis based on the centroid method. The following list provides the centroid distances for the different cluster formations of the experts.

Cluster Formation (underlines indicate connections within clusters)	Centroid Distance
5 9 3 10 4 8 1 2 7 6 11	1.33
5 9 3 10 4 8 1 2 7 6 11	0.27
5 9 1 2 7 6 11 3 10 4 8	0.15
5 9 1 2 7 6 11 3 10 4 8	0.11
5 9 1 2 7 6 11 3 10 4 8	0.09
5 9 1 2 7 6 11 3 10 4 8	0.05
5 9 1 2 7 3 10 6 11 4 8	0.04
5 9 1 2 3 10 7 6 11 4 8	0.04
5 9 1 2 3 10 7 6 11 4 8	0.02
5 9 1 2 3 10 7 6 11 4 8	0.00

The greatest change in the distance measure is from the single-cluster formation to the two-cluster formation. This change delineates the major division between the high-answer experts and the low-answer experts. All other cluster formations differ very little in the distance measure and do not have any better interpretative value.

III: Cluster Analysis of the Experts Using Three Variables

A cluster analysis on the answers was performed including the three variables that describe years of items. Again, the centroid method was used, and the distance measures were used for cluster formation determination.

Cluster Formation (underlines indicate connections within clusters)	Centroid Distance
5 4 8 7 6 11 2 1 9 10 3	1.14
5 4 8 7 6 11 2 9 10 3	0.84
5 4 8 7 6 11 2 9 10 3 1	0.70
5 4 8 7 6 11 2 9 10 1 3	0.63
5 4 8 7 6 11 2 9 10 1 3	0.59
5 8 7 6 11 2 9 10 1 3 4	0.51
5 8 7 6 11 2 1 3 4 9 10	0.34
5 8 6 11 2 1 3 4 7 9 10	0.31
5 8 6 11 1 2 3 4 7 9 10	0.23
5 8 1 2 3 4 6 7 9 10 11	0.05

These cluster formations differ from the ones in Part II of this example. The reason for this difference is that the information from the other three variables is being used to determine clusterings. The new information forms clusters differently from those formed using just the answers.

Again, a substantial change occurs in values from the single-cluster formation to the two-cluster formation. This change indicates that the two-cluster formation shows promise. However, the experts in these two clusters are different from those in the two clusters in part II. This discrepancy indicates that the three added variables make a difference in the results. The results in this example indicate that additional modeling should be done using the other variables in combination with the answer variable. Chapter 15 further discusses model formation and uses. However, valuable information has been gathered from the cluster analyses about variable relationships. Cluster analysis can be a useful tool for such investigation.

■

Factor Analysis

A method related to variable clustering is ***factor analysis***. Factor analysis analyzes the relationships among a set of variables through their correlation and covariance to determine what information is shared among subsets of the variables (common factors) and what information is unique to each variable (unique factors). An example of a common factor is an unobserved, imaginary variable that shares information with (or contributes to the variance of) at least two of the original variables. An example of a unique factor is an unobserved, imaginary variable that represents information from only one of the original variables. Factor analysis assumes that all the unique factors are uncorrelated with each other and that the unique factors are uncorrelated with the common factors.

Factor analysis can be done using many different methods. Some methods rely on the use of principal components (Kshirsagar, 1972). Other methods use least squares methods or ***maximum likelihood*** techniques. Still others formulate scores based on correlations. A process called rotation transforms the variables so that the formation of common and unique factors is both optimized and logical. Example 11.8 illustrates the principal factor method for an unrotated set of variables.

We do not recommend that factor analysis be used in a casual manner. Selecting the proper factor method and understanding how to use rotation requires experience and knowledge of the various factor methods. To interpret the results requires even more knowledge. Only if the variable set has a good underlying structure can a simple factor analysis reveal important information. A successful result can then be used to reduce a large number of variables to a smaller number of variables containing the common factors. However, in most applications, the (common) factors produced are not interpretable. As a result, these factors prove quite useless.

EXAMPLE 11.8: ***Factor analysis***

This example shows a principal components factor analysis using SAS® (using the FACTOR procedure) on the set of six variables from example 11.7. The (common) factor loadings listed below indicate how each variable maps onto each factor. A successful result would be for each variable to map almost completely (e.g., 0.8 or greater) onto one factor and for the set of factors to be much smaller than the original number of variables. Success must also be measured in the interpretation. Can meaning be attached to each of the factors based on the variables that constitute (load onto) them? This is the same problem faced in interpreting the results of a cluster analysis.

The principal components factor analysis yielded six factors for the six variables. This result does not indicate that a successful reduction in the number of variables is possible. Each factor represents a portion of the original variation of the six variables. Examining these portions can help determine which factors are the most important:

	Factor					
	1	2	3	4	5	6
Portion	0.42	0.25	0.18	0.08	0.05	0.02
Cumulative	0.42	0.67	0.85	0.93	0.98	1.00

The first three factors account for 85% of the variation. While 85% of the variation would be considered very good for one factor and good for two factors, it does not represent a good result for three factors.

The six variables map or load onto the following three (common) factors:

	Factor		
	1	2	3
ANSWER	0.88	-0.13	-0.06
YRSEXP	0.56	-0.20	0.79
YRSED	0.90	0.10	0.13
YRSJOB	0.60	-0.43	-0.56
EXPAREA	0.02	0.88	0.16
DEGAREA	0.53	0.71	-0.33

The above mapping is not a successful one. *ANSWER* loads well onto the first factor, as does *YRSED*. *EXPAREA* loads well onto the second factor. However, *YRSEXP*, *YRSJOB*, and *DEGAREA* distribute their loadings over all three factors. If an interpretation was attempted, the first factor would represent some combination of *ANSWER* and *YRSED*, the second factor would represent the two AREA-type variables, and the third factor would represent some combination of *YRSEXP*, *YRSJOB*, and *DEGAREA*. Such an interpretation is not clear. Thus, this example would not represent a successful factor analysis. For a more successful analysis, see chapter 15 on model formation.

■

Discriminant Analysis

In cluster analysis, both variables and observations based on variables could be clustered. Discriminant analysis determines how well other (ancillary or independent) variables predict the classifications or groupings described by the dependent or classification variable. If the ancillary variables successfully classify the dependent variable, then they can be used to predict into which class or group a new observation (a new value of the dependent variable) will fall.

Discriminant analysis calculates a discriminant function from the predictor variables based on the distances between classes and the variation within classes. The theory is based on the data for the variables following a multivariate normal distribution. This assumption is highly restrictive. Very few data from any expert elicitation would be multivariate normal. Therefore, this method is offered as an exploratory data analysis and premodeling tool. It is only a means to other analyses, and not the sole analysis tool. Example 11.9 illustrates how discriminant analysis can be useful.

EXAMPLE 11.9: ***Discriminant analysis***

The eleven experts from examples 11.7 and 11.8 made recommendations based on their answers (scaled 0, 1). They gave the following recommendations (1-4):

Expert	Recommendation
1	1
2	1
3	4
4	3
5	4
6	2
7	1
8	3
9	4
10	4
11	2

We hypothesize that the original answer variable and some of the other variables might be good discriminators in determining the recommendations. To test this hypothesis, we performed a discriminant analysis. The results of this analysis indicate whether the six other variables are good discriminators if they properly classify the recommendations into the four classes. A discriminant function for each class is calculated using the six other variables. These functions are used to determine the existing classifications and are also used to predict classifications of experts who did not make the recommendation but provided information on the six variables. The following results show that the classification was a success:

Expert	Recommendation	Predicted Class Based on Variables
1	1	1
2	1	1
3	4	4
4	3	3
5	4	4
6	2	2
7	1	1
8	3	3
9	4	4
10	4	4
11	2	2

No misclassifications occurred. The discriminant functions for the classes follow:

Recommendation 1 = -109.2 + *ANSWER* • 382.6 + *EXPAREA* • 9.2 + *DEGAREA* • -14.0 + *YRSED* • 19.2 + *YRSEXP* • 9.0 + *YRSJOB* • 5.0.

Recommendation 2 = -107.8 + *ANSWER* • 483.0 + *EXPAREA* • 7.5 + *DEGAREA* • -14.4 + *YRSED* • 20.2 + *YRSEX* • 8.3 + *YRSJOB* • 1.5.

Recommendation 3 = -611.1 + *ANSWER* • 1145.9 + *EXPAREA* • 15.2 + *DEGAREA* • -32.0 + *YRSED* • 43.3 + *YRSEXP* • 17.7 + *YRSJOB* • -1.4.

Recommendation 4 = -908.1 + *ANSWER* • 1398.8 + *EXPAREA* • 18.4 + *DEGAREA* • -40.6 + *YRSED* • 52.4 + *YRSEXP* • 22.7 + *YRSJOB* • -2.5.

This example yields an important result—the recommendations given are functions of the values of the other six variables. A careful look at the recommendations versus the answer variable reveals a very strong positive correlation between them. The discriminant functions reflect this correlation as well.

Discriminant analysis assumes a multivariate normal distribution for all the variables. However, this assumption may not be valid for expert judgment data. Discriminant analysis also assumes that the discriminating variables are measured without error. In this example, the experts' recommendations might violate that assumption. Therefore, only use this technique as an exploratory analysis tool to gain information about the variate relationships. Results from the discriminant analysis should be validated by the results from other analyses and modeling tools such as the ones described in chapters 13, 15, and 18.

■

Analysis of Variance

Analysis of variance is a broad-based methodology for analyzing data from an experiment in which the dependent or response variables are considered functions of the independent or ancillary variables. Each independent variable is tested (controlled) at specified values (levels) of that variable. In most analysis of variance usage, the experiment is carefully designed using techniques that specify the levels of all the independent variables to be studied (Snedecor & Cochran, 1978). This method helps minimize the number of experimental tests or observations required to determine the significance of all the independent variables on the response (dependent) variable.

Analysis of variance determines the effect on the responses of an independent variable by testing the equality (or lack of it) of the means of the responses for the various levels of the independent variable. If the means for the responses differ from one level to the next of the independent variable, then that independent variable is significant in determining the response.

A single set of experiments can test multiple independent variables. Each variable is individually tested for its influence or effect upon the response. Combinations of two or more variables can be tested for their combined effect upon the response. These combinations are called interactions. Interactions may prove significant while the individual independent variables may not.

Independent variables are often called factors. Analysis of variance factors are the independent variables controlled for the study and are not the same as the factors in a factor analysis.

Because the study's design (experiment) cannot be controlled, avoid using analysis of variance for expert judgment studies. Most of the potentially influential (conditional) variables are gathered during the elicitation and cannot be controlled before the study, precluding experimental design. At most, only single variables (one factor) can be analyzed, each in a separate analysis, as illustrated in example 11.10. We do not recommend performing several one-factor-at-a-time analyses of variance. In doing so, the analyst may lose control over the specified α level or type-I error. Theory states that if multiple tests are performed on the same data set, each at a certain α level, then the total α level of all the tests is not the same as that specified value. Therefore, multiple factors are tested by using a single analysis of variance with a precise and carefully planned experimental design before the elicitation. Because such a good design is not usually possible in expert judgment applications, the book uses analysis of variance as an exploratory tool for examining simple one factor analyses. For example, a section in chapter 14 compares *between* versus *within* experts' values.

EXAMPLE 11.10: ***One factor analysis of variance***

Previous examples in this chapter described the expert's degree area or discipline by using a variable *DEGAREA* which had three values (levels). These values were 1 = mechanical engineering, 2 = nuclear engineering, and 3 = physics. Therefore, *DEGAREA* represents a single variable (also called a factor) with three levels that could be analyzed using analysis of variance.

Three experts qualify under *DEGAREA* = 1, six experts qualify under *DEGAREA* = 2, and two experts qualify under *DEGAREA* = 3. For a single factor analysis of variance (AOV), this imbalance of the number of experts across levels does not preclude proceeding with the analysis. However, if two or more factors were to be analyzed, most software programs would require a balance in the numbers of experts in each level for all factors.

For example, suppose the two factors were *DEGAREA* and *EXPAREA*. *DEGAREA* has three levels and *EXPAREA* has five levels. For a balanced design of the experiment, 15 experts would have to be interviewed, one for each combination of all the levels of the two factors. With such a balanced experiment, tests on the effects of *DEGAREA* and *EXPAREA* would be possible. To have a test for the *DEGAREA/EXPAREA* interaction, more than one expert for each combination would be necessary. The problem of finding the required number of experts illustrates the difficulties in designing experiments for expert elicitation.

The data for this example are as follows:

	Factor: *DEGAREA*		
	1	2	3
Answers	0.11	12.00	0.90
	0.88	1.00	0.78
	0.20	0.17	
		0.14	
		0.83	
		1.00	
Sum	1.19	3.26	1.68

To test the differences between the three answer means for the factor, calculate the variances between (across) the three categories and within the three categories. If the variance between the three degree areas is significantly greater (using an F test statistic) than the within degree areas variance (which acts as the noise level), then the three means are not the same. The theory behind this comparison of variations is that the factor levels will influence the response if the answer means for the three levels are different. To test whether the three population means differ, compare the variation between the three levels to some noise level. A between variance larger than the noise level indicates a difference in the three levels. This noise level is determined as the variance within the three levels (the within variance). Thus, the name analysis of variance implies what is actually tested. Comparing and testing the variances determine if the means of the factor levels are different.

The following analysis of variance table outlines the steps of the variance calculations and the significance test (based on the F distribution) used to compare the variances:

Term	df	Sum of Squares	Mean Square	F Statistic
Between	*dfB*	*SSB*	*MSB*	*F*
Within	*dfE*	*SSE*	*MSE*	
Total	*dfT*	*SST*		

MSB represents the variance between the three levels of the factor. *MSE* represents the error or noise level calculated using the within-levels' variance. The F statistic measures the significance of the between variance relative to the within or noise variance. If the *F* value is larger than the critical value for an F distribution with parameters of *dfB* and *dfE*, then the between variance is significantly larger than the noise. We can then conclude that the means of the factor levels differ and that the factor significantly affects the experts' answers.

The following formulas indicate how to calculate the quantities in the analysis of variance table:

n = number of observations = 11
T = grand total = sum of all data = 1.19 + 3.26 + 1.68 = 6.13
SST = sum of squares of all 11 answers = $0.11^2 + 0.88^2 + \ldots + 0.78^2 = 5.00$
SSB = sum of squares of categories (between) = $1.19^2/3 + 3.26^2/6 + 1.68^2/2 = 0.47 + 1.77 + 1.41 = 3.65$
C = correction factor = the grand mean = $T^2/n = 6.13^2/11 = 3.42$
SST = total sums of squares = SST - C = 1.58
SSB = between sums of squares = SSB - C = 0.23
SSE = within (error) sums of squares = $SST - SSB = 1.35$
dfB = ***degrees of freedom*** for between = number of categories - 1 = 2
dfT = degrees of freedom for total = $n - 1 = 10$
dfE = degrees of freedom for error = $dfT - dfB = 10 - 2 = 8$
MSB = ***mean square*** between = $SSB/dfB = 0.23/2 = 0.12$
MSE = mean square error = $SSE/dfE = 1.35/8 = 0.17$
F = F test statistic = $MSB/MSE = 0.12/0.17 = 0.71$

To determine if the *F* value is larger than the critical value for an F distribution with 2 and 8 degrees of freedom requires a table or program of F distributions. Statistical packages and textbooks (Snedecor & Cochran, 1978) provide such tables. To use the tables, the analyst must specify a significance level. The level represents the chance that the analyst is willing to accept for making the following error: declaring that between variance is larger than the within variance when in truth they are the same. A 5% value is a common choice for the chance of making this type-I error. Sometimes a conservative value of 1% is chosen; other times a liberal value of 10% is chosen.

If a significance level is chosen at 5%, then the critical (cutoff) value for this F distribution (with degrees of freedom two and eight) is 4.46.

$$
\begin{aligned}
F(2, 8, 0.05) &= 4.46 \\
F &= 0.71 \\
F(2, 8, 0.05) &< F \quad \text{(no factor effect indicated)}
\end{aligned}
$$

The F value calculated in the table above must be larger than 4.46 for the factor to have a significant effect. In this case, the F value is only 0.71; therefore, the factor means are considered the same, and the factor itself has no effect on the experts' answers.

■

Saaty's Technique for Pairwise Data Analysis

Rather than compare multiple items simultaneously either by numerical or ***qualitative*** evaluations, experts can make relative comparisons and evaluations. Humans find it very difficult to simultaneously examine and evaluate many items. However, ***pairwise comparisons*** require examinations of only two items at a time. Paired comparisons can be made by evaluating items relative to each other in a qualitative evaluation (e.g., better, worse, or equal). Comparisons can also be made in a ***quantitative*** evaluation using specified numerical scales. Either way, the comparisons the experts make are then quantified using a matrix algebraic approach. The results are a set of relative numerical weights for all the items being compared. The paired comparisons technique and one of the scales designed for this technique are part of the Saaty Analytical Hierarchy Process (AHP) (Saaty, 1980).

Many ***decision analysis*** problems apply the AHP. This method appeals to experts because it is easy to use, and it easily quantifies qualitative evaluations. A simple example illustrates the usefulness of the technique. An expert is asked to determine which of the following meteorological conditions would most likely cause a loss of off-site power in a power plant:

1. Flash flooding at plant site with 0.5 to 2 inches of water
2. Flash flooding with 2 to 4 inches
3. Flash flooding with more than 4 inches
4. Lightning (direct hit to power lines)
5. Direct hit by a tornado

6. Winds between 20 to 40 mph
7. Winds higher than 40 mph

The expert first thoroughly defines and clarifies the seven conditions. The expert then chooses an evaluation scheme. If a qualitative evaluation is required, the expert only needs to compare all possible pairs of the seven items (making 21 comparisons) using the terms *better*, *worse*, or *equal*. If a quantitative evaluation is required, then the choice of scale (e.g., the Saaty scale shown in Table 1) is made. The Saaty scale (1980) describes the numerical evaluations for the comparisons:

TABLE 1
Saaty scale

Number	Description
1	The two items are either of equal importance or equally likely
3	A slight favoring of the first item over the second
5	A strong favoring of the first item over the second
7	A demonstrated dominance of the first over the second
9	An absolute affirmation of the first over the second
2, 4, 6, 8	These are used when compromise is needed
1/3, 1/5 1/7, 1/9	These values indicate the above relationships when the first item is worse or less likely to occur than the second

The expert begins comparing all possible pairs of conditions. The analyst records the numerical comparisons in a 7 x 7 matrix of values, with 1s on the diagonals and the comparisons in the upper triangular portion of the matrix. The lower portion is filled later with the reciprocal values of the upper portion. In other words, if condition 3 is strongly favored over condition 4, row 3, column 4 is assigned a value of 5 (from the scale). Then, row 4, column 3 is assigned a value of 1/5.

With the comparisons made and the matrix filled, matrix theory then obtains the set of relative weights of the seven conditions. Specifically, these weights are the normalized eigenvectors of the maximum eigenvalue of the 7 x 7 matrix. The reason why these weights are formulated in such a fashion may not be obvious; however, the mathematical theory behind it is sound.

The Saaty method also provides the ability to monitor the consistency of the expert's evaluations. For instance, if an expert evaluates condition 1 versus condition 2 as a 4, and evaluates condition 2 versus condition 6 as a 3, and evaluates condition 1 versus 6 as a 1, then his or her three evaluations indicate an inconsistency. Using matrix theory, the Saaty technique provides an index of consistency for the comparisons made in a single matrix. The expert is warned when his or her consistency lacks a high value for this index of consistency. When this happens, the experts should reexamine the definitions and evaluations and resolve the inconsistencies. Example 11.11 illustrates the pairwise comparisons, the relative weights, and the inconsistency measures for the meteorological conditions described above:

EXAMPLE 11.11: ***Saaty's pairwise comparison method or AHP***

The following represents seven meteorological conditions important for affecting loss of off-site power (LOSP) in a reactor:

1. Flash flooding at plant site with 0.5 to 2 inches of water
2. Flash flooding with 2 to 4 inches
3. Flash flooding with more than 4 inches
4. Lightning (direct hit to power lines)
5. Direct hit by a tornado
6. Winds between 20 to 40 mph
7. Winds higher than 40 mph.

To find a set of relative weights, compare these seven meteorological conditions using all possible pairs [number of pairs = 7(7-1)/2 = 21]. The weights are interpreted according to the pairwise comparisons. In this case, comparisons are made by determining which of the pairs is more likely to cause LOSP. The pairs are evaluated using the Saaty scale listed above as follows:

1 vs 2 -------------------- 1/3
1 vs 3 -------------------- 1/4
1 vs 4 -------------------- 1/5
1 vs 5 -------------------- 1/5
1 vs 6 -------------------- 1/3
1 vs 7 -------------------- 1/4

2 vs 3 ------------------ 1/2
2 vs 4 ------------------ 1/3
2 vs 5 ------------------ 1/3
2 vs 6 ------------------ 1/2
2 vs 7 ------------------ 1/3

3 vs 4 -------------------- 1/2
3 vs 5 -------------------- 1/2
3 vs 6 -------------------- 1
3 vs 7 -------------------- 1/2

4 vs 5 ------------------ 1/3
4 vs 6 ------------------ 1/5
4 vs 7 ------------------ 1/4

5 vs 6 -------------------- 5
5 vs 7 -------------------- 4

6 vs 7 ------------------ 1/2

These comparisons form the upper triangle of a matrix. The diagonal terms are 1s, and the lower triangle contains the reciprocals of the upper triangle:

	1	2	3	4	5	6	7
1	1	1/3	1/4	1/5	1/5	1/3	1/4
2	3	1	1/2	1/3	1/3	1/2	1/3
3	4	2	1	1/2	1/2	1	1/2
4	5	3	2	1	1/3	1/5	1/4
5	5	3	2	3	1	5	4
6	3	2	1	5	1/5	1	1/2
7	4	3	2	4	1/4	2	1

The principal eigenvalue of this matrix equals 8.001. The weights for the seven factors are formed by normalizing (so that they sum to 1.0) the seven terms in the eigenvector for this eigenvalue. These normalized weights are

$$(0.03, 0.06, 0.11, 0.11, 0.35, 0.15, 0.19)\,.$$

The Saaty method provides a consistency check in the form of a ratio value (called the consistency ratio) that indicates the deviation of the principal eigenvalue from the theoretical eigenvalue of a perfectly consistent matrix. The ratio is also adjusted for the number of factors (the dimension of the matrix). A consistency ratio greater than 0.10 indicates inconsistency.

In this example, the consistency ratio is 0.13, indicating some problems. A closer examination of the meteorological conditions comparisons reveals the following results:

1 vs 4 is the same as 1 vs 5
2 vs 4 is the same as 2 vs 5
3 vs 4 is the same as 3 vs 5
6 vs 4 is the same as 6 vs 5
7 vs 4 is the same as 7 vs 5

These results imply that 4 and 5 are the same; however, 4 vs 5 is given as 1/3. Also, $6 < 7$, but examining 4 vs 6 and 4 vs 7 indicates that $6 > 7$. There may be other minor inconsistencies in the magnitudes of the relationships; however, three major corrections are made as follows:

Comparisons	Correction	Objective
4 vs 5	1	To make 4 and 5 the same
4 vs 6	5	To match 5 vs 6
4 vs 7	4	To match 5 vs 7

The consistency ratio now becomes much more acceptable at 0.06. The weights become

$$(0.03, 0.07, 0.11, 0.28, 0.28, 0.08, 0.15).$$

We recommend a relative comparison interpretation of these weights indicating that a direct hit lightning (4) and a tornado (5) are the most likely to cause LOSP and that flooding with 0.5 to 2 inches of water (1) is the least likely. Because the original scale relies on a ratio relationship among the comparisons, Saaty interprets the resulting weights in terms of ratios.

■

The example problem above consisted of a single matrix evaluation. As the name AHP implies, most applications that use this technique have a hierarchical structure. The above example examines only one of many plant conditions that are of critical concern to operations. Another matrix could be formed by comparing all such critical concerns. A matrix of meteorological conditions could be attached below each of those concerns. These condition matrices do not have to be identical to the one for the loss of off-site power concern. Thus, analysts can construct an entire hierarchy of as many levels as required. The hierarchy is usually constructed from the top down, with the top levels being the more general environmental or scenario factors. The middle levels consist of more specific criteria or characteristics under consideration. The bottom level consists of a list of competing alternative decisions, actions, or choices. The decision among these alternatives must be made to answer the question.

Each matrix is evaluated at each level, resulting in a set of relative weights. The weights are multiplied down the levels to form a final set of weights for the bottom level items. This final set of weights is then used to make the decisions regarding the choices of the bottom level items. The higher the weights, the more desirable the item.

Many codes and packages perform this technique at various levels of user interaction. Some codes merely provide the algorithms for the technique; others take the user through the entire problem, from the initial building of the hierarchy to the final set of weights (Booker, Bryson & McWilliams, 1984). For general references regarding this technique, consult Saaty's two books which provide codes and instructions (Saaty, 1980, 1982). Appendix A provides a FORTRAN user-interactive code for a single matrix evaluation. This code is adapted from the Saaty's second book (Saaty, 1982).

Advantages and Disadvantages of Saaty's Method

This technique has several advantages: experts find it easy to use, it easily monitors the consistency of the expert's evaluations, and it easily quantifies qualitative information. These advantages make it suitable for use in expert judgment problems.

The problem must be formulated in a well-structured hierarchy. Criteria in a given level must be independent such that each addresses different aspects of the problem. In other words, a high rating on one criterion should not imply a high rating of another criterion. Feedback cycles and complex pathways are difficult to implement in the hierarchy. A single level structured problem can be used to avoid the hierarchy; however, a single level formation usually oversimplifies the problem.

Uses for Saaty's Method

Decision analysis problems commonly have a hierarchical structure. In these cases, the weights help a decision maker choose among the competing alternatives at the bottom level of the hierarchy. Therefore, the major disadvantage in applying this technique in expert judgment problems is that usually expert judgment problems cannot be neatly formulated into a hierarchical structure. In addition, this type of forced structure formation is not consistent with the data analysis philosophy advocated in this book. We believe that the data should determine analyses and models, not that the data should be forced to fit to the analyses or models.

In expert judgment applications, the Saaty scale and the ***quantification*** feature of this method can be used as a ***response mode*** and as a quantification technique, respectively. We apply Saaty's method for these limited uses.

Descriptions and Uses of Bayesian Techniques

The Bayesian Philosophy

The second section in this chapter discussed the two different statistical philosophies for analyzing data and for interpreting the roles of probability distributions. These two approaches are the classical (or frequentist) approach and the Bayesian approach.

The data or sample represents the population for the random variable. Characterizing the population as a probability distribution with certain parameters (e.g. mean, variance, maximum value, and degrees of freedom) is a common practice. Frequentists consider parameters fixed but unknown quantities. The estimators (statistics) of these parameters are values calculated from the sample data. For example, if a normal distribution represents the population, then a sample of 20 values randomly chosen from that normal will represent that population. The population mean was 2.0, but this is not known. The sample has a mean of 1.82, which is the best available estimate for that unknown population of 2.0. Chapter 18 further discusses how to use the sample mean (statistic) to draw conclusions about the population mean (parameter).

The Bayesian approach or philosophy interprets the parameters of the distribution representing the population. The population parameters are interpreted as uncertain quantities. Probability distributions represent the state of knowledge or information about the parameter before the sample is taken. These probability distributions are called ***prior distributions***. The sample (data) also forms a distribution called a likelihood. The likelihood distribution represents how likely it was for that sample to be taken from the population. After a sample is taken, the ***likelihood*** distribution can be combined with the prior distribution to form a final combined distribution called the ***posterior distribution***. The posterior represents the combined state of knowledge or information from before and after sample data is taken. The analytical tool (equation) used to perform this combination is Bayes theorem. Hence, the mathematics, the approach, and the philosophy are all labeled Bayesian.

The philosophy is a logical one. Most researchers tend to have information about the problem before completing an experiment or gathering data. Using all available information to draw conclusions makes sense. The Bayesian approach provides a method for combining different sources of information. Application of the technique involves representing the previously known information as a prior distribution, gathering the sample data, and using Bayes theorem to combine the distributions into the resulting posterior distribution. Bayes theorem is as follows:

$$g(\theta|x) = f(x|\theta)\, g(\theta) \,/\, f(x),$$

where

$g(\theta|x)$ is the posterior distribution,
$f(x|\theta)$ is the likelihood or data distribution,
$g(\theta)$ is the prior distribution for the parameter θ, and
$f(x) = \int f(x|\theta)\, g(\theta)\, d\theta$ is the marginal distribution that can be considered as a normalizing constant in the denominator of the above theorem equation.

Therefore, the theorem can be stated as follows: the posterior equals the prior times the likelihood divided by the marginal, or the posterior is proportional to the prior times the likelihood (Martz & Waller, 1982). Example 11.12 illustrates how to easily combine distributions in an interpretable fashion.

EXAMPLE 11.12: ***Combining prior and experimental data in a Bayesian analysis***

A component fails only once in 30 tests. A binomial distribution characterizes the experimental data:

$$f(x|p) = \frac{n!x!}{(n-x)!}\, p^{x}\, (1-p)^{(n-x)} = \frac{30!1!}{(29)!}\, p^{1}\, (1-p)^{(29)}.$$

An identical component operates in a different system where it has never failed in 50 demands. This information is the prior data and can be characterized by a beta distribution:

$$g(p) = \frac{1}{B(x_0, n_0 - x_0)}\, p^{(x_0-1)}(1-p)^{(n_0-x_0-1)} = \frac{1}{B(0,50)}\, p^{(-1)}(1-p)^{(49)}.$$

The beta function works well for the distribution of the binomial parameter (p) because that parameter represents the probability of a single failure for a trial, and the beta distribution can only have values in the range from 0 to 1.

Using Bayes theorem, a binomial likelihood combined with a beta prior forms a beta posterior with the parameters

$$x_0 + x = 0 + 1 = 1$$

and

$$n_0 + n = 20 + 50 = 70.$$

The posterior then has a mean failure rate of

$$\frac{(x + x_0)}{(n + n_0)} = \frac{1}{70} = 0.014,$$

instead of a failure rate of 0.0333 from the test data only. The posterior variance is

$$\frac{(x + x_0)(n + n_0 - x - x_0)}{(n + n_0)^2(n + n_0 + 1)} = \frac{69}{347{,}900} = 0.0002.$$

The test data alone indicates that the component is somewhat reliable. However, combining the test data with the operational data on the component indicates that the component is very reliable.

■

Advantages, Disadvantages, and Uses of Bayesian Methods

The Bayesian approach provides a coherent, consistent methodology for combining or pooling information from different sources. The philosophy of using all the available information seems logical and reasonable, especially when information from a single source is sparse or lacking, which tends to occur in most reliability applications.

Some short examples illustrate the uses of Bayesian methods as a pooling mechanism: expert estimates could provide the information for the prior, and that could be combined with sparse data; expert estimates could be combined, one at a time, to form an aggregation estimate; an expert aggregation estimate (prior) could be combined with information from a decision maker; older information from an expert (prior) could be combined with his or her new assessment to update his or her judgment in view of different conditions or information; generic data such as an overall failure rate of all check valves (prior) could be combined with data on a specific check valve; and uncertainties (prior) could be modeled with the data. Bayesian methods are suggested and discussed in more detail for aggregation (chapter 16), for characterizing uncertainties (chapter 17), and for updating (chapter 7).

A disadvantage of Bayesian methods lies in transforming all the available information, regardless of its source or form, into probability distributions. Transformation is a difficult task for qualitative data. (Chapter 12 discusses ways of handling qualitative data.) Transformations may not be any easier for quantitative data. Once distributions are formed, the second disadvantage of Bayesian methods emerges. Bayes theorem combines these various distributions. These combinations may prove mathematically difficult to construct. However, with modern simulation techniques, using the theorem for combining distributions is not as difficult as it was a decade or so ago.

12

Initial Look at the Data—The First Analyses

The information gathered from the elicitation will appear as a large, complex mass of words and numbers. To become familiar with this mass of information, first examine the data that has been gathered. Second, focus on some important data features, investigate transforming the data or quantifying it, and formulate a data base for further analysis.

What Data Has Been Gathered?

In this chapter, we assume that the reader has completed the following steps:

- selected the components of elicitation, which include the elicitation situation, mode of communication, elicitation technique, response mode, type of aggregation, and method of documentation, as described in chapter 7,
- tailored these components for use in a particular application, as described in chapter 8,
- and conducted the elicitation, as described in chapter 10.

The post-elicitation information base will likely consist of large amounts of ***qualitative*** and quantitative information from each expert on every question. Information documentation (as suggested in chapters 7 and 8) helps reduce some of the volume of information to a more compact and efficient form at the chosen levels of detail. However, even the documentation guidelines will not eliminate the difficulty of the analyzing a qualitative/quantitative data mixture of a potentially large number of variables.

These variables consist of two major types: the answers to the questions, and the ancillary information. This ancillary information is also in two groups: the information about the expert (e.g., his or her background and experience), and the information called ***expert data*** in chapter 1. Expert data consists of the expert's problem-solving technique and other related information (e.g., how long since he or she had seen such a problem).

The qualitative or quantitative structure of the answer data and the ancillary data depends on the choices of the response mode and the documentation. In many cases, the answer data is quantitative and the ancillary data is a mixture. In either case, some quantification of some of the qualitative information becomes necessary for analysis. To determine the need for and type of quantification, consider the level of generality or granularity of all the data and of the analyses.

Overview of the Data

In the process of analyzing the data, the analyst often forgets two important overall features of the data. The analyst should be aware that these two important features of the data set play a key role in all stages of the analysis, and that they critically affect the final conclusions. The two features are ***granularity*** and ***conditionality***.

Granularity is the level of detail defined or chosen for the data, the analysis, and the conclusions. Two examples of the information recorded on an expert's problem-solving may be (1) information in the form of detailed steps, equations, heuristics, definitions, and descriptions, or (2) information in the form of a general categorization of this problem-solving stating simply that the expert used a pessimistic approach. These two recording methods represent two different granularities for the information regarding the expert's problem-solving process. Each method records at a different level of detail.

Conditionality refers to the inescapable fact that all the elicited data is conditioned on many other factors. Some of these factors are controlled, some are not controlled, and some are unknown.

The *Pitfalls* section in chapter 2 provides a more detailed description of the importance of these two features for expert judgment applications. In addition, all the chapters dealing with the analysis of the data have sections on granularity and conditionality.

Establishing Granularity

To some extent, selecting the response mode and the documentation recording schemes establishes the granularity. The granularities chosen for these two could be different. Then the analyses can be performed at a third or more level. However, for interpreting the results and drawing conclusions, one level, the most general of all, must prevail. This level is the only one that applies to the results and conclusions. Therefore, it is wise to establish one desired level of detail in the initial planning phase of the study before the elicitation. If that is not possible, then at least establish the level at the preanalysis phase and use it throughout all the analysis steps. Otherwise, analyses may require repeating at the proper granularity.

Mixing levels during the analysis can change the conclusions. For example, comparing variables at the granularity gathered in raw form could reveal some significant correlations among the variables that vanish if they are compared after combinations or reductions to a more general level. Of course, such a combining or collapsing process might not change the significant results among the variables from the raw form, but it could produce significance where none existed in raw form, or it could lose significance where it existed in raw form. The effects of changing granularity on the results is not predictable.

The chapters that follow describe many stages of data analysis. In each, the granularity is important. At each stage, the results can change if the granularities are changed. Examining how results can change with different granularities within each stage and across stages is an exercise consistent with the spirit of investigative data analysis advocated by this book. However, great care must be taken not to confuse such an academic exercise with the goal of determining conclusions for the problem at hand. Another important point in examining granularities is that detailed information can be made more general, but not the reverse. To avoid confusion and problems, we recommend that one granularity be chosen and used throughout all analysis stages.

Establishing Conditionality

Analysts often ignore the conditional structure of the data, especially the answer data, in the analysis and in the conclusions. Disregarding the conditional structure of the data produces conclusions that are a mixture of differing effects (i.e., a mixture of apples and oranges). Most analysts would agree that such conclusions are worthless. Indeed, this lack of care in analysis may be why many do not trust expert judgment data or claim that expert information cannot be analyzed.

A recent example will help illustrate the problems in dealing with conditionality. In an effort to revise the probabilistic risk assessment (PRA) methodology for nuclear reactors, the Nuclear Regulatory Commission invested time and money in the NUREG-1150 project (U.S. NRC, 1989). As part of this task, several panels of the world's top experts were gathered to elicit their data on many rare and undefined events affecting reactor safety. These events were decomposed into decision-type trees. (Chapter 15, as well as Raiffa [1970], describe these decomposition and decision trees.) The tree structures and probabilities for each branch were elicited from the experts. The final answers came from multiplying these probabilities through the tree. Each answer is therefore conditioned on the tree and its estimates—this conditioning cannot be ignored. Two experts could arrive at exactly the same final answer but for very different reasons, or two experts could arrive at different answers for exactly the same reasons.

Analyses, such as the ones described in this chapter and in chapters 13-15, investigate what effects, if any, such conditioning has on the final answers. If the answers do not depend on the conditions, then conditioning can be ignored; however, this determination is necessary before setting conditionality aside.

How to Quantify

Quantification helps prepare the complex post-elicitation data set for the data base. Both the qualitative data and the quantitative data may require transformations and cleanup for the data base. Statistical and other analysis techniques often require that word descriptions or preference scales (*worst, worse, bad, neutral, good, better,* and best) be transformed into numerical values. In addition, some raw data already in numerical form may require additional numerical transformation to more convenient scales or to a chosen granularity. In both cases, the transformation process is quantification—transforming the raw information into a desired numerical form.

Several commonly used methods of quantification are described in the sections below. The major problem with quantification is not to impose additional assumptions about the information to *fit* the data into the desired form. In most instances, this imposition is difficult or impossible to avoid. However, sometimes other available information elicited from the experts can aid in the quantification process, as illustrated in example 12.1.

EXAMPLE 12.1: ***Using definitions to quantify***

The analyst attempts to convert a statement of the expert's preference about how good a reactor system design might be. The analyst can refer to the expert's elicited definition of *good* and use that definition to compare to definitions from other experts. This comparison can form a consistent numerical scale across experts as follows:

Expert's Definition of Good	Numerical Scale Value
System functions outside specifications at all times	5
System functions within specifications at all times	3
System functions within specifications 90% of the time	3

At this stage of the analysis, the reasons for eliciting information from the experts about definitions, assumptions, and problem-solving processes become obvious.

■

When Is It Necessary to Quantify?

According to theory, all the information gathered during the elicitation should be quantified to a common numerical scale for comparisons using statistical analyses. However, this complete quantification is usually not feasible, nor is it entirely necessary. Many times gathered information is redundant. The expert will state the same information in different forms, as illustrated in example 12.2.

EXAMPLE 12.2: ***Detecting redundant information***

An expert gives a lengthy explanation of a physical phenomenon. Five minutes later, the expert realizes that he or she simply applied a basic principle or law. The expert's information and its usage remain unchanged. The expert has merely given the same information in two different forms. Many times the expert does not realize this redundancy, but the analyst can find it in the course of his or her analysis, with proper documentation.

■

Answers to the technical questions usually require quantification first. In most instances the answers will already have the desired numerical form from the chosen response mode designed in the elicitation. (See chapter 7, *Selecting from Response Modes* and *Selecting from Dispersion Measures.*)

Initial quantification of assumptions, definitions, and problem-solving processes may prove difficult. However, some assumptions about physical quantities such as temperature are easily converted to a scale of values or ranges of values. Do not reduce ranges of values to a single value. The process of such a reduction imposes the analyst's assumptions on the original data, and these assumptions might not have been those of the expert. In general, the assumptions, definitions, and problem-solving information should be kept in raw form until the modeling stage of analysis (chapter 15).

The ancillary information or data, such as data on the expert's background and experience, also need quantification. These data can have many different forms and usually have been elicited without much advanced planning or designing of the analysis. Therefore, these data may range from completely descriptive information to strictly numeric values, with everything in between.

Quantification Schemes

We advise extreme caution when applying the following quantification schemes. The analyst can easily impose assumptions on the data to make it *fit* into the desired quantification scheme. The higher the degree of qualitative structure, the more such assumptions are required to transform the raw information to numbers useful for standard analysis techniques. The *Pitfalls* section in chapter 2 discusses in more detail the interviewers, analysts, and knowledge engineers as sources of this bias.

The following methods of quantification cover ways of transforming qualitative information to quantitative information. Each method is illustrated with an example.

Dummy Variable

This method is widely used when dealing with quantification. The raw data are transformed into artificial indicators or dummy numerical values. Many times the transformations are done without proper logic or thought. Examples of this transformation follow:

1. Transforming information that has only two options into dichotomous values such as (0, 1) or (-1, 1).
2. Transforming information to integer values, such as in the use of scales or ranks (1, 2, 3, etc.).
3. Transforming information to the real number line, which covers a continuum of values for a specified interval (e.g., 0.0, 0.1, 0.15, 0.27, 0.96 in the [0,1] interval). The choice depends on the information gathered, its potential use, and its precision (the number of significant digits).

Transformations to numerical values are often necessary for the data base formation (discussed later in this chapter) and for the modeling process. Most statistical and analytic procedures require numerical variables, making numerical values desirable. The following sections provide more detail on how to effectively form the various types of dummy variables.

Dichotomous Quantification

Most information, whether qualitative or quantitative, can be transformed to a (0, 1) or (-1, 1) dichotomy with little or no assumptions required for the transformation. Example 12.3 illustrates dichotomous quantification for cases involving problem-solving information, experts' background information, and expert answers.

EXAMPLE 12.3: ***Dichotomous quantification***

Experts are asked if they applied the first law of thermodynamics. To quantify the simple *yes* or *no* responses, set *yes* = 1 and *no* = 0.

Experts are asked if they consider themselves engineers. To quantify the responses, set *engineer* = 1 and *nonengineer* = 0.

Experts are asked if the probability of an event is greater than 0.001 or less than 0.001. To quantify the responses, set values greater than 0.001 = 1 and values less than 0.001 = -1.

■

Rank or Rating Quantification

If more than two choices are required to represent some information, multiple values can be used. The values can be in ascending or descending order (ranks), or the values can be chosen from a specified scale. The ordering can be of two different types: ordinal and cardinal. Cardinal ordering is based on a linear, numerical interpretation implying equal spacings between the ranks and relationships, such as a rank of 4 is twice a rank of 2. Ordinal ranks or ***ratings*** only imply a logical ordering that is not necessarily linked to the number line. In either case, the ranks reflect a logical ordering of the information. Example 12.4 illustrates the proper use of ranks, and examples 12.5 and 12.6 contain more detailed uses of ordinal ranks.

EXAMPLE 12.4: ***Rank quantification***

As part of the expert's background information, the interviewer asks the experts if they have any reactor operator experience. The experts responded to that question in verbal terms such as none, some, and extensive. The ordinal ranks 0, 1, and 2 can be assigned to the answers none, some, and extensive. The ascending order implied by the ranks is logical.

The decision maker rates his 5 experts on their experience in a particular problem area. His or her ratings are (2, 1, 4, 2, 2). This is a cardinal rating. His or her interpretations are that experts 1, 4, and 5 are twice as good as expert 2, and expert 3 is 4 times as good as expert 2.

The experts are also asked to describe their major discipline area. The responses are given in terms such as nuclear engineer, civil engineer, mechanical engineer, physicist, and mathematician. The ranks 1, 2, 3, 4, and 5 should *not* be assigned to these answers because they do not have a logical order; they also should not receive an order through the use of ranks.

■

Nonnumeric information that has some logical or relative ordering can be transformed to ordinal. The rank values assigned are also descriptive (words) in nature. The variables formed using this quantification can be incorporated into the data base and can be used in many analytic procedures such as correlation detection and understanding the conditional nature of the data base.

Because the ranks are relative comparisons, exercise care when using definitions to achieve and maintain consistency of application. Assuming that the information from the experts was elicited in the proper fashion, clarifications are available in the documentation to help the analyst or decision maker assign relative comparisons for certain quantities.

EXAMPLE 12.5: ***Assigning ordinal ranks***

A group of engineering experts was asked to solve a problem. Most of them used a basic principle from thermodynamics but in varying degrees of emphasis of use. This information can be added to the data base using the following ordinal rank variable:

Expert No.	Descriptive Use	Rank
1	Did not use the principle at all	1
2	Extensively used the principle	6
3	Only mentioned the principle	2
4	Used the principle a few times	4
5	Used the principle once	3
6	Used the principle several times	5

■

The Saaty pairwise comparison technique (chapter 11) is a convenient quantification technique for making relative comparisons using a numerical scale. However, these resulting numbers or weights can only be interpreted in a relative sense because the information used in the method is only relatively compared. A resulting relative weight from this technique of 0.25 cannot be interpreted as half as good as a weight of 0.50. The relative interpretation is that the value 0.25 is less important or less preferred than 0.50. This technique works best as an elicitation method for obtaining responses from the experts. However, this technique can also be used as a quantification scheme for the post-elicitation data. One major advantage of this method is that the relative comparisons are made in pairs and do not have to be made simultaneously.

EXAMPLE 12.6: ***Ordinal ranks from pairwise comparisons***

Of the five experts solving a problem, some used a rule of thumb or a modification of that rule. The analyst was able to evaluate the experts using pairwise comparisons by answering the following question: Did expert i apply the rule more completely than expert j?

Expert i	Expert j	Comparison
1	2	Same
1	3	Yes
1	4	No
1	5	No
2	3	Yes
2	4	No
2	5	No
3	4	No
3	5	No
4	5	Yes

Using Saaty's method, the resulting relative weights for these 5 experts follow:

Expert	Weight
1	0.138
2	0.138
3	0.079
4	0.387
5	0.257

These weights indicate that expert 4 applied the rule more completely than the others. Expert 3 applied the rule less completely than any of the others. Experts 1 and 2 applied the rule in the same manner. No further interpretation is possible with such a relative comparison.

■

Number Line Quantification

As mentioned earlier, precision of information content is important in the transformation process, especially for transformations to the number line. These transformations to numbers are not limited to integer values but can be equal to any value on the continuous number line. The precision of the numbers is an issue related to granularity. Numerical precision is considered when determining how many significant digits accurately represent the information. The number of significant digits used in the analysis reflects how precisely that value is known and relates to the level of detail (granularity).

EXAMPLE 12.7: ***Significant digits***

Experts are asked to provide experience in a field in terms of the number of years that they have worked. Experts will likely differ in the number of digits used to answer this question. Some say *about* 2 years and mean greater than 1 year but less than 3 years. This is only one significant digit. Some say 2 years and

6 months, or 2 1/2 years. This is 2 significant digits. Some say 2 years and mean exactly 2.0 years (2 digits), or mean *about* 2 years (1 digit). Because the number of digits differs across experts, the most general level of detail (the least precise) must be used for all experts. In this case, that means the lowest number of significant digits (1) would be used for all experts.

Using only 1 significant digit results in a loss of information at the finer level of detail offered by some of the experts. To avoid this loss, the elicitation of the information should be more thorough. The first and second experts should be queried for the desired level of detail using the verbal probe or ethnographic methods described in chapter 7. The elicitation process can help guarantee that the level of information content is consistent among experts, thereby minimizing quantification problems.

■

Proper elicitation planning and execution also involves understanding why the ancillary information is being gathered and what the potential use it will have in the analysis. Knowing this, the analyst should make certain that the information is being elicited at the desired granularity (e.g., number of significant digits) rather than getting mixed levels of detail. If mixed levels occur, then the analyst has the difficult problem of transforming all the information to a common (usually very general) level in the post-elicitation phase.

Number line quantification can be used to combine information from two or more variables, provided these variables have a common basis of accuracy. Example 12.8 illustrates how this is done.

EXAMPLE 12.8: ***Combining number line quantifications***

Experts were each asked how much training and experience they had in thermodynamics. The answers follow:

Expert No.	School Training (years)	Job Experience (years)
1	1.50	0.00
2	0.50	5.50
3	0.00	2.33
4	2.25	3.10

If training is only half as valued as job experience, then the results would be

1	$0.50(1.50) + 1.00(0.00) = 0.75$ years
2	$0.50(0.50) + 1.00(5.50) = 5.75$ years
3	$0.50(0.00) + 1.00(2.33) = 2.33$ years
4	$0.50(2.25) + 1.00(3.10) = 4.23$ years

We do not recommend that information of a more descriptive structure (words) be transformed into values on the real, continuous number line. Only information gathered in a continuous, numerical form should be treated in this manner.

■

Categorical Variables

In cases where qualitative information cannot be ranked or ordered in preference, the information can be transformed and stored into groups or categories, usually according to verbal descriptions. We do not recommend that these verbal categories be coded into an arbitrary numerical code (a dummy variable). When this is done the analyst is tempted to analyze the numeric data as if the numbers reflected ranks or orderings. Most modern software can easily handle character (word) information as a part of the information data base.

Educational information on the experts is usually put into the data base as categorical or classification variables. Degree titles such as BS, MS, MBA, etc., are examples of these categories. Degree disciplines such as civil engineering, mechanical engineering, nuclear physics, and thermodynamics are other examples.

If there are too many classes, or if there are few experts (less than 3) in each class, the classes may have to be consolidated or collapsed to perform some of the later analyses. However, such collapsing does change granularity from specific to general.

EXAMPLE 12.9: ***Collapsing categories***

The following table collapses the disciplines of the highest degree for 10 experts. The original 10 specific categories are collapsed into 3 more general categories:

Expert	Elicited Discipline	Transformed Discipline
1	Mechanical engineering	Engineering
2	Nuclear engineering	Engineering
3	Thermodynamics	Engineering
4	Hydrodynamics	Physics
5	Material science	Physics
6	Nuclear physics	Physics
7	Mechanics	Engineering
8	Computer science	Computation
9	Simulation science	Computation
10	Knowledge engineering	Computation

The definitions and rationale used in making the above transformation should be recorded and then consistently applied throughout the study. The above process changes the granularity of the discipline information from a more detailed description to a broader one. The results of any analysis done on the collapsed version will only be valid for the more general categorization. For this reason, we recommend that collapsing be done only during the later analysis stages and that the categorical data be stored untransformed in the data base.

■

Description Variables

Some information may not be suited for quantification. This information should be condensed to as few words as possible and kept for further analysis applications.

Descriptions of how the experts solved problems can be analyzed as conditioning variables or can be used in model formation (chapter 15). Usually this information is not easily translated into numbers or even categories without making some assumptions or without changing the level of detail. At this initial analysis stage, neither is recommended.

Description variables such as the definitions and assumptions used by experts have another important use. If new information is considered (chapter 10), these variables are needed for purposes of updating the experts' answers. Even if there is no new information, these variables are needed for documentation purposes.

Forming a Data Base of Information

After completing the quantification steps from the previous section, the information can be placed in a data base. This data base could be a computer file or a paper listing containing all the information gathered from each expert including their answers. A suggested list follows:

Expert ID	Name and Number
Interview information	Date, time, place, duration, and environment
Expert's background:	
Education	Degrees, dates, schools, and disciplines
Experience	Years, organizations, colleagues, nature and type of work
Expert's problem-solving technique	Definitions, assumptions, steps, cues, and heuristics
Answers	Values and comments

Each expert will then have many quantities (variables) associated with him or her. Some information could be missing for some experts. A code word or number is needed to denote missing information (e.g., *miss*), and one is also needed for nonapplicable information (e.g., *na*) to distinguish these from 0 values. Many statistical and data-base packages have their own designations for missing information.

If the guidelines for tailoring the elicitation in chapter 8 were followed, the data gathered will reflect the design chosen, and the reasons for the choices are already documented. Therefore, little needs to be done to the data base to conform to the elicitation method.

Because the data base and the elicitation method are so closely connected, the monitoring and use of granularity is important. The results and conclusions that will be found from the data base should be either at the same or at a more general level of detail than the information in the data base. In the analyses suggested in the next chapters, granularity is continuously monitored. Examples are given in which the results change if the granularity changes. Be aware of the detail of the information content in the data at the data base stage. A quick review of the variables and information in the data base is usually sufficient. This review can be aided by listing each answer in ascending order and listing other variables beside the answers for each expert.

During this review, the analyst can get some ideas about model formation (chapter 15). If many variables have missing values, then we suggest the use of more general-level models. If the information is complete in the data base, we suggest forming and analyzing models at the current granularity.

Also during this review the analyst can see which and how many variables require testing for possible conditioning effects on the answers. In addition, some variables may require testing as sources of correlation or bias among the experts (chapter 14).

If the data base contains many variables (more than 20), a strong possibility exists that some variables are repeating or sharing information from other variables. Many statistical techniques (from chapter 11) can be used to monitor for this redundancy of information among the variables. In some cases, variables can be eliminated from the data base. These steps are presented in the next chapter (chapter 13) on understanding the data base.

13

Understanding the Data Base Structure

The analyses in this chapter provide ways of gaining an understanding of the relationships existing among the ancillary variables, among the answer (response) variables, and between these two sets of variables. The information and knowledge gained from the results of these analyses are merely for understanding the data base and are not to be considered as the final results. In addition, conclusions should not be drawn from them. The analyses suggested use standard, statistical techniques, most of which require assumptions about the data that would not necessarily apply in expert judgment problems. Instead, these techniques are only to be used as tools for understanding variable relationships and are not to be used to determine final results. The use of several techniques is for cross-verification of suspected relationships among the variables.

The following are serious words of caution regarding the use of the statistical techniques. We find it difficult to recommend applying a technique when there is reason to believe that assumptions required for its use are being violated. We also find it difficult to recommend using a technique and then strongly urging the reader not to rely on the results. These statistical tools are used to explore variable relationships in the data base. If the results from various analyses do not make sense, or if the results of one test contradict the results from another, the reason might be that the techniques were not used properly. We strongly advise using these techniques with the help of a statistician.

The analyses presented in this chapter begin with the investigation of potential relationships between the ancillary variables and the answers. A separate analysis of the answer data is then presented for two purposes: to present analyses of the answers when no suspected conditional effects exist, and to gain additional information about the answer variables. This chapter also provides the separate analysis of the ancillary data. Finally, analysis techniques are given for analyzing the ancillary data with the answer data.

Many statistical and data analysis software packages contain the analysis techniques. The software used for most of the examples is the SAS® product.

Conditionality—Examining Relationships Between Answers and Ancillary Data

All variables from the ancillary data are potential conditional variables that affect the answers given by the experts. This relationship is called conditionality. Some very basic statistical tests and graphic procedures can be used to begin the investigation of any potential relationships. These tests and procedures are referred to as bivariate analyses, where each ancillary variable is checked against each answer variable. A later section, *Analyzing the Ancillary Data with the Answer Data*, describes multivariate investigative procedures in detail.

Correlations

To begin a bivariate analysis of the ancillary data and the answer data, calculate Pearson pairwise correlation coefficients for all pairs of numerical ancillary variables and numerical answer variables. Most statistical and data analysis packages have correlation routines. If none are available, use the formula in chapter 11.

A significance level is required to determine if any of the correlations indicate potential relationships between the pairs of variables. The significance level for determining correlations among n pairs of variables should be less than $1/n$ for more than 20 pairs, or should be the customary values of 0.05 or 0.01. Example 13.1 illustrates this determination for a large number of experts.

EXAMPLE 13.1: ***Correlations and significance level***

An analyst calculated the following 10 correlation coefficients by comparing two questions (Q_1 and Q_2) with five ancillary variables from the backgrounds of 31 experts. The five ancillary variables are defined as:

Ys = years since they worked on this type of problem;
Ya = years worked as an assessor;
Yp = years worked in applications;
Yd = years worked as a developer; and
Yn = years worked on documentation.

The correlation coefficients for the 10 pairs of variables are as follows:

		Ancillary	Variables		
Correlation (Level)	Ys	Ya	Yp	Yd	Yn
Q_1	0.095	-0.484	0.167	0.153	0.035
	(0.61)	(0.006)	(0.37)	(0.41)	(0.85)
Q_2	0.322	-0.452	-0.095	-0.089	-0.309
	(0.08)	(0.011)	(0.61)	(0.63)	(0.09)

The actual significance levels are listed in parentheses below the correlation coefficients. Because there are less than 20 pairs of variables, the analyst would specify a significance level of 0.05 or 0.01 for significance testing. A 0.01 specified level represents a conservative value, which means that the chances of a type-I error are 0.01. Any correlation whose level in parentheses is less than 0.01 would be considered significant, and a relationship would be suspected between those two variables. In this case, only the Ya / Q_1 relationship is indicated, but not Ya / Q_2 because the significance level is 0.011. Because the correlation coefficient is a negative value, large values of Ya correspond to small values of Q_1. This conclusion is relevant for the population of the type of experts and the type of questions in this example.

■

At this point in the analysis, record all the significant relationships along with any possible reasons or explanations. The ancillary variables that are significantly correlated to the answers are potential conditional variables. When multivariate analyses are performed later, however, these significant relationships may not hold. Nevertheless, they offer some understanding about the data base, the reasons for the experts' answers, and some directions for future analyses.

Graphs

Correlation coefficients can only be calculated for numerical variables. However, graphs can plot bivariate relationships between the ancillary variables and the answers, where the variables are ordinal. Graphs can also indicate nonlinear relationships, whereas correlation analysis only detects a linear trend.

To graph qualitative or categorical data, equally spaced intervals may be used on the axis using the ordering or ranking inherent in the categories. Unequal spacing should only be used if there is some reason for it. If nonlinear relationships are indicated, transformations of the data by taking logarithms can sometimes produce linear relationships. Example 13.2 illustrates that such a graph of categorical data can be done.

EXAMPLE 13.2: ***Graph of an ancillary variable and an answer variable***

The following graph plots a categorical variable describing the reactor experience of 13 experts versus the answer variable Q1. The three categories of the experience variable indicate an order of importance. That ordering places the categories on the axis.

Q_1			
1.0	X		
			X
0.8		X	X
		XX	X
0.6	X	X	
			X
0.4			X
		XX	
0.2	X		X
0.0			
	None	School	Work

The above graph fails to indicate any linear or nonlinear relationships between the two variables.

■

Plots all possible pairs of the ancillary variables with answer variables should follow the example in 13.2. If the graphs indicate a relationship between the variables, then the ancillary variable is a potential conditional variable. Compile a list of these potential conditional variables. This list should also include any explanations or reasons for the relationship.

As mentioned above, additional analyses (e.g., multivariate analyses) will be used to add and to change the list of the potential conditional variables. Before these analyses, some basic investigation of the answer data is useful, especially if conditionality does not appear to be a problem.

Analyzing the Answer Data

The answers to technical questions constitute the prime reason for eliciting expert judgment. Remember that these responses can be highly conditioned on other (ancillary) information such as the experts' problem-solving processes and response environment. Also remember that any scientific investigation must be done with an understanding of the granularity used at the various stages of the problem: the information gathering, analysis, and conclusions. Granularity and conditionality are considered in the next section on the analysis of the answer data by examining the between/within variance and multimodal structures of the answers.

Investigating Multimodality

Empirical evidence shows that the answers given by multiple experts form a multimodal distribution (Booker & Meyer, 1988*a*; Meyer & Booker, 1987*a*; Baecher, 1979). Multimodality played a partial role in the widespread belief that experts must be correlated or dependent upon one another. Therefore, the explanation of the modes or clusters of their answers reflected membership of the experts in groups based on common backgrounds, education, or experiences. Until recently this belief and the reasons for these clusters had not been investigated (Meyer & Booker, 1987*a*).

In many cases the number of distinctive modes in the data will be obvious to the eye. For example, a bimodal case with one mode at high values and the other at low values with a gap in the midrange obviously splits the data set into two groups or clusters (example 13.3). However, some cases may not indicate such obvious groupings or clusterings. For these cases, a formal cluster analysis can determine the structure of possible groupings.

As described in chapter 11, cluster analysis forms the clusters from a data set according to some distance criterion separating the individual data points. Most cluster analysis programs have several options available for determining this distance criterion; the most commonly used is the centroid method. The results of the cluster analysis can change depending upon the method chosen. In any case, the user or analyst must interpret the cluster analysis results. Most cluster programs print out results according to a hierarchical clustering scheme in which clusters are formed beginning with one cluster containing all the data and ending with clusters containing only one data point each. The user must decide which of these possible cluster formations to use in characterizing the data set.

The analyst can choose from a variety of methods in deciding on which cluster structure to use. The analytically-based method is to choose a cluster formation that shows the largest change in distances between clusters (examples 13.3 and 13.4). Example 13.3 is a frequency data plot of 31 experts' answers to a reactor safety question with a continuous response scale from 0.0 to 1.1 (Meyer & Booker, 1987*a*). At first glance, the data appears bimodal in nature. Example 13.4 provides the results of a formal cluster analysis using the centroid method on the SAS® software. This graph shows the distance between the cluster centers plotted for the different numbers of clusters formed (the different cluster formations). The detailed clusters are also given for all possible cluster formations.

EXAMPLE 13.3: ***The frequency plot of a raw data set***

The following plot shows the responses to one question by 31 experts. The responses were elicited on a continuous scale from 0 to 1.1, where a 1.1 was considered the highest likelihood. Of course, rescaling the raw data could yield a scale from 0 to 1.0.

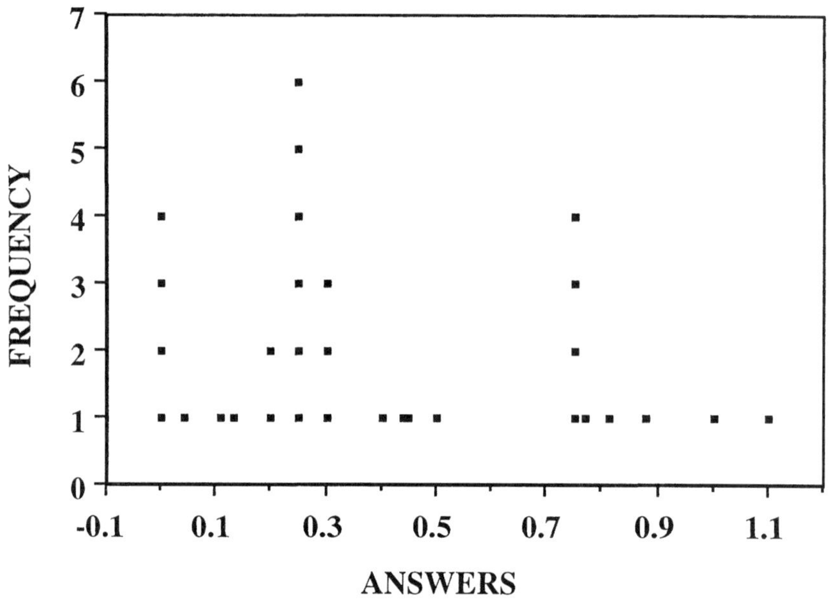

The data form two distinctive groups or distributions. The formal cluster analysis should also indicate these two major clusters.

■

EXAMPLE 13.4: ***Cluster analysis graph***

The following graph indicates the distances between all possible cluster formations for the 31 responses to one question. The distances are based on the centroid method of cluster analysis.

Cluster formations range from 17 clusters where each value given by an expert forms its own cluster, to 1 cluster, where all 31 observations form a cluster. (Some experts gave the same value, making only 17 distinctive values in the data set.) The plot indicates a dramatic change in cluster distances with the two-cluster formation. The next major breaks in cluster distances occur at the four- and five-cluster formations. By the time 17 clusters are formed, the distance measure is zero.

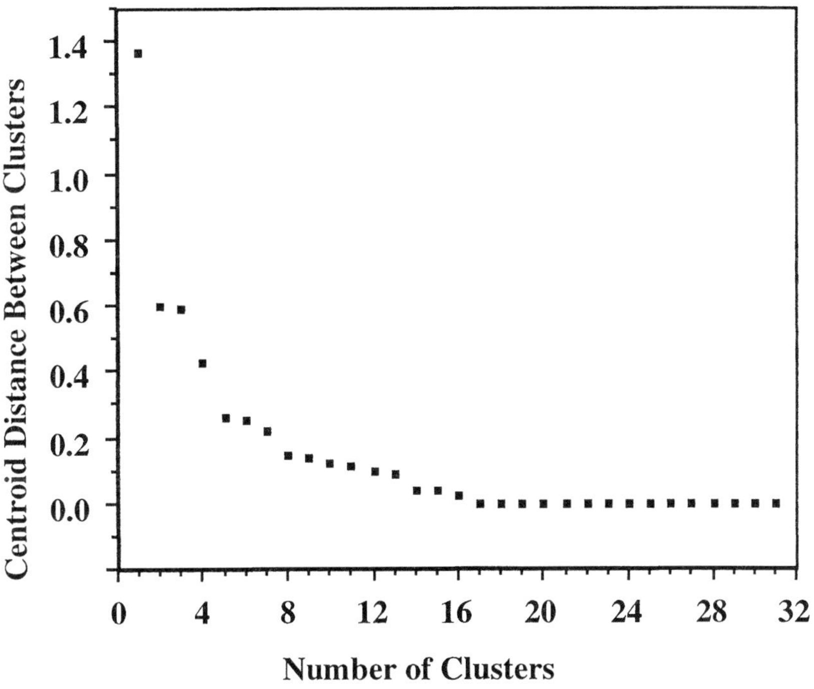

The following table gives the values from the experts belonging to each cluster formation, where individual clusters are marked by square brackets:

No. of Clusters	Cluster Formations with Members
1	[0.25, 0.25, 0.25, 0.25, 0.25, 0.25, 0.30, 0.30, 0.30, 0.40, 0.44, 0.45, 0.50, 0.75, 0.75, 0.75, 0.75, 0.77, 0.81, 0.88, 1.0, 1.1]
2	[0.0, 0.0, 0.0, 0.0, 0.04, 0.11, 0.13, 0.20, 0.20, 0.25, 0.25, 0.25, 0.25, 0.25, 0.25, 0.30, 0.30, 0.30, 0.40, 0.44, 0.45, 0.50] [0.75, 0.75, 0.75, 0.75, 0.77, 0.81, 0.88, 1.0, 1.1]
3	[0.0, 0.0, 0.0, 0.0, 0.04, 0.11, 0.13, 0.20, 0.20, 0.25, 0.25, 0.25, 0.25, 0.25, 0.25, 0.30, 0.30, 0.30, 0.40, 0.44, 0.45, 0.50] [0.75, 0.75, 0.75, 0.75, 0.77, 0.81, 0.88] [1.0, 1.1]

4	[0.0, 0.0, 0.0, 0.0, 0.04, 0.11, 0.13] [0.20, 0.20, 0.25, 0.25, 0.25, 0.25, 0.25, 0.25, 0.30, 0.30, 0.30, 0.40, 0.44, 0.45, 0.50] [0.75, 0.75, 0.75, 0.75, 0.77, 0.81, 0.88] [1.0, 1.1]
5	[0.0, 0.0, 0.0, 0.0, 0.04, 0.11, 0.13] [0.20, 0.20, 0.25, 0.25, 0.25, 0.25, 0.25, 0.25, 0.30, 0.30, 0.30] [0.40, 0.44, 0.45, 0.50] [0.75, 0.75, 0.75, 0.75, 0.77, 0.81, 0.88] [1.0, 1.1]
6	[0.0, 0.0, 0.0, 0.0, 0.04, 0.11, 0.13] [0.20, 0.20, 0.25, 0.25, 0.25, 0.25, 0.25, 0.25, 0.30, 0.30, 0.30] [0.40, 0.44, 0.45, 0.50] [0.75, 0.75, 0.75, 0.75, 0.77, 0.81] [0.88] [1.0, 1.1]
7	[0.0, 0.0, 0.0, 0.0, 0.04] [0.11, 0.13] [0.20, 0.20, 0.25, 0.25, 0.25, 0.25, 0.25, 0.25, 0.30, 0.30, 0.30] [0.40, 0.44, 0.45, 0.50] [0.75, 0.75, 0.75, 0.75, 0.77, 0.81] [0.88] [1.0, 1.1]
8	[0.0, 0.0, 0.0, 0.0, 0.04] [0.11, 0.13] [0.20, 0.20, 0.25, 0.25, 0.25, 0.25, 0.25, 0.25, 0.30, 0.30, 0.30] [0.40, 0.44, 0.45, 0.50] [0.75, 0.75, 0.75, 0.75, 0.77, 0.81] [0.88] [1.0] [1.1]
9	[0.0, 0.0, 0.0, 0.0, 0.04] [0.11, 0.13] [0.20, 0.20, 0.25, 0.25, 0.25, 0.25, 0.25, 0.25, 0.30, 0.30, 0.30] [0.40, 0.44, 0.45] [0.50] [0.75, 0.75, 0.75, 0.75, 0.77, 0.81] [0.88] [1.0] [1.1]

10 [0.0, 0.0, 0.0, 0.0, 0.04] [0.11, 0.13]
[0.20, 0.20, 0.25, 0.25, 0.25, 0.25,
0.25, 0.25] [0.30, 0.30, 0.30]
[0.40, 0.44, 0.45] [0.50]
[0.75, 0.75, 0.75, 0.75, 0.77, 0.81] [0.88]
[1.0] [1.1]

11 [0.0, 0.0, 0.0, 0.0, 0.04] [0.11, 0.13]
[0.20, 0.20, 0.25, 0.25, 0.25, 0.25,
0.25, 0.25] [0.30, 0.30, 0.30]
[0.40, 0.44, 0.45] [0.50]
[0.75, 0.75, 0.75, 0.75, 0.77]
[0.81] [0.88] [1.0] [1.1]

12 [0.25, 0.25, 0.25, 0.25, 0.25, 0.25] [0.30,
0.30, 0.30]
[0.40, 0.44, 0.45] [0.50]
[0.75, 0.75, 0.75, 0.75, 0.77]
[0.81] [0.88] [1.0] [1.1]

13 [0.0, 0.0, 0.0, 0.0, 0.04] [0.11, 0.13]
[0.20, 0.20]
[0.25, 0.25, 0.25, 0.25, 0.25, 0.25] [0.30,
0.30, 0.30]
[0.40] [0.44, 0.45] [0.50]
[0.75, 0.75, 0.75, 0.75, 0.77]
[0.81] [0.88] [1.0] [1.1]

14 [0.0, 0.0, 0.0, 0.0] [0.04] [0.11, 0.13]
[0.20, 0.20]
[0.25, 0.25, 0.25, 0.25, 0.25, 0.25] [0.30,
0.30, 0.30]
[0.40] [0.44, 0.45] [0.50]
[0.75, 0.75, 0.75, 0.75, 0.77]
[0.81] [0.88] [1.0] [1.1]

15 (The distance from 14 to 15 clusters was the same.)

16 [0.0, 0.0, 0.0, 0.0] [0.04] [0.11] [0.13]
[0.20, 0.20]
[0.25, 0.25, 0.25, 0.25, 0.25, 0.25] [0.30,
0.30, 0.30]

[0.40] [0.44, 0.45] [0.50]
[0.75, 0.75, 0.75, 0.75]
[0.77] [0.81] [0.88] [1.0] [1.1]

17 [0.0, 0.0, 0.0, 0.0] [0.04] [0.11] [0.13]
[0.20, 0.20]
[0.25, 0.25, 0.25, 0.25, 0.25, 0.25] [0.30, 0.30, 0.30]
[0.40] [0.44] [0.45] [0.50]
[0.75, 0.75, 0.75, 0.75]
[0.77] [0.80] [0.88] [1.0] [1.1]

■

The centroid distance measure can help determine which clusters are reasonable. The strongest cluster separation splits the data set into two cluster sizes of 9 and 22 experts. This corresponds to the bimodal structure indicated in example 13.1. The next cluster structure suggested by the analysis forms four clusters, breaking off the two largest answers and the seven smallest answers for the original two clusters. The graph 13.3 also shows this separation.

A second way to make a decision on clusters is to use the ancillary information gathered on the experts. For example, if two clusters appear in the data, which ancillary information corresponds to the experts in each cluster? Perhaps all the experts in the first cluster made very optimistic assumptions when answering the question, while the experts in the second cluster made very pessimistic assumptions. The cluster dividing lines could then be drawn based on the experts' assumption making in conjunction with the statistical clustering results from the formal cluster analysis. The next section further discusses investigating the relationships between the answer data and the ancillary data. However, keep potential conditional relationships in mind and be watchful for them.

Determining the number of modes and clusters will be useful later in the bootstrap simulation applications for investigating correlation and bias and for forming aggregation estimates. The cluster formations that look reasonable and any possible ancillary information or explanations relating to the clusters should be documented for these later investigations.

Investigating Between/Within Variation Structure

In most problems in which expert judgment is elicited, several experts are asked more than one technical question. The multiple technical questions may be either totally different, or some could be quite similar in content and structure. In either case, examining the variation *between* and *within* the experts can help gain information regarding any differences in the experts. A commonly used technique for analyzing between variation versus within variation is analysis of variance (see chapter 11). Example 13.5 illustrates the mechanics of the analysis of variance technique for calculating these two sources of variation. (For further details, consult introductory statistics or analysis of variance textbooks, including Snedecor & Cochran, 1978.)

EXAMPLE 13.5: ***Between and within response variation calculation***

Eight experts were asked four technical questions on scales from [0, 1]. The between expert variance is estimated by *MSB*, and the within expert variance is estimated by *MSE*. The response of the ith expert to the jth question is x_{ij}. The overall mean of all 32 responses is C, the number of experts is t (= 8); the total number of responses is n (= 32); and n_i is the number of questions asked of each expert (= 4).

	Expert							
Question	1	2	3	4	5	6	7	8
1	0.90	0.50	0.75	0.65	0.80	1.00	0.22	0.44
2	0.95	0.50	0.75	0.65	0.80	1.00	0.22	0.40
3	0.80	0.50	0.75	0.75	0.30	0.38	0.06	0.40
4	0.94	0.48	0.75	0.75	0.65	0.65	0.06	0.38
Total, $x_{i.}$	3.59	1.98	3.00	2.80	2.55	3.03	0.56	1.62
Mean, $\overline{x}_i$	0.90	0.50	0.75	0.70	0.64	0.76	0.14	0.41

$$MSB = \sum_i n_i(\overline{x}_{i.} - C)^2/(t-1)$$
$$= 1.63/7 = 0.23$$
$$MSE = \sum_i \sum_j (x_{ij} - \overline{x}_{i.})^2/(n-5)$$
$$= 0.49/24 = 0.02$$

$$MSB/MSE = 0.23/0.02 = 11.50$$

The MSB/MSE is a ratio of variances that measures the relative difference of between experts versus within experts. This ratio is also an F statistic that follows an F probability distribution. If the ratio is large, then the differences of the between-experts values are large relative to those of the within experts. In this example, 11.50 is large because it is in the far upper-right-hand tail of the F distribution, with (t-1) and (n-t) degrees of freedom. Therefore, the variation of between experts is significantly larger than the variation of within experts.

■

If two or more questions are similar, then the variation of the responses to these questions can be used as a source of random or background variation for the experts. This random source provides a gauge that measures the variability between experts. The section on *Using Analysis of Variance* in chapter 14 describes how this variation comparison can be used to investigate interexpert correlation. Basically, if the variance between experts for the similar questions is much larger (e.g., four times larger) than the variance within experts, then the experts are giving quite a variety of responses, and positive correlation among them is not suspected. If not, then the variation from one expert to another is similar to the individual experts' variation, and correlation among experts might present a problem.

If all the technical questions differ in either content or structure, the within-expert variability would be larger than it would be if the questions were similar. The within-expert variability could be larger than the between-expert variability (Meyer & Booker, 1987*a*). This case requires an investigation on why such large within-expert variation is present. Such variations could result for many reasons: the experts may be familiar with some questions, unfamiliar with others; the experts may have difficulty in using the response mode; the experts may use an anchoring/adjustment heuristic on some questions while simply guessing on others; the experts may experience fatigue; or there may be some inconsistency in the elicitation process. To trace and understand these possibilities, properly record the experts' rationale and properly monitor the elicitation.

Even though the multimodality and between/within investigations were done only on the answer data, explanations and reasons for the results found incorporated the ancillary information. At this point the analyst should investigate the ancillary data.

Analyzing the Ancillary Data

The ancillary variables and information in the data base can be separately analyzed from the answer data. However, the conditional relationships between the answers and this data should not be ignored. The primary purpose for the separate analysis is to investigate any redundancies in the ancillary variables, thereby reducing the number of variables in the analysis with the answers later on. A secondary purpose is to gain an understanding into the structures and relationships among these variables.

The suggested analysis of the ancillary data is given in a series of steps using standard statistical multivariate techniques. Chapter 11 provides more detailed descriptions of these techniques.

Step 1. Factor analysis of the ancillary variables. Factor analysis produces a new set of factors from the original set of variables. The original variables are mapped onto the new factors according to their common information content (based on variability). The mappings are sometimes difficult to interpret, and thus the results may not prove very useful. However, if the ancillary variables map well into a set of new factors, either use the new factors as the ancillary variables or reduce the original number of variables.

Example 13.6 illustrates two cases of factor analyses. The first case shows how to apply a successful factor analysis. The second case illustrates a factor analysis that does not effectively reduce the set of ancillary variables.

EXAMPLE 13.6: ***Use of factor analysis for ancillary variables***

Case I: Successful factor analysis

The following table contains the factor loadings from a factor analysis of 12 numerical ancillary variables gathered from an elicitation of 20 experts. The factor analysis results reduced the 12 original variables to a set of 4 factors:

	Factors			
Ancillary Variable	1	2	3	4
A_1	0.054	0.112	0.802	0.032
A_2	0.556	0.236	0.080	0.128
A_3	0.754	0.001	0.026	0.219
A_4	0.011	0.218	0.099	0.672
A_5	0.832	0.064	0.002	0.102
A_6	0.000	0.000	0.347	0.653
A_7	0.107	0.883	0.000	0.001
A_8	0.256	0.495	0.202	0.017
A_9	0.109	0.001	0.389	0.501
A_{10}	0.000	0.049	0.076	0.875
A_{11}	0.672	0.256	0.003	0.069
A_{12}	0.000	0.837	0.006	0.157

The loadings are interpreted as follows: variables A_2, A_3, A_5, and A_{11} comprise (load onto) factor 1; variables A_7, A_8, and A_{12} load onto factor 2; variables A_4, A_6, A_9, and A_{10} load onto factor 4; and only variable A_1 loads onto factor 3. In addition, A_2, A_3, A_5, and A_{11} are the only set of variables containing information on the experts' education. Variables A_7, A_8, and A_{12} refer to the experts' recent work experience. Variables A_2, A_3, A_5, and A_{11} refer to the years of work on various related projects. Variable A_1 indicates how long the experts took to interview. With this clean breakdown of variables, the 12 original ancillary variables can be combined into the four factors using the factor loadings.

However, the four new factors have interpretations that are a little more general than the original 12 variables. As a result, the granularity has changed.

Case II: Unsuccessful factor analysis

	Factors			
Ancillary Variable	1	2	3	4
A_1	0.054	0.112	0.802	0.032
A_2	0.256	0.236	0.080	0.328
A_3	0.754	0.001	0.026	0.219
A_4	0.011	0.218	0.099	0.672
A_5	0.832	0.064	0.002	0.102
A_6	0.000	0.300	0.347	0.353
A_7	0.107	0.883	0.000	0.000
A_8	0.256	0.495	0.202	0.017
A_9	0.109	0.001	0.389	0.501
A_{10}	0.000	0.049	0.875	0.076
A_{11}	0.672	0.256	0.003	0.069
A_{12}	0.400	0.437	0.006	0.157

These factor loadings do not indicate to which factors variables A_2, A_6, and A_{12} belong. Also, in factor 3, the variables A_1 and A_{10} have nothing in common, making the interpretation for factor 3 difficult. Furthermore, factor 1 does not include all the educational variables, factor 2 does not include all the experience variables, and factor 4 does not include all the work variables. These results do not help in gaining an understanding about the ancillary variable relationships or structure.

■

Step 2. Graphical analysis. Factor analysis can only be used on the numeric ancillary variables. To examine relationships among qualitative variables or among mixed qualitative/quantitative variables, use plots or graphs as suggested in the section on conditionality.

Graphs that show all the data points falling on or near a line or curve indicate possible redundant information among the two variables. Therefore, these plots help the analyst search for redundant information and for interesting variable relationships. A list of such variables should be made and checked against the correlation analysis suggested in the next step (3). The pairs of variables on this list should be either positively correlated (for a line indicating positive slope) or negatively correlated (for a line indicating a negative slope).

Step 3. Correlation analysis of the ancillary variables. Another useful step in understanding relationships among the ancillary variables is the Pearson pairwise correlation coefficient for all possible pairs of ancillary variables. Deciding whether any correlation is significant requires specifying a level of significance. Guidance for specifying a level is based on the number of pairs, *n*, for which correlations are calculated. If *n* is less than 20, then use the standard 0.05 or 0.01 levels. If *n* is greater than 20, then use a level that is less than $1/n$.

Although correlation analysis is generally considered a bivariate analysis technique, it can help explain relationships among all the ancillary variables. It is also useful in verifying graphic results for numeric variables. If any correlation is highly significant (a significance level of less than 0.0001 or a correlation greater than 0.90), then redundancy is suspected and only one of the pair of variables needs to be considered for further analysis.

Step 4. Categorical analysis. The ancillary variables can be modeled among each other. One way to do this is by modeling all variables that have integers (ranks, dummy variables, or quantifiable categories) as dependent variables, with the numerical ancillary variables as independent variables. Categorical analysis techniques will indicate any potential relationships among the dependent and independent variables. This analysis technique is based on linear modeling (Grizzle, Starmer & Koch, 1969), but is not discussed or used extensively in this handbook. Any significant relationships indicated by categorical analysis should be noted and added to the list for potential redundancies.

Step 5. Cluster analysis for variables. Like factor analysis, cluster analysis can be used to examine how much information is shared among the numerical ancillary variables. If the variables cluster in distinct and tight groups, then information is shared among the variables in the group. The variables from any tightly formed groups should be added to the list of potentially redundant variables.

Following some or all of the above steps yields a list of potentially redundant ancillary variables. The following flow chart indicates how the steps can be used, and example 13.7 indicates how results can be interpreted and used. Only results indicating the strongest information redundancies should be used for trimming the set of ancillary variables. The next section provides additional similar tests on the combined ancillary/answer variables data.

Summary of Steps for Ancillary Data Analysis

For Numeric Data:	For Descriptive Data:
(1) Factor analysis indicates shared information	(2) Graphs indicate possible relationships
(3) Correlation analysis indicates possible relationships	**For Integer (Ranks & Categories) Data:**
	(4) Categorical analysis indicates possible relationships
(5) Cluster analysis indicates shared information	

EXAMPLE 13.7: ***Ancillary variables analysis***

The 12 ancillary variables from example 13.6, case II, plus 5 more ($A_{13\text{-}18}$) descriptive variables were analyzed using the five steps. The results follow:

Step 1. Factor analysis, case II, example 13.6: Indicated no clear redundancies of information; no new factors could replace other variables.

Step 2. Graphs of all pairwise combinations of the 18 variables: Indicated a strong relationship between A_1 and descriptor A_{13} and between A_3 and A_5.

Step 3. Correlations of all 12 pairwise numerical variables: Indicated one pair (A_3, A_5) strongly correlated with a significance level of 0.0001 and that several other pairs were barely significant.

Step 4. Categorical models of all integer variables (A_7, A_8, A_9) with the other numerical variables: A_2, A_8, and A_{12} have some influence over A_7.

Step 5. Cluster analysis of the numerical variables: Indicated a weak clustering of A_7, A_8, A_2, and A_{12}, and a weak clustering of A_3, A_5, and A_9.

Three interpretations can be made from the results:

1. Weak clusterings and no clear factor analysis results indicate little shared information.
2. The only strong result is the A_3, A_4 correlation. Either of those two could be eliminated from the ancillary variables set.
3. All other variables should be kept for further analysis.

■

Analyzing the Ancillary Data with the Answer Data

Many of the steps presented in this section are identical to those described in the previous section, *Analyzing the Ancillary Data.* The section's main objective is to compile lists of possible multivariate relationships; specifically, which answer variables are related to which ancillary variables. Ancillary variables related to the answers are called conditional variables. The model formations and analyses use multivariate techniques. Results from these analyses should be consistent with results found in the previous sections of this chapter.

Step 1. General linear models (GLMs). The descriptor may tempt the analyst to formulate one giant model of all the numerical variables; such a model would consist of the ancillary variables as the independent variables and the answer variables as the dependent variables. Such a general linear model would describe relationships existing between the answer and ancillary variables and determine which of the answer variables were conditioned on which ancillary variables. However, this temptation should be resisted for several reasons. First, many variables may share the same information. Including all of them in a single model results in erroneous variable relationships. Second, the variables may consist of different types (dummy variables, etc.) and different granularities. The model would then be a mixture of levels of detail in the information. Finally, some variables may have missing values. Many computer packages cannot perform the analysis or will give erroneous results.

Using information obtained from bivariate analyses and ancillary data analysis can help formulate some logical general linear models. For example, if any of the graphs of the answer data versus the ancillary data indicate a trend either in the positive or negative direction, these variables should be analyzed using the GLM procedure known as regression. The answer variables are the *Y*s, or dependent variables, and the ancillary variables are the *X*s, or independent variables. Likewise, models can be formed using variables indicating significant correlations with one variable as the dependent (the one measured with error) and the other variable as the independent (measured without error). In addition, models can be formed by using other information (e.g., suspected relationships between the variables).

To assist in regression model formulations, many software packages have procedures called stepwise procedures that indicate which models produce significant relationships between the independent and dependent variables. However, stepwise procedures do not make model choices—they only give a set of models. Model choices should still be based on the information already gathered from previous analyses and logical variable selections.

Once the models have been run on the regression analysis code or a linear model package, a list of significant models and relationships among the variables should be compiled. These significant results should be consistent with the information already gathered in the previous analyses; however, some differences will result because bivariate results do not necessarily hold for multivariate models, and weak significant relationships can change to no significance or to stronger significance, depending upon the procedure used.

Model formations are not restricted to using only ancillary variables for the *X*s. If the answer variables are related or correlated to each other, then one answer variable may be modeled in terms of the other answer variables. However, this model may be subject to a violation of the assumption that the independent variables (other answer variables) must be measured without error for general linear models.

The model formations at this stage of the analysis should only be used for investigating variable relationships. Chapter 15 describes how to formulate models for the purpose of obtaining final results and interpretations.

Step 2. Discriminant analysis. This procedure can be used in conjunction with or in place of the GLM for investigating variable relationships. The objective of this modeling is to find a set of ancillary variables that best discriminates among the values of an answer variable.

Studies have indicated that answer variable values tend to be distributed with multiple modes (Booker & Meyer, 1988*a*; Meyer & Booker, 1987a; Baecher, 1979). To determine which, if any, ancillary variables are responsible for this clumping of values, use discriminant analysis. The results of a discriminant analysis indicate a list of variables that best discriminate among the values of the chosen answer (dependent) variable. To prepare a discriminant analysis, group or classify the values of the dependent variable into categories. If there are several modes, this grouping is obvious. If no value groupings are evident, then discriminant analysis is not indicated. If a discriminant analysis is done for each answer variable, then any significant (effective) discriminating ancillary variable indicates a potential variable relationship. New ancillary/answer relationships should be added to the list, and any old relationships should be noted as confirmation of a previous result. Chapter 14, which deals with correlation and bias detection, will use the variable relationships found from the discriminant analyses.

Step 3. Multivariate correlation. We suggest using pairwise correlations for analyzing the ancillary data, the answer data, and the combined set of both. The information from all these pairwise correlations can be combined graphically to form an ad hoc, multivariate correlation structure. Such a graph depicts many intertwined relationships among the variables.

One way of constructing this graph is by forming a distance between the pairs of variables based on their correlations and plotting the variables according to their distance. Distance measures based on variance and covariance (related to correlation) are used in cluster and discriminant procedures. A simple distance based on the correlation is calculated by $1-r$, where r is the correlation coefficient. The higher the value of r, the closer the distance will be for any pair of variables.

For the graph, distances should only be calculated for pairs of variables that have significant correlations. The bivariate correlation step and example 11.3 outline the rules for determining significance.

Example 13.8 is a graph from the Meyer and Booker study (1987*a*).

EXAMPLE 13.8: ***Multivariate correlation analysis***

Three separate conglomerates of variables are evident from the significant correlations among the variables. The first group of variables comes from five different ancillary variables describing the experts' backgrounds. The second group of variables comes from four different ancillary variables describing the experts' experiences. The third group of variables comes from all the answer variables (labeled A), 10 of the variables describing the experts' problem-solving processes (labeled R), and the single variable which described the experts' evaluation of the complexity of the technical question (labeled C).

Many interesting results can be seen from this diagram:

1. No connection exits between the background variables, the experience variables, and the answer/problem-solving variables. The separation of the answer variables from the experience and background variables implies little evidence for conditionality.
2. The answer variables, the problem-solving variables, and the complexity variable are closely connected. This connectivity implies strong evidence for the answers being conditioned on the problem-solving variables and the complexity variables.
3. The answer variables are closely interconnected among themselves. The problem-solving variables are also closely interconnected.

■

In general, performing GLMs, discriminant analysis, and multivariate correlation steps for ancillary and answer data helps search for strong, consistent relationships among the variables in general. In particular, these techniques help search for answers conditioned on ancillary variables. Chapter 15 discusses how to construct an expert judgment model from the results of these analyses.

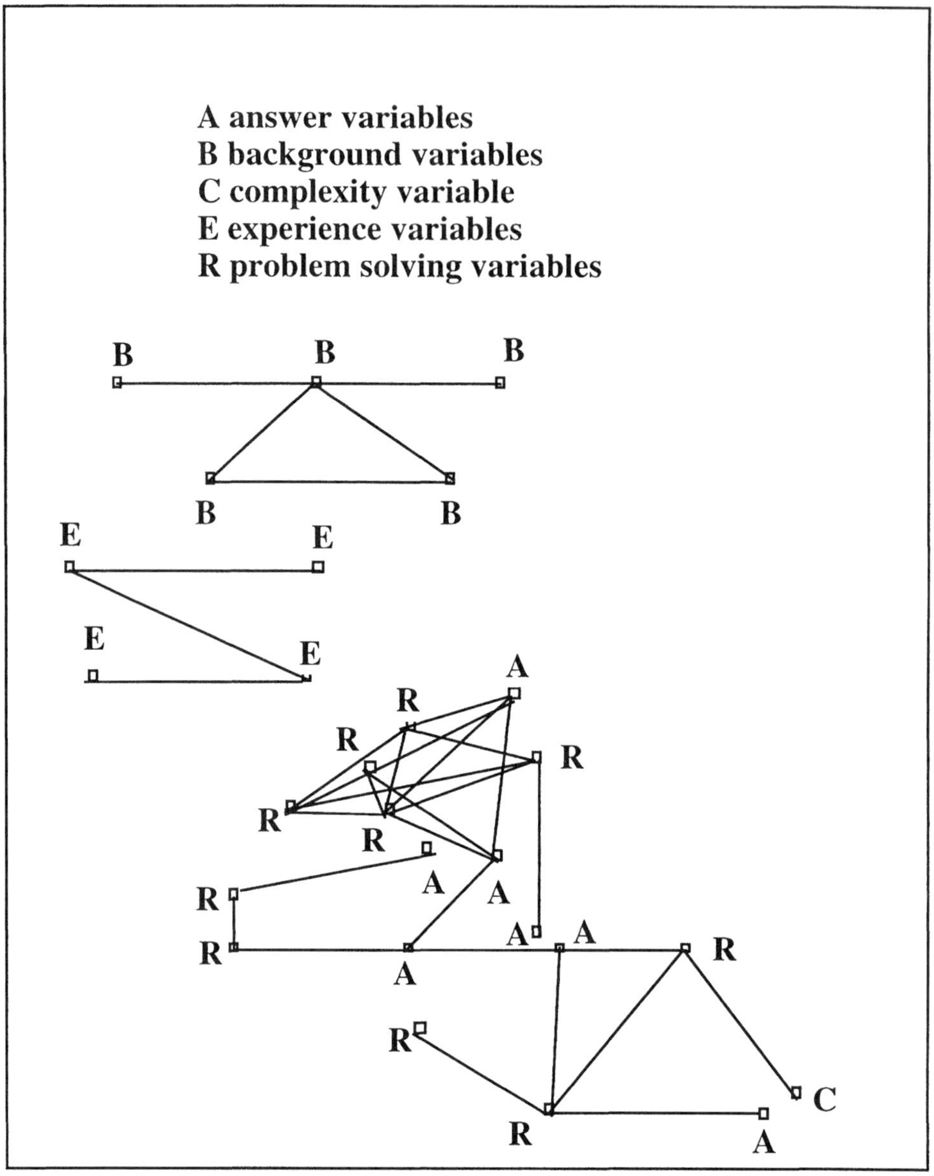
A answer variables
B background variables
C complexity variable
E experience variables
R problem solving variables
B
B
B
B
B
E
E
E
E
A
R
R
R
R
R
R
A
A
R
R
A
A
A
R
R
C
R
A

14

Correlation and Bias Detection

This chapter defines the concept of correlation and shows how it relates to expert judgment applications.

Correlation among experts is closely related to the concept of dependence. The section in this chapter on *Defining Correlation and Dependence* delves into the distinctions and similarities between both concepts.

Correlation is also closely related to the various forms of bias. Part II discusses bias in detail. The affinity of these two concepts is addressed in *Bias and Correlation Relationships.*

Because correlation among experts is often considered a problem area in analysis of expert judgment data, *Detecting Correlation in the Analysis* focuses on various detection methods. This section is organized as a series of 14 steps. Each step relates to the use of several different analysis techniques. These steps and techniques may appear redundant. Indeed, they are meant to be redundant. Comparing results from different techniques is the only way to verify conclusions about correlation. In *Analysis Summary and Conclusions*, we summarize the 14 steps and the results from the examples in the steps.

Defining Correlation and Dependence

Correlation among expert judgment answers has typically meant dependence or lack of independence among expert answers. In effect, to discuss correlation requires the clarification of the concepts of independence and dependence.

One concept of independence comes from a mathematical definition and is called probabilistic independence. The mathematical procedures for combining data from multiple experts require that the data have this type of independence. In probabilistic terms, two events, A and B, are considered independent if the probability of A is unaffected by the probability of B. In other words, the unconditional probability of A $[P(A)]$ is unaffected by B such that $P(A) = P(A|B)$, which represents the probability of A given B (Feller, 1957).

The same definition applies to examining independence in data. However, the process of identifying independence is not so straightforward. Analysts tend to think of two pieces of data as being independent if the occurrence of one datum is unaffected by the other. One way to define *unaffected* is to examine the data-gathering techniques. This examination involves investigating conditionality. If data are collected under various conditions, they may be unconditionally dependent because those conditions affect the data which, in turn, causes the data to affect each other. Thus, the observation of one datum is affected by another such that $P(A) \neq P(A|B)$. However, if the data are collected under conditions (C) mutually affecting both A and B, they could be conditionally independent such that $P(A|C)$ is independent of $P(B|C)$. In either case, conditionality becomes the focal point of investigating independence or dependence.

The terms correlation and dependence have been used interchangeably and synonymously in expert judgment problem settings. This usage is mathematically valid and is consistent with the probabilistic definition. However, the terms zero correlation (uncorrelated) and independence are only interchangeable when the data are normally distributed. Only in normally distributed cases does zero correlation guarantee independence. In nonnormal cases, zero correlation could imply either independence or dependence. Thus, for most of the analyses in this book, the dependence/independence problem is discussed from the dependence (or correlated) viewpoint.

In collecting expert judgment data, analysts have historically speculated that the data were not independent (Baecher, 1979; Winkler, 1981). The reasoning for such speculation was that dependence is likely to occur because the experts had many conditions in common that would affect their estimates. Analysts considered such conditions would be shared training, common work experiences, and exposure to the same data bases. Through time the speculation about the effects of these conditions became identified as sources of correlation among the experts.

A simple example can illustrate how this line of reasoning developed. The following sample contains probability estimates from five different experts for an event: (0.1, 0.15, 0.1, 0.6, 0.65). The bicluster structure of the estimates is a common occurrence (Baecher, 1979; Booker & Meyer, 1985; Meyer & Booker, 1987*b*). The analyst who studies the clustering of the answers tends to arrive at several conclusions: the first three experts give the same answer, the last two give the same answer, only two independent answers exist, and this sample does not contain five independent pieces of information. The data appear to have a correlation structure with the first three experts correlated and the last two experts correlated. If the analyst assumes that experts should be correlated, then the clustering of the data supports that assumption. However, note that the clustering is the only reason for suspecting dependence. No conditions have been examined to support the dependence idea, nor has any reason been given for dependence.

The reasoning used in the above example helped develop a new body of literature on how to deal with data from dependent sources (Winkler, 1981; Lindley & Singpurwalla, 1984). The focus of this research was to establish methods for handling dependent data, assuming that the dependencies existed, and in many cases assuming that the correlation structures were known. The analytical focus concentrated on these assumptions and not on the real issues. First, the correlation structures are generally unknown. Second, correlation among the experts may not exist, and if it does, it may not be an analytical problem. Third, conditionality and granularity require consideration in determining and interpreting correlation.

Two studies (Booker & Meyer, 1985; Meyer & Booker, 1987*b*) had the goal of identifying possible sources of correlation among experts. Neither study assumed that experts were correlated or that clusters of answers implied correlation. The approach instead was to investigate possible sources of correlation using the definition of dependence which is based on finding conditions responsible for correlation, including the effect of granularity. Sources (conditions) were sought from many different forms of information gathered in intensive interviews of the experts.

The granularities of the model formations in each study differed. The model granularity in the first study was more detailed than in the second. For the chosen granularity, the results of each study indicated that the answers were conditioned on the problem-solving processes of the experts. These conditions appeared to be a reasonable source for possible correlation or dependence among answers. However, an equally valid result would be that the answers were conditionally independent, having been conditioned on the sources in the same way. Therefore, discovering conditions affecting answers does not automatically imply dependence.

Also consider granularity when interpreting the results. In both studies, there was a granularity finer than the ones chosen for analysis in which the experts had nothing in common in background, experiences, or problem solving (i.e., there were no conditions that could induce dependence among them). By applying the conditioning argument at a different granularity, the analysts reached the conclusion that there is no dependence among the experts. Of course, another equally valid possibility is that sources of correlation could exist at an even finer level of detail than was gathered. Gathering information at such an extremely fine granularity might be difficult or impossible because the subjects might not be capable of providing information at such a level. Even if the expert could provide such fine detail, the elicitation session would be prohibitively long.

Therefore, in defining correlation or dependence among experts, the conditioning argument from the definition of probabilistic independence can be used as a guide. Therefore, use the predecided granularity for the entire problem because conclusions about correlation and dependence can change if levels change.

Bias and Correlation Relationships

In the broadest definition, bias can be related to a number of sources. Bias can be induced from the interview environment through factors such as the interviewing technique, the question phrasing, and the interactions with others, including the interviewer present at the interview (motivational bias). Bias can be related to the internal consistency of the experts' reactions, conditioning, and thinking (cognitive bias). This bias occurs when the expert is not consistent in his or her own reasoning or when the expert is not consistent with the fundamental rules of logic. Faulty memory retrieval can induce bias (availability bias). Therefore, bias can be considered a conditional phenomenon. In other words, certain external or internal conditions during the elicitation can initiate bias.

Because bias can be found in any of the above forms, it becomes difficult to monitor and to control. However, monitors and controls such as those covered in Part II should be used and considered an integral part of the experimental design. Chapters 3, 7, 8, and 10 discuss the various ways of handling and minimizing bias in conjunction with the elicitation techniques.

The bias in the experts' answers can often be traced to the conditional variables. These variables are also responsible for correlated data. Therefore, application of bias-minimizing methods is important for minimizing possible dependence. Of course, answer data can be correlated for reasons other than bias; however, the original postulated sources of correlation can all be traced to the different types of bias.

The correlation in the data appears in the forms of interexpert correlation or between-expert correlation. If an expert has a motivational bias that drives him or her to be consistently optimistic, then his or her answers will reflect that bias by being more optimistic than the other experts. This expert's answers could indicate interexpert correlation but not between-expert correlation. If all the experts shared a common goal for a project, then their answers might exhibit between-expert correlation and little or no interexpert correlation. The remainder of this chapter describes how to detect and analyze correlated data in both forms, regardless of whether a bias or some other condition is the source of correlation.

Detecting Correlation in the Analysis

In this section, several steps describe several methods to investigate any potential correlation among the experts' answers. Each step/method will focus on detecting correlation or dependence in the answers given by the experts using the answer data itself and the conditionality definition of dependence. In other words, dependence is sought in terms of possible conditions that may influence the answers of multiple experts or that may bias a single expert's answers. Do not assume, a priori, that dependencies exist. The investigation uses only the data (except for step 14) and minimizes the assumptions necessary for using the analysis techniques.

The methods for detection include using the results from correlation analysis, multivariate analysis, analysis of variance, and simulation techniques (e.g., the bootstrap). Detection is also done by using the features of the chosen elicitation method and using assumed correlation structures and distribution forms. Any one of these steps or methods can be used individually. However, to maximize the information inherent in the data, apply all steps (except step 14) for a data analysis approach. Because the last step is an assumptional approach, it may not always be appropriate to use, but it can be done in conjunction with the others or by itself.

The following 14 steps fall into various categories according to usage. These categories are listed as subsection headings. One or more steps may appear under each heading.

Using Granularity

Step 1. Define a level of detail to use in the analysis and in the interpretation of the results. After having assumed and established a definite granularity for the problem, the definition of dependence is applied under that granularity. Before fixing a granularity, a common question would be, *What does it really mean to say the experts are correlated?* With a fixed granularity, the now qualified question becomes, *What does it really mean to say the experts are correlated for this granularity of data and analysis?* The answer to this question can be different if the granularities are allowed to change. Therefore, the first and most important step is to fix the level in the analysis and in the interpretation of the results. The results of the rest of the steps in the correlation investigation are given in terms of the granularity chosen in this first crucial step.

Using Hypothesized Sources of Correlation

The next few steps prepare the data for some of the remaining methods of correlation detection. They do offer the analyst an opportunity to hypothesize what conditions in the ancillary data might be potential sources of correlation. In addition, these steps offer an opportunity to do some hands-on data analysis, which in turn leads to a better understanding of the data set for formal model formation as discussed in chapter 15.

Step 2. Compile a list of conditions from the ancillary information that are suspected or hypothesized as being potential sources of correlation. This list could include suspected sources from the literature (Booker & Meyer, 1985), such as the experts' educational background, commonly shared work experiences, how recently they worked on a similar problem, assumptions made in the problem-solving process, and heuristics or rules used in solving the problem. Likely candidates for this list can come from the results of the multivariate analyses done on the answer data and from the ancillary data found in chapter 13. This list should also include any ancillary variables that were significantly correlated to answer data.

Step 3. Examine the raw data clusterings. Chapter 13, which dealt with understanding the data-base structure, also investigated the multimodal structure of the data. A raw data frequency plot, such as in example 13.3, helped determine the raw data clusters. The important point here is that these clusters of the raw data do not necessarily imply a correlation or dependence structure. For one reason, the data may be conditionally independent on one or more ancillary variables. Also, there may not be any rationale for the clusterings. Attempts to find reasons for the raw clusterings are encouraged, however. Also, do any of the ancillary variables group the data according to these clusters? Chapter 13's ancillary data analysis partially answered this question, but a more complete answer to this question is needed in the correlation investigation. The steps below pursue this answer.

Step 4. Form data clusters using the suspected conditions. Use the conditional variables list from step 2 to form new data clusterings. The data clusters are formed by splitting the values of the conditional variables into categories or clusters and listing the corresponding answer data for each category. If the answers within these clusters are numerically similar, then the question posed in step 3 is answered affirmatively. An alternative formation can be done by plotting the categories of the conditional variable against the numerical answers to see if clusters form. In either case, if no cluster formations are evident, information is still gained regarding the relationships of the data and the ancillary information. This information should be consistent with the results from the analysis in chapter 13.

Step 5. Compile a list of unsuspected variables. This list should consist of randomly selected ancillary variables which are not hypothesized as sources of correlation. Ideally, this list should have the same length as the list in step 2. However, such a long list may not be possible because there may not be enough ancillary variables available or because such a list would be too long for the analyses in the following steps.

This unsuspected list has a twofold purpose. First, a random selection from this list offers the chance of discovering conditions that might be sources but originally were not considered possible sources. Second, this list provides a comparison to the list of suspected sources from step 2.

The lists compiled in steps 2 and 5 should not be considered final at this point in the investigation. The results from the remaining steps/methods will help determine if these hypothesized lists are correct.

Step 6. Form clusters of the data using the unsuspected variables. Graphs or hand listings of the unsuspected ancillary variables and the answers should be examined to see if the answer data clusters are similar to the raw clustering (from step 3). The clustering for an unsuspected variable could be better (closer to the raw clustering) than the clustering for the suspected variables (from step 4).

Steps 2 through 6 will give three different clustering mechanisms to interpret and compare: (1) the raw data clusters, (2) the data clustered by the various conditions suspected of inducing dependence, and (3) the data clustered by other unsuspected conditions. If any of the variables forms clusters of approximately the same size and of similar values to the raw data clusters, then there is reason to suspect that the variables are sources of correlation. The remaining steps will help verify this conclusion. If none of the variables produces good clusterings, a correlation may still exist among the experts, making the other steps necessary.

Example 14.1 illustrates hand listings of cluster formations from ancillary variables.

EXAMPLE 14.1: *Clusterings using different variables*

Analysts selected four variables to test as potential sources of correlation. For eight experts, two background variables were selected from the list of suspected sources: years of work experience in the technical area (*YRWORK*) and one of the variables describing the expert's problem-solving process (*PSRATE*). Two other variables, not suspected, were also chosen: discipline of highest degree (*DEGREE*) and whether the expert had any experience in a nuclear experimental facility (*EXPFAC*). The answers to the technical question are listed as *P*.

Expert	*P*	*DEGREE*	*EXPFAC*	*YRWORK*	*PSRATE*
1	0.90	Engineer	Yes	1.5	1
2	0.50	Engineer	Yes	0.5	-1
3	0.75	Engineer	Yes	3.0	0
4	0.65	Physics	No	3.0	-3
5	0.80	Engineer	Yes	1.2	-1
6	1.00	Engineer	No	10.5	-1
7	0.22	Physics	No	7.6	-2
8	0.44	Engineer	Yes	4.9	2

The following table shows how two clusters formed for each of the four variables:

Variable	Clustering Description	Cluster 1	Cluster 2
P	[0, 0.50] & [0.50, 1.0]	0.22, 0.44, 0.50	0.65, 0.75, 0.80, 0.90, 1.0
DEGREE	Physics & engineer	0.65, 0.22	0.90, 0.50, 0.80, 0.75, 0.44, 1.00
EXPFAC	Yes & No	0.65, 1.00, 0.22	0.90, 0.50, 0.75, 0.80, 0.44
YRWORK	> 5.0 & [0, 5.0]	1.00, 0.22, 0.44	0.90, 5.00, 0.65, 0.75, 0.80
PSRATE	[0, 2] & [-3, 0)	0.90, 0.44, 0.75	0.50, 0.65, 0.80, 1.00, 0.22

This table shows that none of the ancillary variables cluster the answer data in a raw data cluster (line *P*) format, even though the cluster sizes (2 or 3 for cluster 1 and 5 or 6 for cluster 2) are similar for the ancillary variables. ■

Using Correlation Analysis

Step 7. Calculate the Pearson correlation matrices of the answers. Calculating the Pearson pairwise correlation matrix (chapter 11) for the experts' answers to all questions helps find experts who are numerically correlated in their answers. This correlation is strictly a numerical correlation; it does not necessarily imply dependence of the experts under the definition given in the first section. Knowing which pairs of experts' answers are significantly correlated is important because significant correlation coefficients can indicate potential expert correlation. These expert correlations can be further tested under the dependence definition by using the remaining steps/methods.

The correlation coefficients are calculated for pairs of experts using the numerical answers to more than two technical questions. Note the implicit assumption made regarding the similarity of the technical questions asked. The coefficient calculation assumes that the answers to these different questions represent repeated measures of each expert's knowledge and problem-solving processes. This assumption makes sense if the technical questions have a similar structure (form and response mode) and a similar content. To verify similarity among the questions, check the results of the analysis done in chapter 13. If the answer variables are related to each other from Pearson correlation coefficients, from graphs, from GLMs, or from cluster analysis, then similarity among the questions is plausible.

Example 14.2 illustrates two different correlation matrices for the data from example 13.1. The first matrix is the expert correlation matrix of coefficients calculated across answers to the questions. The second is the answer matrix of coefficients calculated across experts to help verify question similarity through correlation of the answers.

EXAMPLE 14.2: ***Correlation matrices of experts and answers***

Eight experts answered four questions with similar subject matter by using the same response mode: a continuous scale from 0.0 to 1.1. The Pearson correlation coefficients for each pair of experts form the matrix as follows:

	E_1	E_2	E_3	E_4	E_5	E_6	E_7	E_8
E_1	1.00	0.41	0.00	-0.46	0.87	0.74	0.46	-0.15
E_2		1.00	0.00	-0.58	-0.03	0.24	0.58	0.66
E_3			1.00	0.00	0.00	0.00	0.00	0.00
E_4				1.00	-0.80	-0.93	-1.00	-0.69
E_5					1.00	0.96	0.80	0.35
E_6						1.00	0.93	0.52
E_7							1.00	0.69
E_8								1.00

The only significant correlations in this matrix (using a 5% significance level) have been underlined. The lower triangular values are not listed because they are identical to the upper triangle values folded over at the diagonal of 1.00 values.

Note the perfect negative correlation of expert number 4 and expert number 7. Analysts usually focus on the effect of underestimating the expert variation associated with positive correlations among experts. However, the effect of negatively correlated experts could result in overestimating the variance. Neither effect is desirable and both produce statistically biased estimates of the expert variation. In any case, correlations (positive or negative) should be investigated.

The correlation matrix for the answers follows:

	Q_1	Q_2	Q_3	Q_4
Q_1	1.00	0.99	0.51	0.86
Q_2		1.00	0.52	0.87
Q_3			1.00	0.85
Q_4				1.00

The underlined values are significant correlation coefficients (using a 5% significance level). This table shows how well the answers to the questions are connected into a structure. This structure indicates that using the answers could serve as repeated measures of the experts' processes of answering similar technical questions. Therefore, the correlation results from the first matrix (experts) are also usable. The answer structure can be diagramed as follows, with the connecting lines representing significant correlations:

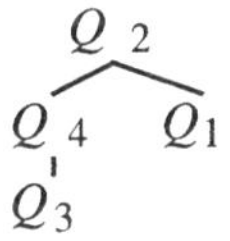

Using Multivariate Analysis

Step 8. Determine which clusterings match the raw clusters. This step helps determine which of the ancillary variables cluster the data in a manner similar to the raw data clusters by using multivariate analysis methods (e.g., discriminant analysis) and the results from the analyses in chapter 13 for understanding the data base. If an ancillary variable clusters the answers well, and this clustering has a basis in terms of cognitive theory, then a potential source of correlation is identified (Meyer & Booker, 1987*b*). The word *potential* points out that not all the raw clusters of the answer data indicate the existence of a correlation structure. Additional analysis, such as using simulation techniques and examining variances (in steps 9 and 10), can help make such a determination.

Discriminant analysis could be used to complete the investigations done in steps 3 and 6, where the answer data are clustered according to the ancillary variables from the lists compiled in steps 2 and 5. Discriminant analysis can be used in two different ways. First, it can determine which of the ancillary variables best predicts to which raw cluster each datum belongs. Second, it can determine which ancillary variables predict the clustering behavior of any other ancillary variable, such as the variables from the lists in steps 2 and 5.

To implement the first application, choose a variable from either the list in step 2 or in step 5. Categories of values for the clustering are formed using the values of that variable. For example, the variable *PRSOLV* in example 14.3 could be categorized as a positive (1) or a negative (-1) value. *PRSOLV* is the variable describing the expert's problem-solving score accumulated over several problem-solving characteristics. This variable came from the list of suspected correlation sources (step 2). Another variable, *DEGREE,* categorizes the discipline of the highest degree earned by the experts. It is not suspected as being a source of correlation (from the list in step 5).

EXAMPLE 14.3: ***Using ancillary variables as discriminators***

The answers of 8 experts are listed below as Q_1. Two other ancillary variables, *PRSOLV* and *DEGREE,* represent choices from the lists in steps 2 and 5, respectively. To determine if any of the ancillary variables might be potential discriminators for the answers, use discriminant analysis. The results given below indicate how many answers would be misclassified if the ancillary variable was used as the discriminator.

Expert ID	*PRSOLV*	*DEGREE*	Q_1
1	1	1	0.90
2	-1	1	0.50
3	1	1	0.75
4	-1	0	0.65
5	-1	1	0.80
6	-1	1	1.00
7	-1	0	0.22
8	1	1	0.44

Discriminant analysis results where the asterisks indicate misclassification by *PRSOLV*:

Expert ID	From *PRSOLV* Class	Classified Into *PRSOLV* Class	
1	1	1	
2	-1	-1	
3	1	1	
4	-1	-1	
5	-1	1	***
6	-1	1	***
7	-1	-1	
8	1	-1	***

Discriminant analysis results where the asterisks indicate misclassification by *DEGREE*:

Expert ID	From *DEGREE* Class	Classified Into *DEGREE* Class	
1	1	1	
2	1	0	***
3	1	1	
4	0	1	***
5	1	1	
6	1	1	
7	0	0	
8	1	0	***

With so few experts (8), *any* misclassifications indicate poor discrimination by the ancillary variable. As a result, neither of these variables works well as a clustering mechanism for the answers.

■

The second use of discriminant analysis can be applied in two ways. To determine if ancillary variables are also discriminators, either analyze the variables with each other or compare their discriminating abilities by direct examination of the misclassification output. Example 14.3 contains two ancillary variables with the same misclassification pattern. This difference in patterns indicates that these two variables had little in common. This result is consistent with the results from the analysis of the ancillary variables done in chapter 13.

Using Analysis of Variance

Step 9. Examine between and within cluster variations. The basic philosophy of analysis of variance procedures in statistics (chapter 11) is to compare between-expert to within-expert variations. Chapter 13 suggests the formal analysis of between and within answer variation to help understand the data-base structure. This step provides a method for interpreting the results from this analysis in the context of correlation.

Analyses in the previous steps have concentrated on data clusterings. How do clusterings relate to correlation? To determine if data within a given cluster are correlated, compare the variance structures of between-expert variation to within-expert variation. Formal analysis methods such as discriminant analysis are based on complex but similar variation comparisons. If interexpert correlation is suspected, then within-expert variation can be used as a measuring standard. Within-expert variation is how closely each expert repeats him or herself on the answers to various but similar questions.

For example, if experts are asked multiple but similar questions, then the variation between experts can be compared to the variation within experts. If the two variations are the same, the analyst would interpret no difference in the expert's answers. If between-expert variation is much larger than within-expert variation, then this result indicates that the experts act more independently of each other. The variation comparison approach only works if multiple questions are asked of the experts, and these questions must be very similar in content, structure, and response mode.

If the multiple questions asked vary in content or structure, then the within-expert variation comparison measures a combination of question variation and within-expert correlation. Even so, the between versus within variance comparison can still be useful. If each expert answers the various questions differently, indicating no consistent bias, such as always answering with low values, then the within-expert variation would be large. In this case, if the between variance is of the same size as or larger than the within variance, correlation among the experts might not be suspected. If the between-expert variance is small relative to the within-expert variance, correlation might well be suspected.

Example 13.3 illustrates how to calculate these variations for the eight experts answering four very similar questions. The between-expert variation is 0.23 and the within-expert variation is 0.02. An F significance test reveals the within-expert variation is significantly smaller than the between-expert variation. An examination of the raw data unearths a suspicion of a systematic bias among experts 1, 2, 3, 4, 7, and 8. Each of these experts consistently answers the four questions as low or high. Therefore, expect the within-expert variation to be a small value, making it a good measuring standard for the between-expert variation. In this case, the between variation is much larger, indicating that the eight experts may not be correlated.

Using Simulation Techniques

A major concern with correlated data occurs when forming an aggregation or pooled estimate. Most rules for aggregation require the independence of the data (not just uncorrelated data). However, the chosen aggregation estimator can be used on the existing data to help determine if correlation presents a problem. To make such a determination, investigate the behavior of the chosen estimator using correlated versus using uncorrelated data sets. A convenient way of determining this behavior is by using simulation techniques.

Chapter 11 discusses two such simulation techniques, Monte Carlo and bootstrap. Using the median as an example of a chosen aggregation estimator, steps 10 through 12 illustrate how to use simulation for investigating correlation. The bootstrap technique is heavily used because it does not require any distributional assumptions for the data and because it relies solely on the information content of the raw data.

Step 10. Compare different stratified bootstrap sample results. The bootstrap simulation technique can be used to investigate correlation in conjunction with ancillary variables listed in steps 2 and 5. The resulting bootstrapped distributions of a chosen estimator (such as the median) are formed for each clustering using stratified sampling techniques. This sampling differs from the ordinary bootstrap (as described in chapter 11) because the clusters of the data act as the strata. Each bootstrap sample is formed by randomly selecting one datum from each strata. The resulting distribution of the median from stratified bootstrapping for a cluster formation from one variable can then be compared to the results from other clustering variables.

Specifically, this comparison can be made by examining the dispersion of the resulting bootstrapped distribution of the medians. The dispersion of highly correlated data (r = 0.9) is smaller than for uncorrelated data (r = 0.0). Dispersions can be measured using the variances of the bootstrapped distributions, the ranges, or some central probability coverage interval such as the central 90% putative interval (the difference between the 95th and 5th percentiles).

If the central 90% putative intervals of the distribution of the median are used as the measure of dispersion, then these intervals are at least three times wider for uncorrelated data than for highly correlated data (Booker and Meyer, 1988*b*). Thus, comparing the 90% putative intervals for the bootstrapped medians from different cluster formations reveals a relative correlation structure. If one cluster formation from an unsuspected clustering variable results in an interval three or more times wider than another suspected or good clustering variable, then the data in the first case is uncorrelated relative to the second case. Again, the goal of the correlation investigation is to find conditional variables that comply with the definition of dependence.

To form these bootstrapped distributions, the bootstrap technique (chapter 11) requires modification. The first step is to classify the data into clusters using the clustering variables from the lists in steps 2 and 5. From these clusters, form stratified bootstrap samples (e.g., 1000) by randomly sampling 1 datum from each of the m clusters. Calculate the median from the m values. One thousand or so samples form a distribution of the median. The corresponding percentiles (e.g., 5th and 95th), variance, or range can be taken from this distribution as estimates of dispersion.

Example 14.4 illustrates the results of stratified bootstrapped median distributions for 4 ancillary variables and for the raw data itself. Two of the ancillary variables were from the suspected list (step 2), and two were from the unsuspected list (step 5). The central 90% putative intervals plotted in the example indicate relative correlations that result from the various clustering mechanisms. The raw cluster formation is the most narrow. *DEG* is the only other clustering variable that indicates possible relative correlation. The other variables appear to have little potential as sources of correlation.

EXAMPLE 14.4: ***Ninety percent putative intervals for bootstrap medians using different variables as strata***

Each of the five variables formed six clusters; from the suspected variable list came *PSP* and *YRS* and from the unsuspected list came *DEG* and *ASR*. Stratified bootstrap sampling was done by forming 1000 samples. The medians of each sample were calculated and sorted. The vertical lines (|---|) mark the 5th and 95th percentiles of these medians. Each sample was of size six, with one datum randomly selected from each of the six clusters (strata) in the manner of simple random stratified sampling (Cochran, 1963). The variables chosen follow:

RAW	-	the raw data as it clustered numerically
PSP	-	a variable describing the expert's problem-solving process
YRS	-	the number of years since the expert had worked on that type of problem
DEG	-	the discipline of the expert's highest degree
ASR	-	the number of years the expert worked as a code assessor

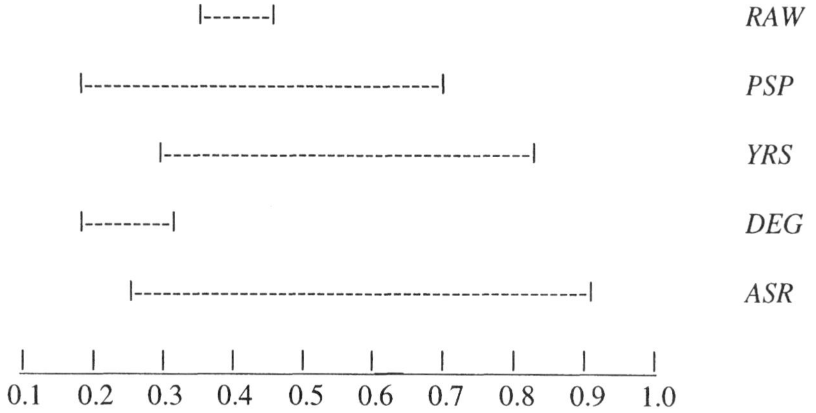

■

Step 11. Compare bootstrap distributions using pairwise correlation. The bootstrap distribution of the chosen estimator (median) for the entire, unclustered data set can be used in conjunction with the information from the pairwise correlation analysis of the answer data (from chapter 13). First, bootstrap the entire data set obtaining the distribution of the chosen estimator. Second, remove from the sample the experts' answers that were correlated in a pairwise manner from the correlation analysis. Next, find the bootstrap distribution of the estimator for this reduced but perhaps more independent data set. By comparing the dispersions of the two resulting distributions for the median, the effect of including those highly correlated experts can be readily seen.

Example 14.5 shows some results of this analysis for a data set of 31 experts (example 13.3) from the Meyer & Booker study (1987*b*). This example shows how the removal of the five correlated experts significantly changes the dispersion of the estimator, indicating that the inclusion of those correlated experts does affect the dispersion of the aggregation estimator (the median). The results from this example would indicate the removal of those five experts' answers from the data set, replacing them with an average of their values. This average represents a single but more independent expert in conjunction with the remaining, uncorrelated experts. This new average is analogous to finding the equivalent number of experts, as described in step 14 below.

EXAMPLE 14.5: ***Using the bootstrap with pairwise correlation***

Thirty-one experts were asked a question with a response mode scaled from 0.0 to 1.1 (Meyer & Booker, 1987*b*). The median was chosen as the estimator for aggregating the 31 answers. A bootstrap sampling procedure was done for 1000 randomly formed samples for this data.

In this set of 31 experts, a pairwise correlation of the experts over eight similar questions resulted in the following significant correlation structure:

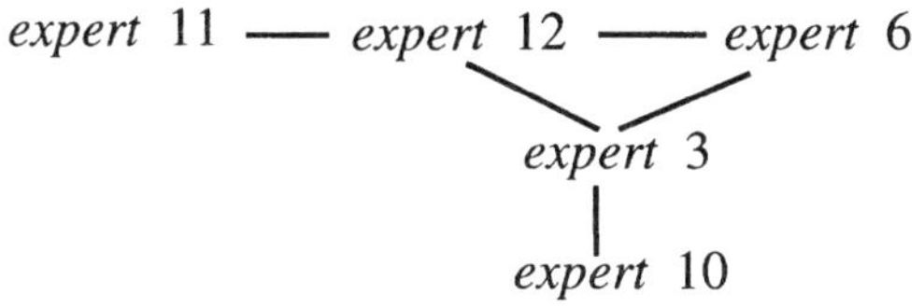

A bootstrap distribution for the median was then calculated from the data with the above five experts' answers deleted. However, because sample size affects dispersion, the results for a sample of 31 might differ from a sample of 26. This effect would be more pronounced if the samples were 15 versus 10. To account for sample size changes in comparing these two bootstrapped distributions, a third bootstrap was done on the original data set, deleting five experts' answers from five randomly selected experts that were not at all pairwise correlated.

The following plot shows 5th and 95th percentiles for the bootstrapped median:

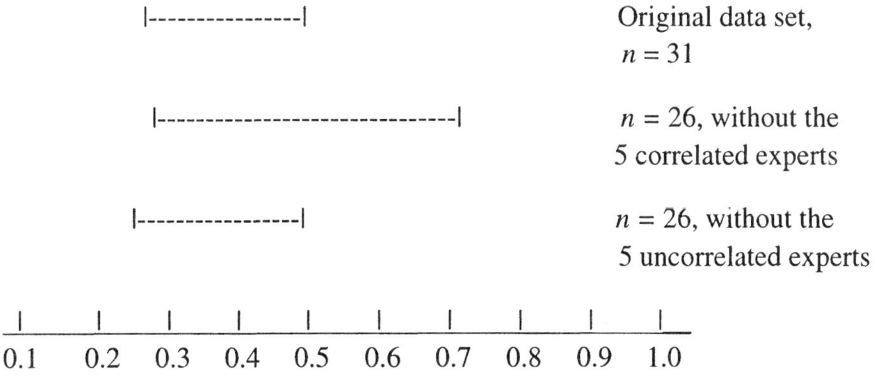

The results of the three comparisons indicate that the sample size difference between 31 versus 26 has little effect on the median dispersions; however, removing the five correlated answers indicates a significant increase in the dispersion. Therefore, we recommend replacing those 5 experts with a single average value, making a data set of 27 uncorrelated expert answers.

■

Step 12. Compare simulation results of the data to known distributions. Relative comparisons using bootstrap simulation do not provide definitive information on the correlation structure of the data. To gain this information, compare the bootstrap results of the data to bootstrap results of data sets with known correlation structures. Comparisons can be made by forming random samples of data from distributions with known correlations. These dispersions of the chosen estimator (e.g., median) are then compared to dispersions of the estimator for the data.

The major difficulty in this approach is deciding on which known distributions and correlation structures to compare with the raw data bootstrap results. For example, if the data are assumed to be normally distributed, then the values of the parameters of the normal distribution need to be chosen. If the mean and variance of the raw data are used to estimate the mean and variance of the assumed normal distribution, then the resulting dispersion of the normal sample with zero correlation will be the same as the dispersion of the original data with unknown correlation. Such a comparison fails to provide any information about the unknown correlation in the raw data. However, if many different correlation structures are bootstrapped, the effects in dispersion can be seen for changes in correlations. For example, the central 90% putative interval of the median estimator for a normal sample with correlation structure of 0.5 is about twice as large as the central 90% putative interval for a normal sample with a correlation structure of 0.9.

Using a mixture of distributions rather than a single distribution can improve the bootstrap comparison of the data set with distributions of known correlation structures. This method has two advantages. First, the new data set can be modeled into clusters, with means and variances reflecting the raw data set clusters. Second, the new data set can be mixed according to the results of the correlation analysis on the raw data set. Example 14.6 shows such a mixture. In this case, the correlations and clustering structure of the 31 experts (Meyer & Booker, 1987*b*) were used to form a three-distribution mixture using the normal family. The three normal distributions for the mixture come from the two distinctive raw data clusters and the set of the five correlated experts (see example 14.5). The raw data bootstrap compares favorably to the bootstrap results for this three-normal mixture. Of course, other goodness-of-fit techniques (Conover, 1971) could have been used in making this comparison, but this section focuses on the use of simulation techniques.

EXAMPLE 14.6: ***Using the bootstrap for a normal mixture***

The following answers from 31 experts (Meyer & Booker, 1987*b*) are cleanly grouped into two distinctive clusters. In addition, a correlation analysis indicates that experts 6, 3, 10, 11, and 12 are significantly correlated with correlations larger than 0.90. The data can be divided into three groups using the correlation and raw clustering information. These three groups are as follows:

Group Size	Mean	Variance	Correlation
17	0.24	0.03	0.00
5	0.18	0.01	0.99
9	0.84	0.02	0.00

Using the normal distributions with the above parameters for each group, a combined sample of 31 was formed. This sample was bootstrapped and compared to the bootstrap distribution for the median of the raw data. The resulting 5th and 95th percentile values of the medians follow. The close correspondence implies that the normal mixture is a reasonable fit for this data set.

```
         |------------------|                          n = 31, raw data

       |--------------------|                          n = 31, mixture
                                                       normal sample

 |    |    |    |    |    |    |    |    |    |
0.1  0.2  0.3  0.4  0.5  0.6  0.7  0.8  0.9  1.0
```

■

Using Elicitation Methods

The extent of control and monitoring of the resulting biases and correlations depends on the elicitation technique used. Chapter 3 describes the biases and some methods used to counter, reduce, or handle them. Bias consideration plays a crucial role in the design of the expert judgment elicitation, as presented in chapter 8. Chapter 10 explores bias control and monitoring as part of conducting the elicitation.

Step 13. Examine the correlation or bias relative to the elicitation method. Of the three basic elicitation situations (interactive group, Delphi, and individual interview), the first two can easily induce correlation among experts. These two situations allow the use of analysis steps 1 through 12, but the results should indicate much more correlation among the experts than the examples taken from individual interview situations.

In the group and Delphi situations, the ancillary variables should indicate strong clustering or discriminating ability on the experts' answers. In other words, experts should reach similar answers for similar reasons, especially for the group situation. In the Delphi situation, the analysis steps 1 through 12 can be performed at the various stages of the process. The results at each Delphi iteration will indicate how the correlation structure among the experts becomes stronger. The answers of the first stage should be no more correlated than those from an individual interview situation.

The final section of this chapter summarizes the results of the correlation analyses steps for all three elicitation situations. Chapter 16 will apply the conclusions reached from these steps.

Using Assumptions

Step 14. Make assumptions about the correlation structure based on the analysis or on any additional information. Example 14.6 illustrates how to make such assumptions about the correlation structure. The results of the correlation and bootstrap analyses indicated a potential correlation structure for the data set. These results also indicated the presence of 26 uncorrelated experts and 5 correlated experts. One possible conclusion based on this result would have been to form a 27th uncorrelated expert by using the average of the 5 correlated experts' answers.

The concept of forming an equivalent number of experts (Clemen & Winkler, 1985) led to the idea of combining the five correlated experts. For k normally distributed experts with a common population correlation (ρ) and variance (σ^2) the asymptotic equivalent number of independent experts is

$$n(\sigma^2, \rho) = k\,[1 + (k\text{-}1)\rho]^{-1}.$$

Applying this formula may be impractical because $n(\sigma^2, \rho)$ is much smaller than k for large values of ρ, making the number of independent experts extremely small or equal to one. Another problem in using this concept is the assumption that the experts are normally distributed with a common correlation and a common variance.

Correlation structures are assumed using a mathematically convenient model (e.g., a multivariate normal model) requiring values for correlation coefficients or values defining a covariance structure for the data. This model is commonly assumed and can easily handle the dependence problem. Because normality is assumed, the concepts of zero correlation and independence are interchangeable. Example 14.7 shows how such an assumed model and correlation structure can be used (Winkler, 1981). The major problem with this approach is the restrictive assumptions. Because very little experimental evidence is available on distributions and correlations of expert judgment data in most applications, many analysts consider these assumptions too unrealistic to make. However, applications such as blue-chip and weather forecasting provide immediate feedback which in turn provides adequate estimation of the correlation and variance.

EXAMPLE 14.7: ***Dependent experts with assumed normal distribution***

Three experts ($k = 3$) estimate the average cost and standard deviation (in thousands of dollars) for a newly designed earthquake-proof valve. If each expert's estimate is normally distributed with means and standard deviations as $\mu_1 = 60$, $\sigma_1 = 6$; $\mu_2 = 62$, $\sigma_2 = 5$; $\mu_3 = 70$, $\sigma_3 = 7$, then the distribution for all three experts is a multivariate normal with mean vector $\mu = (60, 62, 70)^t$ and covariance matrix Σ. The elements of Σ are $\sigma_{ij} = \rho_{ij}\sigma_i\sigma_j$, where ρ_{ij} is the correlation coefficient between the ith and jth experts. In this example the values for these correlations were based on previous experience with these experts. These correlations are $\rho_{12} = 0.6$, $\rho_{13} = 0.5$, and $\rho_{23} = 0.6$, making the resulting matrix Σ:

$$\Sigma = \begin{bmatrix} 36 & 18 & 21 \\ 18 & 25 & 21 \\ 21 & 21 & 49 \end{bmatrix}.$$

This example does not use estimated values for the means, variances, and correlations. Instead, these values are considered known, and the standard Greek letters for the parameters are used.

One way of formulating a single normal distribution from this multivariate (three-variate) normal distribution is to use a Bayesian approach. Chapter 16 describes this approach to aggregation in more detail. The way to reduce this multivariate normal to a single normal is by combining the multivariate normal with a prior for the parameter of interest. In this case, the parameter of interest is the aggregation estimator for the combined mean responses. A prior that has little influence on the data would be a diffuse prior that does not involve Σ. Combining the above multivariate normal with such a prior gives a posterior distribution (the aggregation distribution) that is normal with a mean μ^* and variance σ^{*2}, where

$$\mu^* = e^t\Sigma^{-1}\mu / e^t\Sigma^{-1}e,$$

and

$$\sigma^{*2} = 1/e^t\Sigma^{-1}e,$$

and where $e = (1, 1, 1)^t$ for three experts. This formulation gives weights to the three experts according to the following:

$$w_i = \sum_{j=1}^{3} \sigma_{ij} / \sum_{j=1}^{3} \sum_{m=1}^{3} \sigma_{mj}.$$

The resulting values from this data are

$$\mu^* = 62.0$$

and

$$\sigma^{*2} = 22.8,$$

with the experts' weights being $w_1 = 0.26$, $w_2 = 0.67$, and $w_3 = 0.07$.
If the experts were considered independent, then the values for r_{ij} would be 0.0. The resulting values for the mean and variance would be

$$\mu^* = 63.2$$

and

$$\sigma^{*2} = 11.3,$$

with the experts' weights being $w_1 = 0.32$, $w_2 = 0.45$, and $w_3 = 0.23$.

Ignoring the dependence results in a slightly larger mean and a much smaller variance.

■

The mathematical formulations in the above example are not trivial. However, the multivariate normal model is the simplest and most convenient model that allows closed form calculations. If the data do not indicate that a multivariate normal distribution is appropriate, or if the correlation structures cannot be assumed as known, then even these formulas are not useful. In such cases, the results from the other steps that rely on the evidence from the raw data itself must be used to draw conclusions.

Analysis Summary and Conclusions

Investigating the possible existence of correlated data is an important step before the aggregation analysis (chapter 16). Many aggregation methods assume that the experts' answers are independent. The steps (10-12) involving the bootstrap simulation as a tool for investigating correlation demonstrated the use of aggregation and its relationship with correlated answers. If dependencies can be identified or controlled, then the aggregation schemes can be used without violating the independence. The reason for investigating ancillary variables that might affect the answers was to identify and control dependency. If such conditions were found, the experts might be conditionally independent. Under conditional independence, aggregating on a conditional basis will not violate the independence assumption.

Conclusions from the analyses done in the above steps (1-14) can be drawn by using the results collectively and relying on results that are consistently indicated by several steps. After performing the above steps, summarize the results from each step to determine consistency. In most cases, the same results are indicated by several of the steps, making conclusions obvious.

For instance, example 14.8 shows the summary of the results for a set of 31 experts answering a question (Meyer & Booker, 1987*b*). Most of the results for the steps can be found in the examples in this chapter. The conclusion from all the steps is that five of the experts are correlated numerically, but that the source of this correlation is not known. Averaging the answers of those five and using that average as a 27th expert can solve the numerical correlation problem. The normality fit in step 12 indicates that the data can be considered a mixed normal distribution. For a single normal distribution, zero correlation and independence are identical. This equivalence supports the conclusion that the data could be considered a set of 27 independent experts for the purposes of aggregation (chapter 16).

EXAMPLE 14.8: ***Summary of the correlation detection steps***

The following summarizes the series of 14 steps outlined in this chapter for the detection of correlation and bias. The data set used in these steps is from example 13.3 and consists of 31 experts' answers to 8 technical questions. The ancillary information gathered was reduced to 17 variables relating to the experts' background and problem-solving processes (Meyer & Booker, 1987*b*).

Step	Step Summary
1.	The analyst chose the granularity at the detail level of the analyzed ancillary variables. This level was more general than the original data gathered because a problem-solving score variable was formed for each of the eight answer variables from combinations of the original problem-solving characteristics. ***Results or conclusions:*** *Results will be valid if interpreted at the level of the problem-solving score variables.*

2. The analyst formed a list of nine variables suspected as sources of correlation.
Results or conclusions: *These variables represented the experts' recent background and problem-solving processes.*

3. The analyst performed a cluster analysis on the answer data for each answer variable.
Results or conclusions: *The data formed two or three major clusters.*

4. The analyst clustered the data using the suspected ancillary variables.
Results or conclusions: *The problem-solving scores were the only variables clustering the answer variables in a way similar to step 3.*

5. The analyst formed a list of 11 variables not suspected as sources.
Results or conclusions: *These variables represented the experts' earlier history.*

6. The analyst clustered the data for each answer variable using the not-suspected ancillary variables.
Results or conclusions: *None of these ancillary variables clustered the answer data in a way similar to step 3.*

7. The analyst calculated the pairwise correlation matrices for the 31 experts and the 8 questions.
Results or conclusions: *All 8 questions had highly correlated answers. Of the 31 experts, 5 were highly mutually correlated.*

8. The analyst tried a discriminant analysis on three of the suspected ancillary variables and on three of the not-suspected ancillary variables.
Results or conclusions: *Only the problem-solving score variables discriminated between the answers for some of the 8 questions.*

9. The analyst performed an analysis of variance on the 8 answers for the 31 experts.
Results or conclusions: *The within-expert variation was much smaller than the between-expert variation, indicating interexpert consistency and possibly bias, but also indicating a potential for little or no correlation between the experts.*

10. The analyst tried stratified bootstrap simulations for the median using the 6 variables from step 8 for strata on each of the 8 answer variables. The descriptions and results for the remaining steps are only listed for the one answer variable that indicated potential expert correlation. For this one answer variable, each ancillary variable was

used to cluster the answer data into 6 clusters of nearly equal cluster sizes.
Results or conclusions: *Some of the results appear in the figure in example 14.4. Of the answer variables analyzed, only one, DEGREE, indicated any potential as a source of correlation. This result may have been because one cluster of the DEGREE variable was of size one, which induces a bias into the stratified sampling.*

11. The analyst bootstrapped the raw data for the answer variable in step 10 in three ways: with the five correlated experts deleted, with the five uncorrelated experts deleted, and with no experts deleted.
Results or conclusions: *Removing the five correlated experts resulted in a significant increase in dispersion compared to the bootstrap results with no experts removed (example 14.5). Removing five uncorrelated experts produced the same results as did the case with no experts removed. Therefore, the five correlated experts affect the variance of the median, and their estimates should be combined into a single value to represent a single expert that is uncorrelated to the others.*

12. The analyst formed a three-normal distribution mixture from the raw data and compared the bootstrap results for the median of the answer variable in step 10 with the bootstrap results for the median from the mixture.
Results or conclusions: *The bootstrap results for this mixture modeled the data well (example 14.6). The correlation structure of the mixture could represent the raw data. If the data can be modeled using normals, then the concepts of zero correlation and independence can be interchanged.*

13. The analyst used the ethnographic method with verbal report in an individual interview situation to minimize bias.
Results or conclusions: *Biases and correlations during the interview process were controlled and minimized. The analyst did not suspect that other biases would affect the results of the data analysis.*

14. The analyst made no additional assumptions about the correlation structure.
Results or conclusions: *An analysis of a mixture of three normals could have been developed similar to the one in example 14.7. However, such development would be useful only as a special case for this problem. The information already gained in the previous steps is enough to make some conclusions about correlation for this data set.*

The results from steps 2-6, 8, 10, and 13 reveal two possible conditional variables. Steps 6 and 8 indicate a rather strong conditioning effect from the problem-solving score. However, the effect of the *DEGREE* variable from step 10 is doubtful. Step 9 also indicates that there is little numerical interexpert correlation. However, some numerical correlation is indicated from the results in steps 7, 11, and 12. This result led to reducing the five correlated experts to one expert by using the concept of equivalent number of independent. This expert would be added to the remaining 26, for a total set of 27 pseudoindependent experts. Therefore, the final conclusion would be to use the 27 experts and consider them conditionally independent upon the problem-solving score variable formed at a more general level of detail than the raw information gathered. ■

15

Model Formation

Chapters 12 through 14 emphasized investigation and preliminary analyses for the purpose of gaining familiarity with the information gathered from the elicitation. This chapter and chapter 16 focus on what might be termed final analysis procedures that establish interpretable conclusions. This chapter concentrates on forming models whose results provide inferences. Chapter 16 concentrates on forming aggregations or combinations of the experts' judgments also for inference purposes.

This chapter presents modeling techniques and suggestions for describing the experts' answers in terms of other variables. These models are chosen based upon information already gained from the exploratory analyses done on the data base. Some modeling techniques use standard statistical procedures (e.g., least squares regression) as their foundation. These techniques fall under the heading of general linear models (GLMs). Applying multivariate methods such as factor, discriminant, and cluster analyses appropriately model the multivariate structure of the data base. However, as stated in previous chapters, we do not recommend the use of these techniques for final conclusions because of the assumptions required to use them. Other, more applicable modeling techniques are based on decision analysis methods and can be described as conditional models.

In general, the model is a functional relationship, h, between the answer variables, Y, and the ancillary variables, X, and is expressed as

$$Y = h(X).$$

GLMs define the expected value of the answer variable as a function of the ancillary variables:

$$E(Y) = h(X).$$

GLMs also define the observed values of Y as the function, h, of the observed values of the Xs plus some error ***residual***, ε:

$$Y = h(X) + \varepsilon.$$

The above model also applies to discriminant analysis, where h represents the linear discriminant function.

Model formation based on conditional relationships between the answer variable and the ancillary variables can be written as

$$h(Y/X).$$

Cluster analysis can be written in terms of conditional relationships between any and all variables of *X* and *Y*:

$$h(Y/X),\ h(X/X'),\ h(Y/Y').$$

Factor analysis also can be modeled in terms of defining new variables, *Z*, from *X* and *Y*:

$$Z= h(X,Y).$$

Model formation can create special problems from granularity and the structure of the available data base. The chosen granularity and the granularity in the variables determine whether models have general or specific granularities. Model selections are also integrally linked to the elicitation technique used in gathering the information. As with any analysis technique, the assumptions required for the selected modeling procedure (e.g., cluster analysis) must be examined and/or tested before application.

General Linear Models

GLMs are models formed using the statistical method of least squares. The following explains how the least squares method got its name: the method determines the model coefficients (*b*s in the equation below) such that the squared distances between the values of *Y* and the *E*(*Y*) are minimized. Regression and analysis of variance techniques use the least squares approach. Analysis of variance usually refers to cases in which the *Y*s are continuous, numerical variables and the *X*s are categorical or rank variables. The analysis of variance technique is not applicable for modeling expert judgment data when the *X*s are a mixture of categorical variables and continuous variables or when the categorical variables are not properly designed so that the categories are configured in a specified, balanced way. Regression usually refers to the cases where both *Ys* and *X*s are continuous, numerical variables. Chapter 11 discusses their uses in more detail. Because the *X*s may be dummy variables, the term GLM, rather than regression, better describes the models discussed in this section.

Full-Scale General Linear Models

A full-scale GLM models the experts' answers as functions of the ancillary variables. The form of the GLM is

$$Y= b_0 + b_1X_1 + b_2X_2 + \ldots + b_mX_m + \varepsilon,$$

where

Y is the dependent variable and is one of the answer variables,
X_i (i = 1, m) are the independent variables including many, if not all, ancillary variables,
b_i (i = 0, m) are the coefficients estimated in the GLM analysis that determine the impact of the X_i variables on Y, and
ε is the random error in the model. This error represents the residual that cannot be modeled by the other variables.

A full-scale GLM includes all the independent variables (the Xs) in the model in the same form as they appear in the data base. Certain procedures can screen out redundant variables and insignificant variables to form a streamlined model. Redundant variables contain overlapping information allowing one variable to substitute for another without changing the total information content in the model. Insignificant variables do not predict the values of Y. Therefore, their b values are not statistically different from zero.

The screening procedures for both types of variables are called stepwise regression and are available on most statistical packages that have GLMs (Snedecor & Cochran, 1978; chapter 13). A step-up procedure formulates a model by adding *Xs* one at a time. The *Xs* are added to the model according to which ones best predict the dependent variable. A step-down procedure formulates a model by starting with the full set of *Xs* and eliminating, one at a time, those *Xs* that contribute least to the model. A stepwise procedure applies both the step-up and step-down techniques simultaneously to select the best sets of *Xs* that predict the Y. Stepwise procedures conveniently find the best model for Y for any selected number of *Xs*. However, cautions and assumptions are necessary for using the stepwise procedure. The last section of this chapter discusses some of the cautions. The regression section in chapter 11 discusses the assumptions for any regression or GLM:

1. All the *Xs* are variables with no measurement errors.
2. The εs are independent, normally distributed random variables with a mean of 0 and a variance of σ^2.

Because the answers are considered variables measured with error, violation of the first assumption occurs when one or more of the answer variables are included in the model as *Xs* . Other *Xs* may not be considered as measured without error. To avoid violating the first assumption, include only the variables that are measured without error as *Xs*. This limits model formation possibilities, which limits discovering some conditional relationships among the variables. Modeling *Xs* with errors leads to an underestimation of the variance of the dependent variable. Such underestimation impacts the variance associated with predictions made by the model. Approximations for the prediction variances are available (Booker, 1978). However, in expert judgment applications, predictions using GLMs are not usually necessary and are not recommended.

Testing for compliance of the second assumption is relatively easy. A quick test is performed by plotting the residuals (εs) on normal probability paper. If they plot as a straight line, then the normality assumption holds. Many statistical packages have tests for normality, such as the Wilks-Shapiro *W* test, which can also provide easy verification. However, the effects of violating the second assumption are not as severe as violating the first assumption.

Full-scale modeling requires careful use because variables with missing values create poor models. If there are many missing values (from the experts) on many of the variables used in the model, an adequate model fit is not possible using GLM procedures. However, collapsing, combining, or redefining variables can form variables that are appropriate for modeling. Changing the original variables to more general variables changes the granularity of the model to a more general level. The next topic discusses such a model formulation.

Combination Models

Chapter 13 suggested several analysis techniques for analyzing the ancillary data. Analysts can use the knowledge of the relationships among these variables gained in those analyses for possible combinations of the ancillary variables. Combinations of information from many ancillary variables can form new variables called scores or indices.

For example, information (many variables) is elicited regarding the experts' problem-solving processes. These variables may take the form of rules, assumptions, heuristics, and problem-solving steps used by each expert. Each expert may solve a problem differently; therefore, many variables have missing values (as shown in example 15.1). A full-scale GLM using all these different variables would not be possible because these variables would form a sparse matrix of information. If the variables could be combined to form scores with no missing values, then a GLM analysis would be possible, because each expert would have a value for this score. The score would be a new variable for the GLM.

EXAMPLE 15.1: ***Scoring using the anchoring and adjustment model***

Seven experts participated in a face-to-face interview. The information gathered on their problem-solving processes indicated the use of four different assumptions. The matrix below gives the usage of these assumptions by the experts:

Expert No.	Assumption No. 1	2	3	4
1	x		x	
2		x		
3		x		
4			x	
5		x		
6				x
7	x			x

Not all the experts used the same assumptions. We cannot recommend a GLM for these experts' answers using the four assumptions as *Xs* in the model because there are too many missing values in the assumptions.

All experts began solving the problem with some initial impressions. These impressions could be considered anchors and the assumptions used could be considered adjustments made on the impressions, forming an anchoring and adjustment model (see chapter 3, *The Four Cognitive Tasks*). The following information on the experts' initial impressions of the problem helps score the anchoring:

Expert No.	Initial Impression	Value Assigned to Anchor
1	Highly possible	2
2	Possible	1
3	Not likely	-1
4	Not sure about this	0
5	Can never happen	-2
6	Don't believe this	-2
7	Could be true	1

The following evaluation of the assumptions scores the assumptions as adjustments from the anchors:

Assumption No.	Evaluation	Value
1	Assuming this gives a pessimistic view	-1
2	Assuming this does not affect the problem	0
3	Assuming this gives an optimistic view	1
4	Assuming this also gives an optimistic view	1

To produce the final score, replace the assumption matrix entries by the numerical values for the assumption evaluations, then record the initial impression values for each expert. For the final score, sum the values over all assumptions and add that to the value for the anchor as follows:

	Assumption Value					
Expert No.	1	2	3	4	Anchor	Score
1	-1		1		2	-1 + 1+ 2 = 2
2					1	0 + 1 = 1
3		0			-1	0 + -1 = -1
4			1		0	1 + 0 = 1
5		0			-2	0 + -2 = 2
6				1	-2	1 + -2 = -1
7	-1			1	1	-1 + 1 + 1 = 1

■

Because score or index variables are combinations of several variables, the granularity of the model with these new variables is more general than the original variables gathered. Thus, the results and their interpretations must be done at this more general level. As shown in example 15.1, the problem-solving score variable is a significant contributor in modeling the answers using a GLM analysis. Therefore, the answers are conditioned on this general score variable, and not on the specific problem-solving features (assumptions and initial impressions) of the experts.

There are many ways of forming combination variables. Some of the most commonly used methods follow.

Anchoring and Adjustment Scores

This variable combination scheme uses the cognitive theory of problem solving which states that the expert anchors to an initial value or idea and then proceeds with a series of adjustments from that value. To model the expert's problem solving using this theory, formulate the expert's initial impression (*good*, *bad*, *indifferent*) of the problem (as in example 15.1) into a new variable, then formulate the expert's adjustments (e.g., *up*, *down*, *neutral*) into one or more new variables. Any relevant problem-solving information can form adjustment variables. Example 15.1 uses assumptions made by the experts during their problem-solving processes. The new anchor and adjustment variables can be easily quantified into ranks (1, -1, and 0). A final score or index for each expert is then found by adding the new anchoring variable and the new adjustment variables.

Cumulative Scores

Cumulative scoring is a very general variable combination scheme that produces a final score variable at a general granularity. For example, a cumulative score is created by counting all the problem-solving features used by the experts. This accumulation produces a score that reflects how much effort and thought each expert used in solving the problem. Of course, the analyst has the freedom to determine which problem-solving features to count. He or she also has the flexibility of weighting the various features to provide a weighted sum as a score. Example 15.2 uses the information from example 15.1 to illustrate the formation of a cumulative score.

EXAMPLE 15.2: ***Scoring using cumulative scores***

Using the matrix of assumptions from the experts in example 15.1, an analyst forms a cumulative by counting the number of assumptions used by each expert. No evaluation of the assumptions is done to establish different weights.

	Assumption				
Expert No.	1	2	3	4	Cumulative Score
1	x		x		2
2		x			1
3		x			1
4			x		1
5		x			1
6				x	1
7	x			x	2

The interpretation or meaning of the cumulative score can be vague. In this example, the meaning refers only to the number of assumptions used by the experts. This cumulative score variable may not be important for the modeling of the answers because it provides very little and very general information on why the experts chose these particular answers.

■

Collapsing Variables

Combining or collapsing variables forms new variables with different, usually more general, interpretations. Example 15.3 illustrates how information on the expert's education degrees and disciplines can be combined to form a more general variable reflecting all the information. This poses two items worth mentioning. First, the analyst must be careful not to impose his or her own views or interpretations on the original data to successfully collapse variables. Second, the granularity can change through several levels, producing an extremely general variable that may be of little use in conjunction with other variables (at a different granularity) or be of little use in the interpretation of the results.

EXAMPLE 15.3: ***Scoring by collapsing variables***

The following table lists background information on the education of seven experts. To form a new variable, collapse this information into a single composite variable representing the experts' education.

Expert No.	BS	Degree Code* MS	PhD
1	1	3	
2	4	4	4
3	2		
4	6	4	
5	5	5	
6	3	3	3
7	5	4	4

**1 = physics*
2 = mathematics
3 = nuclear engineering
4 = mechanical engineering
5 = civil engineering
6 = electrical engineering

Assigning weights to the degrees (BS, MS, and PhD) and to the degree disciplines (the 6 codes indicated by *) collapses the information into an overall education score. The degree weights are calculated from ranks of importance for the degrees. In other words, a BS degree is not very desirable; an MS degree is; a PhD is only slightly better than an MS. The ranks are normalized so that they sum to 1.0:

Degree	Rank	Normalized Weight w_d
BS	1	0.1
MS	4	0.4
PhD	5	0.5
	10	1.0

The degree codes (the discipline areas) are calculated from the ranks of importance for the disciplines. In this case, the engineering degrees were valued as more relevant, receiving higher ranks and resulting weights.

Degree (Code)	Rank	Normalized Weight w_c
Physics (1)	1	0.05
Mathematics (2)	2	0.09
Electrical engineering (3)	4	0.17
Civil engineering (4)	4	0.17
Mechanical engineering (5)	6	0.26
Nuclear engineering (6)	6	0.26
	23	1.00

As a result, the collapsed variable consists of a combination of the degree weights and the degree code weights:

$$\text{Collapsed variable for the } i\text{-th expert, } V_i = \Sigma w_c w_d .$$

The values for the collapsed variable follow:

Expert	BS w_c •	w_d	+ MS w_c	• w_d	+ PhD w_c	• w_d	=	V_i	V_i *new*
1	0.05	0.10	0.26	0.40	0	0.50	=	0.11	0.42
2	0.26	0.10	0.26	0.40	0.26	0.50	=	0.26	1.00
3	0.09	0.10	0	0.40	0	0.50	=	0.01	0.03
4	0.17	0.10	0.26	0.40	0	0.50	=	0.12	0.47
5	0.17	0.10	0 17	0.40	0	0.50	=	0.09	0.33
6	0.26	0.10	0.26	0.40	0.26	0.50	=	0.26	1.00
7	0.17	0.10	0.26	0.40	0.26	0.50	=	0.25	0.97

Because the maximum possible score for this weighting scheme equals 0.26 and the variation among the experts is small, the V_is can be transformed according to the highest score as follows: $V_i\,new = V_i / 0.26$.

■

Other Multivariate Models

Chapter 14 showed how some multivariate analysis procedures helped detect correlation and bias. Chapter 11 briefly described these procedures. These procedures are used in this section in the model formation process. However, the assumptions and circumstances for applying these techniques make their use limited. Specifically, factors from a successful factor analysis are used as new variables in a GLM. Discriminant analysis models are used for describing relationships between categorical answer variables and ancillary variables. Cluster analysis is used on the variables (rather than on the values of a variable) to model variate relationships.

Factors from Factor Analysis

A successful factor analysis on a set of ancillary variables will produce a set of new variables called factors, which are combinations based upon shared information from the original variables. These factors could be used as new variables in a GLM. The key word here is *successful.* Success implies that the factor analysis produces factors from clear-cut subsets of the original variables and that the factors have an interpretation or meaning that remains consistent with the variables that went into the factors' formation. Example 15.4 illustrates these concepts.

EXAMPLE 15.4: ***Using factor analysis to form new variables***

An elicitation of 15 experts yields 12 numeric, ancillary variables. All 12 are problem-solving variables: variables H_2, H_3, H_5, and H_{11} describe heuristics used; variables A_7, A_8 and A_{12} describe assumptions used; variable R_1 is a rule of thumb used; and variables C_4, C_6, C_9, and C_{10} describe cues used from the problem. The variables, constructed from rating scales, follow.

	Variable											
Expert	H_2	H_3	H_5	H_{11}	A_7	A_8	A_{12}	R_1	C_4	C_6	C_9	C_{10}
1	-1	0	1	0	-1	-2	-1	0	-1	0	-2	-1
2	0	0	0	0	-2	-3	-1	0	-2	-3	-2	-3
3	1	-1	-1	1	-3	-3	-1	0	-1	-1	-3	-3
4	2	1	1	2	-1	-1	-1	1	-1	-2	-3	-2
5	2	2	1	1	-1	-1	0	1	-2	-1	-1	-3
6	1	2	2	2	0	0	-1	1	-1	-1	0	0
7	-2	-1	-1	-2	0	0	-1	0		0	0	0
8	-1	0	0	-1	0	0	0	0	0	-1	0	0
9	-1	0	-1	0	1	0	0	0	0	-1	1	0
10	2	1	2	2	-1	0	0	1	-1	1	1	0
11	2	2	2	2	2	2	3	1	1	3	1	2
12	1	1	2	2	1	1	1	1	2	2	2	2
13	0	0	0	-1	2	2	3	0	3	2	3	2
14	0	-2	-2	0	2	2	3	0	3	2	3	2
15	0	-1	0	0	3	3	3	0	3	3	3	2

A factor analysis performed on the 12 variables by using the principal components method resulted in the following factor loadings on 2 factors:

Ancillary Variable	Factor 1	Factor 2
H_2	-0.23	0.85
H_3	-0.35	0.83
H_5	-0.21	0.87
H_{11}	-0.27	0.86
A_7	0.94	0.15
A_8	0.93	0.27
A_{12}	0.90	0.20
R_1	-0.23	0.93
C_4	0.96	-0.03
C_6	0.87	0.30
C_9	0.93	0.12
C_{10}	0.93	0.20

Factor 1 is a new variable that combines the information from the rule of thumb variable (R_1) and the 4 heuristic variables (H_2, H_3, H_5, and H_{11}). Factor 2 is a new variable that combines the information from the 3 assumption variables (A_7, A_8, and A_{12}) and the 4 cue variables (C_4, C_6, C_9, and C_{10}). Therefore, the original 12 variables can be reduced to only 2: factor 1 and factor 2. The values or scores for each expert for factor 1 and factor 2 follow:

Expert	Factor 1	Factor 2
1	-0.64	-0.64
2	-1.27	-0.87
3	-1.19	-0.84
4	-1.12	0.71
5	-0.99	0.79
6	-0.56	1.07
7	0.10	-1.47
8	-0.02	-0.83
9	0.13	-0.80
10	-0.32	1.17
11	0.78	1.69
12	0.60	1.21
13	1.39	-0.26
14	1.49	-0.76
15	1.64	-0.17

The original 12 variables are now represented by 2 variables. The 2 new variables have a new meaning or interpretation: factor 1 represents the assumptions and problem cues used indicating the problem background items; factor 2 represents the rules used in solving the problem.

Some information is lost by replacing the 12 original variables with the 2 new factors. Most analysis procedures measure the information loss as a function of the percentage of the total variance (from the 12 variables) as explained by the 2 factors. The factors in this example account for 87% of the variation. Therefore, 13% of the variation in the data is lost when using the 2 factors as new variables.

Another way of interpreting this loss of variance or information is by stating that the granularity changed from the original set of 12 variables to the set of 2 factors. If these 2 new variables are used in a GLM, the results from that model would have a more general interpretation than a GLM that used the original 12 variables.

■

Discriminant Analysis

Discriminant analysis determines a linear discriminant function (f) which determines how well the ancillary variables map or classify the categories or groups defined by the answer variable. Two assumptions underline the use of discriminant analysis: first, the variables must follow a multivariate normal distribution; second, the answer (dependent) variable must have a group or class structure (e.g., multiple choice responses, qualitative responses, or naturally occurring numerical groupings). The first assumption is very restrictive and would not hold true for expert judgment data. The second assumption is often true of expert answers. However, the answers should be distinctively and cleanly clustered and the reasons for this clustering should be evident from the exploratory analysis results. Example 15.5 illustrates model formation using this technique under these conditions.

EXAMPLE 15.5: ***Using discriminant analysis in model formation***

Eleven experts estimate the likelihood of a specific event occurring in a nuclear reactor under certain conditions. The response or answer mode given to them was the Sherman-Kent scale (see chapter 7), and their percentage answers follow:

(10, 10, 90, 80, 99, 20, 10, 80, 99, 90).

Each expert provided the type of work environment (listed 1-3 for variable *W*), used certain cues (listed as 1-5 for variable *C*), used certain formula calculations (listed 1-3 for variable *F*), used some rules of thumb (listed 1-5 for variable *R*), provided information on background (listed 1-5 for variable *B*), provided the highest educational degree (listed 1-3 for variable *D*), provided information on work experience (listed 1-5 for variable *E*), and used certain assumptions (listed 1-3 for variable *A*). The data follow:

Expert No.	*ANSWER*	*W*	*C*	*F*	*R*	*B*	*D*	*E*	*A*
1	10	2	3	3	1	1	1	5	1
2	10	2	2	1	3	2	2	5	1
3	90	1	5	3	5	3	3	5	1
4	80	3	3	3	3	4	3	4	2
5	99	2	5	2	3	5	2	5	3
6	20	2	2	3	2	5	2	4	1
7	10	2	1	1	3	4	2	3	3
8	80	3	2	3	3	3	2	3	3
9	90	2	4	2	3	2	2	2	2
10	90	1	5	3	4	1	1	2	3
11	20	2	1	3	2	1	1	1	1

The goal is to determine which variables make good discriminators for the 5 different answer categories (10, 20, 80, 90, and 99). If a discriminant analysis is run on this data set, most packages will either give a warning, an error message, or will not complete the calculation for this data because the data set has a singularity in it. The presence of the singularity means that not all the 8 variables can be used to model the answer because one or more of the variables are exact functions of one or more other variables. In this case, eliminating variables *E* and *W* would allow a solution. This problem is common in data sets with either large numbers of variables or small numbers of experts.

The problem has two solutions: (1) run a cluster analysis or correlation analysis on the variables (as done in chapters 13 and 14) to either learn more about the variable structure or to find the singularity; or (2) use available information to make some discriminant analysis runs on subsets of the variables. In this example, the cluster analysis in example 15.6 will indicate a solution, but a clue already exists on how to categorize the 8 variables from their definitions. Four of them (*A*, *C*, *F*, and *R*) are items that the expert used to solve the problem, and 4 of them (*B*, *D*, *E*, and *W*) are features of the expert's background and work environment. Two discriminant analysis runs were made using the two sets of variables.

I. Variables *A*, *C*, *F*, and *R* as discriminators for the answers.

Most discriminant analysis programs supply a table that both shows how well the discrimination was done and records any observations that were misclassified. The table for this analysis follows:

Expert No.	From *ANSWER*	Classified into *ANSWER*
1	10	10
2	10	10
3	90	90
4	80	80
5	99	99
6	20	20
7	10	10
8	80	80
9	99	99
10	90	90
11	20	20

In other words, no misclassifications imply that the four variables effectively discriminate the 5 answer classes.

The model of the four variables (X_i) and the categories of the answer variable (Y_j) are very similar to the GLM:

$$Y_j = l_{0j} + l_{1j} X_1 + l_{2j} X_2 + l_{3j} X_3 + l_{4j} X_4,$$

where l_{ij} are the linear discriminant function coefficients shown below:

			ANSWER		
Term in Model	10	20	80	90	99
Constant	-241.9	337.4	-538.2	-833.0	-438.0
$A = X_1$	26.1	29.1	38.7	47.7	36.0
$C = X_2$	15.3	12.3	20.1	32.1	27.0
$F = X_3$	105.5	130.3	160.4	191.9	134.2
$R = X_4$	100.2	118.2	149.4	185.4	132.0

This table specifies the model of the experts' answers as classified by these 4 variables. This model could predict the answer (10, 20, 80, 90, or 99) the twelfth expert would give by determining his or her values for the 4 variables. However, this prediction capability is not necessary or important in this case. The goal is to determine which variables are influential in determining the answers. To attempt inferences beyond this goal would exceed the limit of the information contained in the gathered data.

II. Discriminant analysis using *W*, *E*, *B*, and *D*.
The analysis generated the following classification table:

Expert No.	From *ANSWER*	Classified into *ANSWER*
1	10	10
2	10	10
3	90	90
4	80	80
5	99	99
6	20	20
7	10	20*
8	80	80
9	99	90*
10	90	90
11	20	20

The asterisk on experts 7 and 9 indicates that if these 4 variables were good discriminators, then expert 7 should have answered with a 2 and expert 9 should have answered with a 9. While some may argue that this misclassification pattern is not severe, the 4 variables in the first discriminant analysis of this example had no misclassifications. Also, with so few experts, even 2 misclassifications are not considered a good result. Therefore, the results of these discriminant analyses indicate that the first set of variables worked well as discriminators of the answer variable.

■

Because the function determined in discriminant analysis is a linear function, a model from this technique should be similar to a GLM. In other words, if a particular set of ancillary variables is found to be significant in modeling the answer variable from a GLM, that same set should also make good discriminators for the answer variable from a discriminant analysis. If the results do not match, then perhaps violations occurred on the assumptions required for one or both procedures. As a general rule of thumb, the GLM is a more forgiving and a more widely applicable procedure. We recommend GLM over discriminant or even cluster analysis.

Cluster Analysis

Chapter 13 determined that the primary use of cluster analysis was as an exploratory analysis tool in which determinations were made about how the data formed various clusters or groups. To use cluster analysis as a modeling tool is to determine variable relationships according to how the variables are clustered or grouped. In addition, discriminant analysis or GLM use cluster analysis as a premodeling tool. In the discriminant analysis case, a cluster analysis is done on the answer data to determine if the data form clean, definite clusters which, in turn, form the categories for the discriminant analysis (e.g., example 15.5). The results from a cluster analysis on the variables can then be used to determine which variables would be good discriminators, as illustrated in example 15.6. The results from variable clustering can also be used to develop a GLM.

EXAMPLE 15.6: ***Using cluster analysis in model formation***

Using the responses and ancillary information from the 11 experts in example 15.5, an analyst performed a cluster analysis on all nine variables using the centroid method to determine clusterings of the variables. The results of the first cluster formation indicated that the 2 clusters should contain the following variables:

Cluster 1	Cluster 2
(*A*, *F*, *R*, *C*, *ANSWER*)	(*B*, *D*, *E*, *W*)
Interpretation	Interpretation
The background and experience variables	The problem-solving variables and answers

This result indicates a strong relationship between the answer and the problem-solving variables. These 4 ancillary variables could work well as discriminators for the answers. The combination of this cluster analysis result and the discriminant analysis result in example 15.5 imply that these four ancillary variables would make good independent variables for GLM with *ANSWER* as the dependent variable. The 4 variables in cluster 2 make poor candidates for discriminator or independent variables for *ANSWER*.

The previously performed cluster analysis indicates other cluster formations for 3 to 9 clusters. Using different criteria helps determine which formation to use. One such criterion is the measure of the proportion of the original variance as explained by each cluster formation. The following four-cluster formation explains 72% of the variation in the original variable set:

Cluster	Variables
1	(*ANSWER*, *A*, *C*, *R*)
2	(*B*, *D*, *E*)
3	(*F*)
4	(*W*)

Another criterion is using common sense or logical interpretation of the cluster formation. Although the above four-cluster formation explains a sizeable percentage of the variation, it is not desirable to have clusters 3 and 4 with only one variable each. A better clustering is the two-cluster formation where the variables in each cluster have a common definition, even though the two clusters explain only 50% of the variation.

■

Using cluster analysis as a stand-alone modeling tool provides only a conditional model. The model is represented as a cluster formation like the two-cluster formation in example 15.6. This model can be described using a general conditional form of the answer variables (Y) and the ancillary variables (X) and (X') as

$$h(Y|X) \text{ for cluster 1}$$

and

$$h(X/X') \text{ for cluster 2.}$$

The next section describes such conditional models in more detail.

Conditional Models

Techniques from the decision science community can help model the functional relationship between the answer variables and the ancillary variables. These modeling techniques also serve as the response modes for the experts' answers. In other words, the model selection dictates how the interviewers will ask the experts questions. Therefore, because the model selection process becomes an integral part of the elicitation process (chapter 7), it must take place before gathering the data. Part III's analysis philosophy has been to explore the gathered data with the freedom of allowing the data and its information content to direct the analyses. Formulating a model before gathering the data has the disadvantage of losing some of this freedom and runs contrary to the analysis philosophy.

The following sections describe two of the more popular decision analytic methods. Their popularity stems from three advantages: ease of implementation, wide applicability, and a successful track record. The first is based on Saaty's pairwise comparison technique, while the second is a general technique based on problem decomposition using decomposition diagrams.

Saaty

The Saaty pairwise comparison method yields a set of relative weights for a set of competing alternatives. The analyst can use these weights to compare and evaluate the alternatives. Therefore, this method can be used as a decision-making tool. The resulting comparison weights are not probability or likelihood estimates. However, they could represent relative likelihood comparisons. Chapter 11 provides a more detailed description of Saaty's method including its advantages and disadvantages.

The Saaty method acts as both a model formation tool and as a response mode. The model formed is a very qualitative one structured as a hierarchy of descriptive conditions used to compare alternatives. The pairwise comparisons are made at all levels of the hierarchy. Therefore, conditional comparisons are made among all the conditions in all the levels; comparisons are also made of the alternatives, given the various conditions.

The analysis of the hierarchy to determine the weights is a straightforward set of calculations based on the eigenvalues of the matrix of comparisons at each level. For each level, the weights for the items being compared are calculated from the eigenvectors of the principal eigenvalue. The final weights for the bottom level of competing alternatives are determined by multiplying weights through the levels of the hierarchy.

The weights at any level are relative comparisons only; thus, they have no numerical interpretation. For example, if item A has a weight of 0.25 and item B has a weight of 0.50, then A is not half as important (or half as likely or good) as B. The only interpretation is a qualitative one: item A is less important or likely than item B.

Example 15.7 illustrates how to perform the conditioning and calculations for a simple, two-level problem. Most problems have a much more complex structure. Higher levels and items per level in a problem require more relative comparisons and time for evaluation. Each level of m items requires $m(m-1)/2$ comparisons.

EXAMPLE 15.7: ***Using Saaty's pairwise comparison technique for model formation***

In reactor safety analysis, the loss of off-site power (LOSP) can lead to other consequences and important events. Therefore, plant managers want to investigate how LOSP can occur. Meteorological conditions (floods, lightning, and high winds) represent one set of possible events responsible for LOSP. Each of these conditions can occur with varying degrees of impact on the likelihood of LOSP. Different questions facilitate evaluating the impacts: How much flooding? Where does the lightning hit? How high are the winds? To answer these questions, a hierarchy of the impacts on LOSP follows:

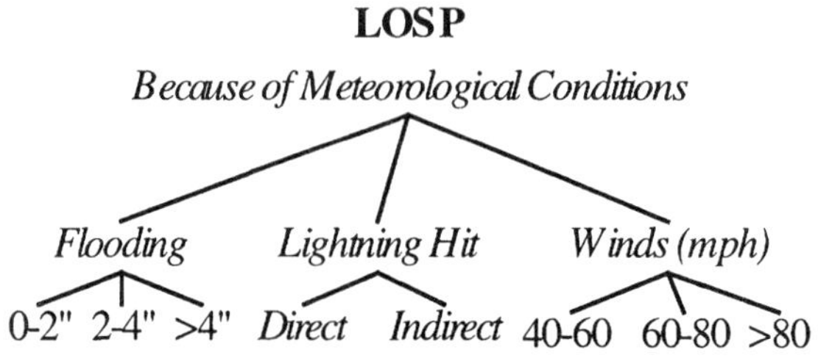

The comparisons of the bottom eight, specific meteorological conditions depend on the levels above them. Although chapter 11 provides a numerical (Saaty) scale to make comparisons on a pairwise basis and although the resulting weights are numerical, weight interpretation is only done qualitatively. This interpretation preserves both the qualitative structure of the input information and its granularity.

The pairwise comparisons follow:

LEVEL 1: LOSP because of meteorological conditions.

LEVEL 2: Which of the 3 general conditions most likely causes LOSP?

Flooding vs lightning? 1/4 (meaning lightning is more likely than flooding using a 4 on the Saaty scale)

Flooding vs winds? 1/3
Lightning vs winds? 2

These three comparisons form the upper triangular portion of the Saaty comparison matrix. To fill the matrix, place 1s on the diagonal and reciprocal answers on the lower triangular portion.

	F	W	L
Flooding	1	1/4	1/3
Lightning	4	1	2
Winds	3	1/2	1

Normalizing the eigenvector of this matrix's principal eigenvalue determines the relative weights for *F*, *L*, and *W*. In this example, the weights are (0.12, 0.56, and 0.32), respectively.

LEVEL 3: Which of the specific conditions is most likely to occur, given the general condition of flooding?

0-2 inches vs 2-4 inches, given a flood?	2
0-2 inches vs more than 4 inches?	6
2-4 inches vs more than 4 inches?	3

The weights for the flooding matrix are (0.67, 0.22, 0.11).

Which of the specific conditions is most likely to occur, given the general condition of lightning?

Direct hit vs indirect hit, given lightning hit?	1/4

The weights for the lightning matrix are (0.20, 0.80).

Which of the specific conditions is most likely to occur, given the general condition of winds?

40-60 mph, given high wind?	2
60-80 mph?	8
Greater than 80 mph?	5

The weights for the winds matrix are (0.75, 0.16, 0.09).

Multiplying the weights of levels 2 and 3 yields the final (unconditional) weights for the eight specific conditions:

(0.08, 0.03, 0.01, 0.11, 0.45, 0.24, 0.05, 0.03).

The results indicate that the most likely cause of LOSP is an indirect lightning hit; the second most likely is winds 20 to 40 mph. These results then would be the important conditions and causes of concern for LOSP.

Statements such as *an indirect lightning hit is almost twice as likely as the 20-to-40-mph winds* are not correct. Such numerically based conclusions are not valid with a relative (nonnumeric) comparison method.

■

This technique can be used as a weighting scheme for aggregating multiple experts (chapter 16). A hierarchy can be constructed to represent a conditional model of information of each expert. Multiple experts can be aggregated by combining their hierarchies. To make such an aggregation, make a new level for the experts and place their hierarchies beneath them.

Decomposition Diagrams

The hierarchical structure of the Saaty method described in the previous section is not the only way to diagram conditions. A more established method is to form a tree structure with branches and nodes (Raiffa, 1970). The tree usually begins as a qualitative description of alternatives or events connected with branches. The tree can be quantified into a decision diagram, (e.g., an event tree) by assigning values or distributions to the branches. If values or distributions are propagated through the branches, then the values or distributions at the end of the tree will have a numerical interpretation. If only qualitative information is propagated through the tree branches, then the results have a qualitative interpretation (e.g., the results from Saaty's method). In either case, the diagram serves as a model describing the conditions relevant to the final results.

Using the diagramming approach as a decomposition tool helps the expert evaluate a complex or large problem one piece at a time (U.S. NRC, 1989). Chapter 5 and Kahneman and Tversky (1982) cite the advantages of using the decomposition principle. Example 15.8 illustrates how to use the diagram method.

EXAMPLE 15.8: ***Using decomposition diagrams for conditional modeling***

An expert determines the probability of failure for an important but never observed event *Prime*. The failure of *Prime* depends on the system's temperature and pressure. The expert is asked to provide a set of potentially fatal temperature values (T_1, T_2, and T_3) and a set of potentially fatal pressure values (P_1, P_2, P_3, and P_4). The expert then diagrams the relationships between these temperatures and pressures as they affect *Prime*. The diagram follows:

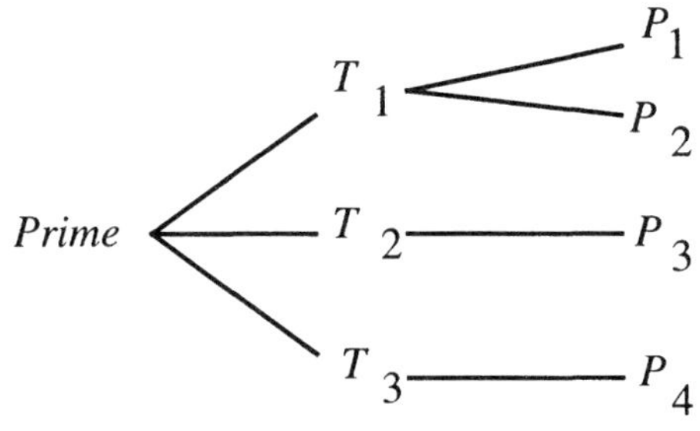

This diagram represents a qualitative model of *Prime*. To quantify the diagram, the expert provides the likelihoods or probabilities of the temperatures and the pressures.

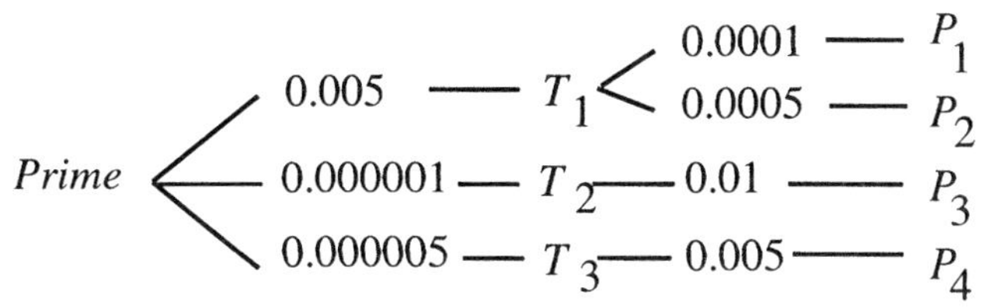

The pressures could be independent of the temperature. The above diagram indicates that the pressures are conditional on temperature because the P_2 values change depending on whether T_1 or T_2 is present.

The resulting probability for *Prime* is determined by multiplying the values across pathways:

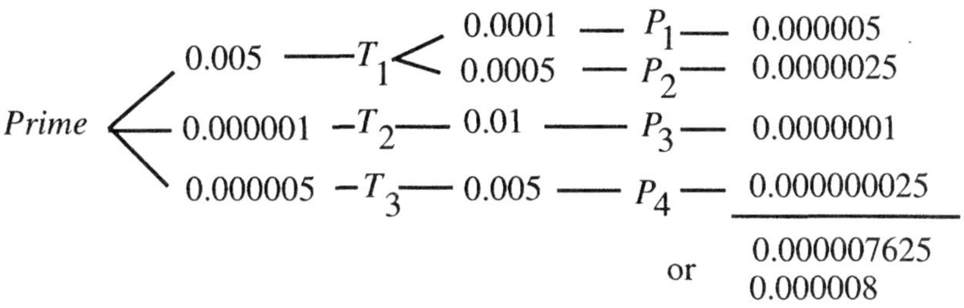

■

Example 15.8 illustrates a conditional model construction for a single expert providing single estimates for the branches. A similar example in chapter 16 illustrates how to combine multiple experts using the diagram approach. Chapter 17 provides a similar example on how to propagate probability distributions, rather than single estimates, along the branches.

Model Selection Suggestions and Cautions

This chapter has described some of the methods used to formulate functional relationships between the answers and the ancillary or conditioning variables. The examples have illustrated some of the pitfalls and shortcomings in these methods. The following page summarizes these cautions and makes some suggestions for model selection.

1. A descriptive model of the answer variable in terms of the ancillary variables (such as GLM) is useful for determining and defining influential variables. Such models help analysts understand the reasons behind the answers given. However, we cannot recommend that these models be used for prediction purposes because the independent variables may not be measured without error. The variable relationships established from these models should be consistent with the results from the exploratory data analysis (chapters 12 and 13).

2. The steps necessary to formulate some models may require that an analyst interpret or quantify variables. Avoid interpretations and keep quantification to a minimum. If the information content or granularity of the elicited or gathered data is insufficient for the desired models, then do not use those models. If all the data are highly qualitative or uncertain in nature, perhaps the model selection should not be done.
3. When using more than one modeling procedure, make sure that these procedures are consistent in their model formations. The examples listed in this chapter indicate how the multivariate procedures are compatible with GLM models. Likewise, the conditional models are compatible among themselves. Only in a very general way are the statistical and conditional types of models compatible.
4. Granularity plays a critical role when collapsing or reformulating variables for modeling. As noted in several examples, the new variables formed are more general in information content, and the conclusions resulting from the analyses must also be stated in more general terms. Granularity is also important when making relative comparisons, such as with the Saaty method. Although numbers are used in the analysis, the results can only be interpreted in relative comparison terms.
5. When using GLMs or multivariate procedures, consult a statistician if necessary. The regression and multivariate methods results are sometimes difficult to interpret. The procedures themselves require computer programs that may be difficult to run, and several different methods may be available for the analysis. For example, the different ways to do cluster, factor, and discriminant analyses produce different results.
6. The conditional models serve as both elicitation and analysis techniques. They are somewhat difficult to use because they consume much time and may require training of the experts.
7. The philosophy of this book is that models should not be selected before gathering the data, nor should they be selected based on popularity or convenience of the analyst. Many choices and changes complicate and compromise the elicitation, and the experts must be motivated to participate in its implementation. Model choice based upon the data and the elicitation makes the experts feel comfortable and provides more flexibility for the analyst.
8. The primary purpose of model formation is to describe the answers in terms of the other variables (information). The results of the modeling effort provide the appropriate conditional interpretations for the answers. Plenty of other information should be available from the elicitation and from the data base so that these relationships can be understood and so that they can be used for aggregating experts (chapter 16), handling the inevitable uncertainties (chapter 17), and making the final inferences (chapter 18).

16

Combining Responses—Aggregation

This chapter consists of three major sections. Section one, *Aggregation Scheme Selection*, presents several recommended and widely accepted aggregation estimators as well as methods used to form aggregation distributions. Section two, *Application Environments*, covers aggregation using the methods in the first section in various problem settings and environments that involve multiple experts, decision makers, and analysts. Section three presents the similarities between solving the aggregation problem and solving the problem of characterizing uncertainties (chapter 17).

Aggregation Scheme Selection

Combining the experts' answers into either a single value (estimator) or a single distribution of values can prove a most difficult analytical problem. This aggregation is a mathematical aggregation. A variety of techniques can perform mathematical aggregation; however, many of these techniques impose restrictions on the data, the experts, the analyst, and on the interpretations of results. The techniques presented in this chapter for both estimators and distributions reflect the general philosophy of the book: from the elicitation side, the aggregation should not require the experts to be force-fitted into unknown or uncomfortable modes of providing data; from the analysis side, the aggregation should not be so complex that an analyst requires a doctorate in mathematics to understand and use it. To achieve both goals, this chapter presents only some of all the available techniques. However, most of the general types of estimators are given in some form. For a complete report on all the different types of mathematical aggregation schemes, see a survey paper by Genest and Zidek (1986).

Aggregation Estimators

The most commonly used method of combining a set of answers is to calculate a single summary value (estimator) based on all the values in the data set. The most popular estimators for central tendency are the mean, median, and geometric mean. As discussed in the following paragraphs, each estimator has its own properties, making it appropriate for different applications. Regardless of the estimator chosen, remember to include an estimate of the variance of each estimator. For example, estimate both the mean and the variance of the mean.

The mean, which is the average of the values, has the advantage of having an easily calculated variance. To estimate the variance for a large sample mean, divide the variance of the sample data by the square root of the sample size (n):

$$var(mean) = \left(\sum_{i=1}^{n} (x_i - mean)^2 \right) / (n(n-1))$$

This variance formula comes from the central limit theorem which states that the formula is a good approximation for the variance of the mean for a sample of sufficient size. However, if the sample size (n) is small, as it is in most expert judgment applications, or if the data set is not unimodal and symmetric, the formula from the central limit theorem does not adequately approximate the variance of the mean. For samples of size less than 10 (even for symmetric, unimodal data), the asymptotic approximation is poor. Therefore, in these cases we recommend an alternative estimate for the variance of the sample mean. One such alternative is to estimate the variance of the sample mean from a simulation (e.g., bootstrap).

The mean possesses a second noteworthy property: it gives equal weight to each datum. This equal weighting has a serious implication: if only a few experts provide answers and one expert gives an answer that is far away in value from the rest, then that extreme value greatly influences the mean value. This result may not be desirable, especially if the extreme value appears questionable or seems unreasonable.

To overcome the influence of extreme values when forming an aggregation estimate, use the median or geometric mean. The central values of the data set tend to influence both these estimators while the extreme values do not.

The median is the 50th percentile value; in other words, half of the data is larger than the median and half of the data is smaller than the median. If the data set has an odd sample size (n is odd), calculate the median by finding the central value of the ordered data points. If the data set has an even sample size, then the median is the average or halfway point between the two center values. No general or convenient formulation of the variance for the median is available. Earlier chapters suggested the use of simulation techniques (e.g., the bootstrap or Monte Carlo methods) to find this variance.

Kahneman and Tversky (1982) provide another interesting reason for using the median in expert judgment applications. These researchers have shown that when experts provide numerical answers, they really estimate the median value rather than the mean. If all values for the answers given by the experts form a distribution that is symmetric in shape (cutting the distribution in half results in one half being the mirror image of the other), then the mean and median are the same. However, most expert judgment data distributions are not symmetric (they are skewed with the mode shifted to one side of the center and the other side having a long tail); consequently, the mean and median are not the same.

Therefore, we recommend the common practice of considering the answers given by experts as median values.

The geometric mean is an average of the data values based on a logarithmic scale, while the simple mean is based on a linear scale. To form the geometric mean, take the product of all the n sample values then raise the product to the $1/n$th power. Simulation methods can determine the variance for the geometric mean. The log-based nature of the geometric mean makes it popular in expert judgment applications because many estimates elicited from experts are small values or probabilities that fit better in a log scale than in a linear scale.

Example 16.1 shows how the three estimators above give different summary results for the data set of seven experts estimating a very low probability value of an event. The variance estimates are bootstrapped from the data using a FORTRAN computer program given in Appendix D.

EXAMPLE 16.1: ***Comparing three aggregation estimators***

Seven experts provided the following probability estimates for a rare event.

Expert	Estimate
1	0.00001
2	0.00010
3	0.00010
4	0.10000
5	0.00001
6	0.00001
7	0.00005

The fourth expert gave an estimate several orders of magnitude larger than the others. The three aggregation estimators (the mean, the median, and the geometric mean) produce the following results:

Mean	=	0.014
Median	=	0.000050
Geometric mean	=	0.000091

The fourth expert's estimate dramatically affects the mean value. Note, however, that the fourth expert's estimate has little influence on the median and geometric mean making them closer to each other in value.

Using the central limit theorem (CLT), the variance of the mean for this data set is

$$var(mean) = var(data)\,/\,7 = 0.0014/7 = 0.000054$$

Bootstrap simulation (BS) can easily determine the variances for the geometric mean and the median. In this case, sampling with replacement from the original data formed 1000 samples of size 7. The mean, median, and geometric mean were calculated for each of the 1000 samples. The following table contains calculations for the variances from the 1000 means, the 1000 medians, and the 1000 geometric means:

$$var(mean)_{BS} = 0.00016$$
$$var(median)_{BS} = 0.000060$$
$$var(geometric\ mean)_{BS} = 0.000000095$$

A large discrepancy appears when comparing the bootstrap variance estimate for the mean with the variance estimate of the mean calculated according to the central limit theorem. This sample is a small data set and is highly skewed (with the extreme value from expert 4). As a result, this estimate is not expected to be accurate. In addition, note that in this case the estimate of the variance for the mean from the CLT is the same as the bootstrap variance estimate for the median.

A comparison of the bootstrap variances for the three estimators yielded the following result: the mean has the largest, the median is smaller, and the geometric mean has the smallest.

In conclusion, the small sample size and extreme value of expert 4 makes the mean estimator and its theoretical variance inappropriate. Both the median and geometric mean estimates work well for this data. In either case, a variance estimate for that estimator must come from a simulation such as the bootstrap. ■

Because the mean estimator weights each datum equally, and the median estimator counts the number of data giving weight to the extreme values, some analysts prefer to use a weighted average or mean. Each datum (expert answer) is given its own individual weight, and the mean is calculated as

$$weighted\ mean = \sum_{i=1}^{n} x_i w_i / \sum_{i=1}^{n} w_i .$$

The advantage of this aggregation method is that the analyst can control which values (or experts) influence the estimator. Theory provides a method to estimate the variance of the weighted mean; however, because of the small sample sizes and potential skewness of most expert judgment data sets, we advise simulation for determining the variance. The biggest disadvantage is that the weights must be determined for each expert.

Determining Weights

Determining weights is not an easy process. The process often requires information about the experts and how they arrived at their answers. It can lead to the dangerous situation in which the analyst imparts his or her knowledge and influence (perhaps erroneously) to the results.

However, a variety of methods can determine weights based on the data itself, on qualitative comparisons of the experts and ancillary data, and on information from model formations. There is also a simple rule of thumb for determining weights: equal weights.

Data-Based Determinations

The data provide information for determining weights. For example, an expert who gives an extreme value relative to the others receives a lower weight than the others. To illustrate this usc of the data to determine weights, example 16.2 shows the effects of different weights on a weighted mean estimator. This example shows that to produce a weighted mean comparable to the median or geometric mean requires that expert 4 be given a weight of 1/1000 of the other six experts. This example does not imply that the goal is to achieve the same value for the weighted mean as for the median or geometric mean. This example merely indicates how the three estimators weight the seven responses.

EXAMPLE 16.2: ***Using the weighted mean estimator***

The data from example 16.1 help determine different sets of weights for the seven experts. In each case expert 4 is given a smaller weight relative to the other six. The other six are given the same larger weight, and expert 4 is given a weight of 1. The effect on the weighted mean estimates is shown in the following table:

$W_{1,2,3,5,6,7}$	W_4	Weighted mean
1	1	0.014
2	1	0.0039
10	1	0.00086
50	1	0.00021
100	1	0.00013
1000	1	0.000055
1	0	0.000047

For cases in which the other six receive weights 1000 times that of expert 4, the weighted mean is comparable to the median and geometric mean values. The mean with expert 4 eliminated ($w_4 = 0$) also compares to the median and geometric mean.

■

If the analyst felt that a weight as low as 1/1000th or even 1/50th that of the others was warranted, why would that person be considered an expert? The expert must have possessed some underlying reason for giving the extreme value. Rather than eliminate him or her from the sample or resolving the issue analytically with outrageous weights, explore the reasons behind the expert's extreme value. Reviewing the documentation of the expert's problem-solving processes could determine the reason for the extreme value. Reviewing the preliminary data analysis results (as prescribed in chapters 12 through 15) may also lead to an explanation. These results should contain information about this expert and his or her answer relative to the others.

If no evidence suggests that this expert used different assumptions, cues, problem-solving methods, or any information different from the others, then do not assign him or her a low weight or eliminate him or her from the sample. The expert's answer is just as valid and reasonable as the others. In this case, use an estimator that has a wide variance to reflect the wide range of data values. The bootstrap results from example 16.1 indicate wide variances for the mean and median.

Using only the data to determine weights can lead to a dangerous situation: the analyst tries different weights to achieve some goal—such as the elimination of a particular expert. Therefore, we recommend using other methods to determine weights. These methods rely more on finding and using the reasons why experts' estimates differ rather than on focusing on mere numerical differences.

Saaty Weight Determinations

Unlike cases in which the gathered data help determine the weights for the experts, some cases permit weight determinations before the answers are given. The Saaty pairwise comparison method helps make such a determination; in addition, the decision maker, the analyst, or even the experts can use this method to determine expert weights.

The determinations are based on other information about the experts. The Saaty method produces weights that are qualitative in nature even though these weights are numerical. However, care must be taken in the use and interpretation of these resulting weights. The process of forming the weight has quantified the original information which, in turn, has changed the granularity from general to specific. The resulting weights should not be used as numerical values. They are only relative comparisons in numerical form. However, relative comparisons can determine other numerical weights for a weighted mean. If the weighted mean value is calculated using the Saaty weights, it should be accompanied by a caveat, such as *this mean is the result of relative weights that are values from qualitative comparisons and are not exact numerical values.* Because even exact numerical weights are highly uncertain in value or fuzzy in nature, such a warning would not be unusual.

A FORTRAN program (see Appendix A) for a single level evaluation of the Saaty method can determine the expert's weights. The user (decision maker, knowledge engineer, analyst, or expert) supplies the comparisons of the experts on a pairwise basis. The resulting relative weights are normalized to sum to 1.0. Example 16.3 illustrates how experts are compared using this method.

EXAMPLE 16.3: ***Using Saaty's method to determine weights***

A decision maker wishes to determine weights for the seven experts in example 16.1. He or she is familiar with their qualifications and will make the comparison according to those criteria before he or she examines the answers that were elicited. Based on his other knowledge of the seven, he or she compares the experts as follows:

1 versus 2: Better	3 versus 4: Same
1 versus 3: Same	3 versus 5: Better
1 versus 4: Same	3 versus 6: Better
1 versus 5: Better	3 versus 7: Same
1 versus 6: Better	
1 versus 7: Better	4 versus 5: Same
	4 versus 6: Better
2 versus 3: Worse	4 versus 7: Same
2 versus 4: Worse	
2 versus 5: Worse	5 versus 6: Same
2 versus 6: Worse	5 versus 7: Worse
2 versus 7: Worse	
	6 versus 7: Worse

The Saaty method (and code in Appendix A) translates the verbal comparisons using the following recommended numerical scale:

better	=	2.72,
same	=	1.00,
worse	=	0.37,

and the resulting relative weights for experts 1-7 are

(0.23, 0.06, 0.19, 0.17, 0.10, 0.08, 0.17).

The consistency ratio for this matrix is 0.06, which is less than the critical value of 0.10. Such a value indicates good consistency among the comparisons.

Using these weights to combine the seven expert answers yields a result of

$$\sum_{i=1}^{n} x_i w_i / \sum_{i=1}^{n} w_i = 0.017.$$

On the other hand, the relative weights can be used to determine other numerical weights. The weights of experts 1, 3, 4 (the extreme-valued expert), and 7 are high relative to those of experts 2, 5, and 6. A decision maker decides on a 2-to-1 weighting of the high group over the low group. This weighting yields the following combined value

$$\sum_{i=1}^{n} x_i w_i / \sum_{i=1}^{n} w_i = 0.018.$$

This value does not differ much from the original mean value of 0.014 or the weighted mean using the Saaty weights (0.017). Note that expert 4 occupies the high-weight group, making the weighted mean estimate even more influenced by his or her answer.

■

The results from example 16.3 may appear disappointing. The decision maker's information about the experts only allowed expert 4's answer to further influence the weighted mean. Thus, the decision maker did not have any information about this expert that would suggest that the experts solved a different problem. In fact, the decision maker's information reinforced the idea that expert 4 should be included in the aggregation.

Model-Based Determinations

The decision maker arrived at the weights in example 16.3 by using a single criterion to compare the experts. Therefore, these comparisons are conditions on the experts' estimates. Other conditions such as the information on the expert's rationale recorded in the elicitation session may be equally important. We need this information to formulate a data base as seen in chapters 12 and 13, and we also use it to formulate models in chapter 15. Weights can be determined from the conditional variables found to be significant in model formation.

Conditional Modeling. The Saaty method represents a conditional model used to formulate relative weights for the experts. Any variables found significant in other models can also be used as conditional variables and put into a hierarchical structure for analysis in the Saaty method. The Saaty method works well even if only a single conditional variable is deemed important, creating a single-leveled hierarchy. In addition, the Saaty method can analyze many significant variables as conditions as long as the variables can be structured into a hierarchy. Recall that complex hierarchies of a many levels with numerous items per level are time consuming and difficult to manage. If such a hierarchy is constructed, consider using other weight-determining methods outlined in this chapter. The Saaty method can easily manage a few levels with a few items per level.

The method has the advantage of making pairwise comparisons among experts either by a simple *better, same,* or *worse* comparison or by a convenient numerical scale such as the Saaty scale (given in chapter 11) or Sherman Kent scale (given in chapter 7). Therefore, this procedure can use quantitative or qualitative conditional variables. Example 16.4 illustrates how a significant variable from a general linear model is used to obtain relative weights for the experts using a simple scale of *better, same,* or *worse.* In this case, the analyst uses the conditional variable to determine the weights rather than using his or her own knowledge or judgment about the experts.

EXAMPLE 16.4: ***Using conditional variables and Saaty's method to determine weights***

Eight experts provided answers to a likelihood comparison question using a linear scale from 0.0 to 1.1. A particular problem-solving variable, ps_1, was created as a cumulative score of several heuristic and cue variables (example 15.1). A general linear model analysis indicated that ps_1 was the best variable (and the most significant) in predicting the answers (v_1). The second best variable for predicting the answers was an experience variable *(yrt)* specifying the number of years that the expert had worked in the particular field of the problem. The data follow:

v_1	ps_1	*yrt*
0.20	-2	5.49
1.00	2	6.00
0.04	-3	2.00
0.00	-2	2.00
1.10	6	1.99
0.45	0	4.50
0.00	-3	3.00
0.75	2	4.90

The following table used the values of ps_1 and *yrt* to compare the experts, where

>	----	is greater (better) than,
=	----	is equal, and
<	----	is less (worse) than.

ps_1	*yrt*
1 < 2	1 < 2
1 > 3	1 > 3
1 = 4	1 > 4
1 < 5	1 > 5
1 < 6	1 > 6
1 > 7	1 > 7
1 < 8	1 > 8
2 > 3	2 > 3
2 > 4	2 > 4
2 < 5	2 > 5
2 > 6	2 > 6
2 > 7	2 > 7
2 = 8	2 > 8

$3 < 4$	$3 = 4$
$3 < 5$	$3 > 5$
$3 < 6$	$3 < 6$
$3 = 7$	$3 < 7$
$3 < 8$	$3 < 8$
$4 < 5$	$4 > 5$
$4 < 6$	$4 < 6$
$4 > 7$	$4 < 7$
$4 < 8$	$4 < 8$
$5 < 6$	$5 < 6$
$5 < 7$	$5 < 7$
$5 < 8$	$5 < 8$
$6 > 7$	$6 > 7$
$6 < 8$	$6 < 8$
$7 < 8$	$7 < 8$

These two variables can be used as two items in a one level hierarchy:

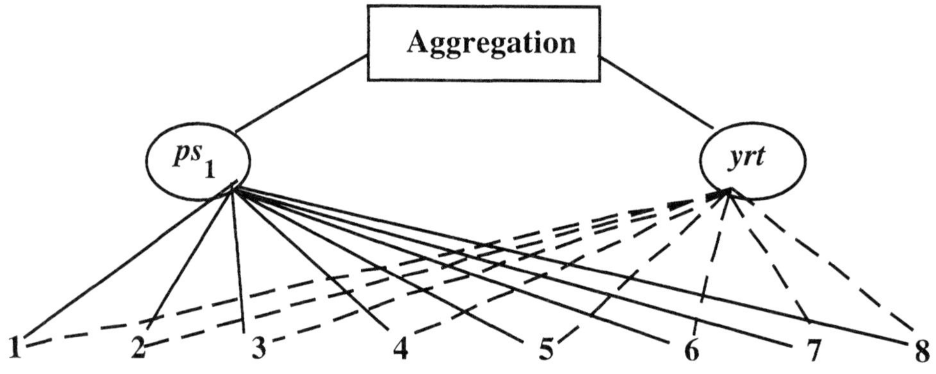

The hierarchical structure helps to combine the influence of the two variables into a single aggregation measure for weight determination. The above pairwise comparisons of the eight experts for each conditional variable form the entries of the two matrices (one for each variable) required for the analysis. Each matrix results in a set of weights for the experts. The weights assigned to the two variables combine the two sets of comparisons. From the GLM done on these two variables, ps_1 accounted for most of the model variation, and *yrt* added very little. A weighting for the two variables might be 0.9 for ps_1 and 0.1 for *yrt*.

The Saaty single matrix code in Appendix A provided the following weights for the eight experts for ps_1 and *yrt*. These two sets of weights are combined using the (0.9, 0.1) split on the two variables.

ps_1 Weights	Yrt Weights	Combination weight $= 0.9 \cdot ps_1 + 0.1 \cdot yrt$
0.08	0.20	0.09
0.18	0.26	0.19
0.05	0.06	0.05
0.08	0.06	0.08
0.15	0.05	0.14
0.16	0.12	0.16
0.08	0.09	0.08
0.22	0.16	0.21

The following table shows the weighted mean value calculation using the combination weights.

V_1	Combination	$v_1 \cdot$ Combination weight
0.20	0.09	0.02
1.00	0.19	0.19
0.04	0.05	0.00
0.00	0.08	0.00
1.10	0.14	0.15
0.45	0.16	0.07
0.00	0.08	0.00
0.75	0.21	0.16
		0.59

This mean value is larger than the original mean of 0.44. The variables chosen for the weight determination create this discrepancy. Both variables have large values for large answers (v_1); therefore, the weights will be large for large values of v_1. The effect causes inflations of the final mean value. This may seem like a weighting scheme with a built-in bias. The scheme is based on conditioning, and conditions can induce bias. The validity of the scheme lies in using conditioning variables that are correlated with the answers.

■

Example 16.4 pointed out a problem inherent in weighting schemes. In some cases, the method for determining the weights induces a bias in the results. The weights in that example in turn biased the resulting mean on the high side simply by the method used to determine the weights. The analyst or decision maker determining weights can also induce biases. The following methods and examples tend to minimize these biases.

GLM Modeling. For example, problem-solving variables play an important role in determining answers (Booker & Meyer, 1988*a*; Meyer & Booker, 1987*b*). Example 16.5 illustrates how one such problem-solving variable can form weights for the experts based upon the residuals from a linear model regression analysis done on that variable and the answer variable. The residuals measure how far away each answer lies from the model (line) predicted by the problem-solving variable. Therefore, one possible set of weights is the inverse of these residuals. This weighting scheme gives larger weights to values with smaller residuals.

Using residuals to determine weights has a distinct advantage: the residuals can come from a model of more than one variable. Therefore, the weight determination reflects the simultaneous impact of several conditional variables. However, as is generally true of multivariate models, the influence of the most important one or two variables has the greatest impact in the model which impacts the residuals and which ultimately impacts the weights.

EXAMPLE 16.5: ***Using residuals to determine weights***

A model of the experts' answers can determine weights for the eight experts in example 16.4. A general linear model analysis indicated that the problem-solving variable *(ps_1)* proved very significant in predicting the answers (v_1). The residuals from a regression of v_1 on ps_1 follow:

v_1	ps_1	Residual	Weight = 1/Residual
0.20	-2	-0.095	10
1.00	2	0.137	7
0.04	-3	0.115	9
0.00	-2	-0.058	17
1.10	6	-0.023	43
0.45	0	-0.045	22
0.00	-3	0.007	141
0.75	2	-0.038	26
			275

Weight	Weight • v_1
10	2.0
7	7.0
9	0.4
17	0.0
43	47.3
22	9.9
141	0.0
26	19.5
	80.06/275 = 0.31 = weighted mean

The mean of the original data is 0.44, and the median is 0.45. Therefore, the residual weighting scheme changed the mean value, but not by much.

■

Direct Estimation. Other conditional variables that were important in model formation can be used to determine weights. Specifically, these variables are either qualitative- or classification-type variables such as those found to be important in cluster analysis.

Weights can be determined from these model variables by direct estimation. This requires knowing how to use qualitative or class variables to form quantitative weights. Also, remember that granularity can change with this type of quantification. Words (e.g., categories or classes) can easily and arbitrarily be transformed into numbers. However, never forget that those numbers have no numerical meaning and cannot be used for interpreting the results. Therefore, any weights formulated from qualitative or class information should be considered relative weights, as in example 16.3.

Direct estimation allows the analyst to use the Saaty method for quantification. On the other hand, the decision maker or analyst can form relative weights without the formal Saaty technique. In either case, the granularity can change. The criteria (variables) chosen for the comparison can be evaluated according to how they cluster the data (see example 16.6). In this case, the classification variable is important because it discriminates between the experts according to a critical assumption made in solving the problem. Such a clear-cut distinction provides a valid reason for eliminating expert 4 from the data set. Expert 4 used an assumption that radically differed from the other experts. As a result, he or she was actually solving a different problem from the one being asked.

EXAMPLE 16.6: ***Using direct estimation from cluster model variables to determine weights***

A decision maker failed to discover one of the critical assumptions made in determining the probability of a rare event until after the completion of the exploratory data analysis and data-base formation. The seven experts made explicit assumptions regarding the constancy of temperatures. Some experts assumed that the temperatures would remain constant (assumption = 1), some assumed it would vary negligibly (assumption = 2), and some assumed it would vary greatly (assumption = 3). The decision maker originally assumed that the temperature variation would not be great and wanted the problem solved without a large temperature variation. At that time the decision maker failed to realize that this assumption might be important to the problem. However, a cluster analysis revealed that the seven answers were clustered according to this assumption.

Estimate	Assumption /Cluster
0.00001	1
0.00010	2
0.00010	2
0.10000	3
0.00001	1
0.00001	1
0.00005	2

Only expert 4 assumed a large temperature variability. Therefore, expert 4 actually solved a different problem from the one intended. The following table shows the weight determination from this assumption:

Estimate	Weight	Estimate • Weight
0.00001	2	0.00002
0.00010	1	0.00010
0.00010	1	0.00010
0.10000	0	0.00000
0.00001	2	0.00002
0.00002	2	0.00002
0.00005	1	0.00005
	9	0.00031/9 = 0.000034

This weighted mean value does not differ much from the unweighted mean value that deleted expert 4 (example 16.2). Therefore, the other weights made little difference.

■

The goal of direct estimation is to effectively determine weights by using variables or conditions that influence the experts' answers (as determined from models). Exercise care in using the methods presented here because these variables are known to be important in the experts' determination of their answers; these variables can induce a bias into the weights that will result in a bias in the weighted mean value. Also, take care in using qualitative information to formulate weights. Such formulation can cross levels of detail, making accurate interpretation of the results tricky at best and incorrect at worst.

One final note: no matter how good the choices for the weights are, most weights make little difference in the value of the weighted means. The question then becomes, why bother with weights? The answer follows.

Equal Weights

The best recommendation to date on the weight determination problem is to use equal weights. This idea is not new (Seaver, 1978); however, several studies indicate that it still remains the best idea (Genest & Zidek, 1986).

Equal weights effectively combine both individual responses from experts and distributions; the remaining sections will deal with combining distributions.

Use equal weights unless some unusual circumstances indicate the use of some different weights. Example 16.7 summarizes the lessons from the first sections of this chapter. These lessons reinforce the idea of equal weights for the two applications used as examples in this chapter.

EXAMPLE 16.7: ***Summary of weight determinations***

In examples 16.1 through 16.3, seven experts estimated the probability of a rare event. Expert number 4 gave a much larger answer than the other six. Those examples calculated the mean (equal weights), median, and geometric mean. Several different weighting schemes were calculated using various weights, and the Saaty method was also used to determine weights. Those results are summarized below.

Estimator	Weighting Scheme	Estimate
Mean	Equal	0.014
Median	Equal	0.000050
Geometric mean	Equal	0.000091
Weighted mean	$w_4 = 2$	0.0039
Weighted mean	$w_4 = 10$	0.00086
Weighted mean	$w_4 = 50$	0.00021
Weighted mean	$w_4 = 100$	0.00013
Weighted mean	$w_4 = 1000$	0.000055
Weighted mean	$w_4 = 0$	0.000047
Weighted mean	Saaty — decision maker	0.017
Weighted mean	Saaty — model	0.000034

All these methods produced only two basic results:

1. Estimates of the order of 0.01 resulted from equal weighting of expert 4 with the rest.
2. Estimates of the order of 0.00001 resulted from deletion of expert 4 from the set, with equal weighting for the rest, or using the geometric mean aggregation method.

The results from simply increasing the weight of expert 4 (w_4) serve only to demonstrate that severe weights are required to reduce the impact of expert 4. The clear decision in this example is either to include or exclude expert 4.

The assumptions used to solve the problem in example 16.6 helped the analyst to conclude that the other six experts made acceptable assumptions for the particular problem but that expert 4 made an unacceptable assumption. Therefore, expert 4 should not participate in the final aggregation for this question.

Note that the results could have been completely reversed. Expert 4 could have provided the sole acceptable assumption, which would make his or her answer different from the rest. Assumptions are only a part of the information gathered in the elicitation. Other conditions under which all experts answered the question must be also be examined and evaluated. Such evaluation may help identify and correct some problem areas during the elicitation such as the expert using questionable assumptions, drifting from the question, or solving a different problem. It is vital that all the experts answer the same question.

In examples 16.4 and 16.5, eight experts estimated the likelihood of an event. In these examples, no experts provided responses different from the rest, but the experts' answers fall into two distinctive clusters (experts 1, 3, 4, and 6; experts 2, 5, and 8). In this case, any centroid type of estimator (mean, median) will yield a result between the two clusters of data (where no actual data exists). A weighting scheme in this case is useful to pull the aggregation estimator toward one cluster or the other. However, the same cautions about choosing and using weights in this case still apply. The following table lists the results for this data set and the chosen estimators.

Estimator	Weight Scheme	Estimate
Mean	Equal	0.44
Median	Equal	0.45
Weighted mean	Saaty, 2-variate	0.59
Weighted mean	GLM residuals	0.31

All these estimates are close in value. The residuals tend to lower the original mean value because the regression model is more influenced by the larger sized cluster, which is the lower valued cluster. The Saaty weightings favor the larger values, making the estimate biased. Neither of these weighting schemes are based on conditions that definitively favor one cluster over the other.

Examining the original data base reveals no other conditions for making such a distinction. In other words, the analyst cannot find an explanation why the two clusters emerged. As a result, neither cluster should be excluded from influencing the final estimate.

In conclusion, the equal weighting (simple mean) method is appropriate in this case. The uncertainty in the data is such that the experts spanned the entire available range of values (response mode) for their answers. The mean estimator will give a value in the center of the range and between the two clusters.

■

We hope that by examining the two problems in example 16.7 the steps are evident for determining weights for the experts. The choices in both these cases led to a simple, equal-weighting mean estimator. By using the data from the data base, the analyst examined many possible explanations for apparent differences in the answers. The first problem produced a valid difference while the second produced no difference. The basic rule for searching and determining such differences is that *all* experts must provide answers to the *same* problem.

Aggregation Distributions

Rather than restricting the expert to providing a single, best estimate for an answer, many analysts prefer that the experts provide either a range of values or an entire distribution of values (or probability distribution function). Many experts prefer to give multiple values. These multiple values do not imply that the experts provide values for different conditions. Giving multiple values for different conditions would be like providing single estimates for multiple problems.

If the problems in handling single-valued estimates from several experts appeared difficult, the problems will prove equally difficult in handling distributions or multiple values from many experts. This section emphasizes the case in which experts provide probability distributions. The next section deals with uncertainties, including the topic of handling ranges (see also chapter 17).

The aggregation methods presented in this chapter reflect the philosophy in this book. In particular, the methods should be easy for the experts to use. The analyst does not have the luxury of deciding upon his or her favorite method and then force-fitting the experts into its use. The methods should also be easy for the analyst to use. These methods are illustrated in the examples; Genest and Zidek (1986) reference others.

Using Bayesian Methods

The philosophy of using Bayesian methods incorporates the idea that all information can and should be used to form final estimations (chapter 11). Bayes theorem provides the mechanics for the combination of information from various sources. The information can appear in several, often different forms, such as (1) information from measured data combined with expert judgments, (2) information from the present that is combined to update previous information, and (3) information from experts combined with information from the decision maker (DM) or analyst.

The first case is the classical reliability problem common in Probabilistic Risk Assessment (PRA) applications (Martz & Waller, 1982). The second case is modified to include combinations of several information sources (experts) without following a time line; this case is examined in example 16.8. The three cases are examined in more detail in the *Application Environments* section.

The major difficulty in applying Bayesian methods is that the information (all information sources) must be quantified into probability distribution forms. Common distributional forms (e.g., the beta distribution) are easy to use for probabilities because they require only two parameters to estimate the entire distribution; they range from 0.0 to 1.0 in value, which is the range of probabilities; and they combine in a mathematically tractable form with other common distributions such as the binomial to form the posterior distribution.

Example 16.8 illustrates how to combine two estimates provided by each expert into a single beta distribution that represents the probability of a single failure (p) of an event.

EXAMPLE 16.8: ***Bayesian-based aggregation of distributions***

Six experts provided estimates of 5th and 95th percentile values for the distribution of a probability (p) for an event. Based on previous studies (Kahneman & Tversky, 1982), these percentiles really only represent a fraction of the true uncertainty. Therefore, the given 5th percentile represents the 30th percentile and the given 95th percentile represents the 70th percentile. Assuming that each expert's percentiles come from beta distributions with parameters x_0 and n_0, the values of those parameters follow:

Percentiles		Beta Parameters	
30th	70th	x_0	n_0
0.0010	0.004	0.82	242.59
0.0001	0.010	0.19	11.04
0.0010	0.100	0.18	1.52
0.0001	0.010	0.41	41.80
0.0050	0.050	0.41	8.64
0.0100	0.050	0.65	15.30

The beta parameter values can be obtained from the beta subroutines in the beta Monte Carlo code in Appendix B. These parameters have specific meaning in a binomial process: x_0 represents the number of failures in n_0 trials, each with a probability p of occurrence.

To combine the six different beta probability distribution functions into a single beta distribution function, $f(p{:}x,n)$, where

$$f(p{:}x,n) = \sum_{i=1}^{6} w_i f_i(p_i{:}x_{0_i}, n_{0_i}),$$

with equal weights for w_i, parameter sets are combined as follows (Winkler, 1968):

$$x = \sum_{i=1}^{6} x_{0_i} = 2.66$$

and

$$n = \sum_{i=1}^{6} n_{0_i} = 320.89.$$

A Bayesian approach uses this new beta as a prior distribution and combines it with a DM's distribution, resulting in a posterior distribution. Example 16.10 illustrates such a combination.

This new resulting beta distribution is the aggregation distribution for the experts. The distribution has a mean and variance of

$$x/n = 0.008$$

and

$$x(n\text{-}x)/(n^2(n+1)) = 0.00003,$$

respectively.

■

Using Assumed Distributions

Other analytical methods combine assumed distribution functions without using Bayesian-based methods. A joint distribution function of all the experts forms the aggregation function. Using this joint or composite distribution as a prior distribution, the decision maker combines the joint distribution of all the experts with his or her distribution [$f_0(x:\theta)$] in a Bayesian manner as follows:

$$f(x:\theta) = c(x:\theta) \bullet f_0(x:\theta) \bullet f_c(x:\theta),$$

where $f_c\ (x:\theta)$ is the joint distribution function with parameter θ for the random variable x of all k experts and where $c\ (x:\theta)$ is a joint calibration function for all k experts. If all the experts are independent, the joint distribution is as follows:

$$f(x:\theta) = c(x:\theta) \bullet f_0(x:\theta) \bullet f_1(x:\theta) \bullet f_2(x:\theta) \bullet \ldots \bullet f_k(x:\theta),$$

where the individual distribution functions of the experts are factored out. In other words, the joint distribution of all the experts is expressed as the products of the individual expert distributions. If the experts' distributions are not statistically independent, then factorization is not possible.

Joint distribution functions are usually assumed, easy-to-work-with distributions. Example 16.8 used a beta distribution to characterize experts' information. Other distributional forms produce mathematically tractable combinations and can be assumed for the experts. One common choice is the normal distribution (Winkler, 1981). If the expert provides very few estimates, or if the expert provides a simple range of values, then the uniform distribution is appropriate. In this case, the posterior or resulting distribution will also be uniform.

If the information each expert provides follows a normal distribution, then a multivariate normal distribution is the combination (joint) distribution. Using a multivariate normal allows a specified correlation structure among the experts. However, the analyst must know this correlation structure.

Example 16.9 illustrates the requirements necessary to apply the multivariate normal. In using the normal, assume that the quantity of interest (quantity being estimated) follows a normal distribution. Most quantities (e.g., failure rates and probabilities) are not symmetric in shape and are not unimodal (Meyer & Booker, 1987*b*). Therefore, they cannot be considered normally distributed. In addition, the analyst rarely knows the values for the required parameters of the multivariate normal, the mean, and the variance. It may also prove difficult to obtain a correlation structure for the experts.

EXAMPLE 16.9: ***Multivariate normal distribution for aggregation***

Three experts estimate a temperature range that is critical to component failure (Winkler, 1981. Numerical values are used with permission.). The experts agree that this temperature range should be distributed as a normal distribution. They understand how to estimate the mean of this normal. The variances of the temperature ranges for each expert are not estimated but are known. In addition, the correlation structure of these three experts is known as

Correlation of experts 1 and 2 = $\rho_{12} = 0.60$
Correlation of experts 1 and 3 = $\rho_{13} = 0.50$
Correlation of experts 2 and 3 = $\rho_{23} = 0.60$

The experts' mean values (m) and corresponding known variances (σ^2) follow:

$m_1 = 60$	$\sigma^2_1 = 36$
$m_2 = 62$	$\sigma^2_2 = 25$
$m_3 = 70$	$\sigma^2_3 = 49$

The multivariate normal mean vector (m) and variance-covariance matrix (Σ) follow:

$$m = (60, 62, 70)$$

and

$$\Sigma = \begin{bmatrix} 36 & 18 & 21 \\ 18 & 25 & 21 \\ 21 & 21 & 49 \end{bmatrix},$$

where the diagonal elements of Σ are the variances and the other elements are $\sigma_i \sigma_j \rho_{ij}$.

To combine these estimates with a Bayesian approach, assume a diffuse prior (a prior that imparts little added information) to combine with the multivariate normal (the joint distribution of the experts). The result of this Bayesian combination produces a posterior distribution for the temperature range. With e being the unit vector of 1s, the posterior mean and variance are

$$m_{post} = e^t\ \Sigma^{-1} m \ /\ e^t\ \Sigma^{-1} e = 62.02$$

and

$$\Sigma_{post} = e^t\ \Sigma^t\ e = 22.83.$$

The m_{post} is actually a weighted mean value with weights that correspond to the following:

$$w_i = \sum_j a_{ij} / \sum_m \sum_j a_{mj},$$

where a_{ij} are the ijth elements of Σ^{-1}. In this case the weights are

$$w_1 = 0.263,\ \ w_2 = 0.669, \text{ and } w_3 = 0.068.$$

The third expert has a small weight because of his or her large variance and the high correlations with the other two.

Rarely are the variances known. If the variances for the experts are estimated, then the posterior variance calculation is much more complicated (Winkler, 1981). ■

The parameters required to use the multivariate normal distribution may be difficult to obtain or may rely on assumptions. However, the normal is a convenient distributional form for a combined distribution. Other distributions could certainly be used to make multivariate combinations; however, these would also require assumptions and estimates of their parameters. Other multivariate distributions are not as easy to form or to work with as the normal, and no more precedence exists for using them than for the normal. If multivariate forms become too complex or difficult to use, then empirical (data-based) distribution forms are suggested in conjunction with simulation techniques.

Using Empirical Distributions

If experts prefer to provide several values or percentiles without specifying distribution forms, then empirical or step distribution functions can be constructed from these estimates. Some cases may require interpretations or rules of thumb to form the distributions. The rules below are based on several studies (Kahneman & Tversky, 1982).

1. When experts provide 5th and 95th percentiles, they really only provide 30-40th and 70-60th percentiles.
2. When experts provide maximum and minimum values, they really only provide 5-10th and 95-90th percentiles.
3. When experts provide their best central estimate, they really give a value that corresponds to a median (50th percentile) rather than a mean.
4. When experts provide a variance, they really represent less than half of the variance.

Percentiles help form empirical distributions (i.e., those percentiles using the rules above rather than what the expert provides directly). The percentiles provide the points for a step cumulative distribution function. The empirical probability distribution function, $f(x)$, is also formed from these percentiles in the shape of a histogram. For example, an expert provides

Best estimate	0.3
5th percentile	0.2
95th percentile	0.4
Minimum value	0.1
Maximum value	0.5

for a random variable (x) with a possible range of values from 0.0 to 1.0. The distribution of x is then a histogram composed of a series of rectangles with starting values at x_l, with ending values at x_u, and with heights of f:

x_l	x_u	f
0.0	0.1	0.5
0.1	0.2	2.5
0.2	0.3	2.0
0.3	0.4	2.0
0.4	0.5	2.5
0.5	1.0	0.1

The following equation calculates the f values:

$$f_i = (\text{percentile level of } x_u)/(x_u - x_l).$$

The areas under these rectangles ($f_i \cdot (x_u - x_l)$) sum to 1.0. A computer program in Appendix C forms empirical distributions for each expert and combines them by a user-specified aggregation function using Monte Carlo simulation.

The aggregation function usually combines the experts' individual empirical distribution functions, $f_i(x)$, and the decision maker's empirical function, $f_0(x)$, by the following weighting scheme (Winkler, 1968):

$$f(x) = w_0 f_0(x) + \sum_{i=1}^{k} w_i f_i(x).$$

The analyst encounters the same difficulties described in the first section of this chapter when determining these weights. The rule of thumb for these weights is also the same: equal weights work best (Seaver, 1978). Distributions can be empirically combined using this equation without specifying distributional forms (such as normals, beta, or uniforms) or parameters. Example 16.10 illustrates weighting methods for empirical distributions from two experts.

EXAMPLE 16.10: ***Empirical distribution aggregation***

Two experts provide the following estimates for a random variable (x) without specifying any distributional form, such as the normal or beta. The rules of thumb listed above help interpret the given values.

Given As	Used As	Expert 1	Expert 2
Best estimate	Median	0.30	0.10
5th percentile	30th	0.20	0.05
95th percentile	70th	0.40	0.15
Minimum value	5th	0.10	0.01
Maximum value	95th	0.50	0.20

In this case, the absolute possible range of values for x is from 0.0 to 1.0. The resulting empirical distributions from the code in Appendix C follow:

Expert 1

X_l	X_u	f
0.0	0.1	0.5
0.1	0.2	2.5
0.2	0.3	2.0
0.3	0.4	2.0
0.4	0.5	2.5
0.5	1.0	0.1

Expert 2

x_l	x_u	F
0.0	0.01	5.00
0.1	0.05	6.25
0.2	0.10	4.00
0.3	0.15	4.00
0.4	0.20	5.00
0.5	1.00	0.06

Using a weighted sum (with equal weights of 0.5) in a Monte Carlo simulation of the two empirical distributions yields the following pooled distribution:

Percentile Level	Value
1st	0.042
5th	0.079
10th	0.099
20th	0.13
30th	0.15
40th	0.18
50th	0.20
60th	0.23
70th	0.25
80th	0.28
90th	0.32
95th	0.43
99th	0.58

Mean = 0.22
Variance = 0.011
Standard deviation = 0.11

■

If the analyst decides to form a pooled (weighted) distribution in a non-Bayesian context (e.g., example 16.10) of only the experts, then the weighting scheme used is very important. If the DM adds or changes the information from the experts in any way, he or she should be treated like any other expert in terms of influence on the final results. Of course, the choice of weights can shift influence slightly toward or away from the DM or any other expert. However, as seen before, results will not differ by much unless drastically different weights are used, and such weights require justification. Lacking strong justification to apply drastically unequal weights, we recommend equal weight assignments.

Using Monte Carlo Simulation

Monte Carlo simulation works well aggregating empirical distributions of experts as demonstrated in the previous section. Additional applications for aggregation also use Monte Carlo simulation.

One commonly used method of eliciting and modeling expert information is through a series of distributions, each representing a phase or characteristic of the problem. The method decomposes the problem into simpler parts. Information is then gathered on each part. The information can take the form of distributions of interest variables in the various parts. The difficulty is how to recombine the information from all the parts for each expert and how to combine all the information from the experts.

Many applications use a tree diagram to structure the parts. Decision analysis applications commonly used this type of diagram. Some diagrams could be hierarchical in structure while others could be quite complex with feedback loops and ill-defined connections. Whatever the structure, logic must be used to determine how the parts should fit together. Accurately diagramming a complex problem can prove a most difficult task; like any model formation process, however, it must be done correctly.

Combining different structures for several experts is another difficult task. Each expert will have a structure that is uniquely his or her own. These structures can be viewed as conditional models, $f(x|c)$, for the final result or answer, x. Because the structures are complex, one way of combining them is to determine $f(x|c)$ through simulation. Assumed or empirical distributions can characterize the information provided for the various parts of the structure. To connect the various parts (however complex) requires arithmetic expressions similar to Boolean expressions for the failure or reliability of a system fault tree. Simulation is performed on each expert's structure by sampling from the distributions in the structure and then combining the sampled values using the arithmetic expression. A final or resulting distribution of the whole structure is formed by sampling and calculating the expression many times. The resulting distributions for each expert can then be combined using any of the above distribution aggregation methods, including Monte Carlo simulation. Example 16.11 illustrates how this is done for two experts solving a problem by decomposition.

EXAMPLE 16.11: ***Decomposition and aggregation by simulation***

Two experts evaluate the probability of an event E. The question posed to them provides a limited set of conditions for determining the probability. These conditions involve establishing specific values: temperature (T), pressure (P), and flow rate (F). Both experts agree that these are the three most important conditions; however, each differs in his or her assessment of how the three conditions affect the event and what probabilities are associated with them.

To find the probability of the event $P(E)$, the analyst sums the products of the conditional probabilities times the probabilities of those conditions:

$$P(E) = P(E|T)\bullet P(T) + P(E|P)\bullet P(P) + P(E|F)\bullet P(F).$$

Expert 1 claims that T, P, and F all follow a uniform distribution with a probability value of 0.01. Expert 2 claims that all three follow a unimodal distribution between the values of 0.0 and 1.0, with a reasonable range of 0.1 to 0.9. The rules of thumb for ranges and percentiles characterize this information as a beta distribution with 40th and 60th percentiles of 0.1 and 0.9, respectively.

The experts then estimate the probability of the event given these values of T, P, and F. The probabilities associated with the ranges of the three conditions given follow:

Conditional Probability of E	Expert 1	Expert 2
$P(E\|T_{hi})$	0.1	0.01
$P(E\|T_{lo})$	0.01	0.001
$P(E\|P_{hi})$	0.01	0.001
$P(E\|P_{lo})$	0.001	0.0001
$P(E\|F_{hi})$	0.01	0.01
$P(E\|F_{lo})$	0.001	0.001

The provided ranges for each condition are used to estimate the 40th and 60th percentiles of a probability distribution. Again, a useful distribution type for probabilities is the beta distribution. If these values are used to form betas, the following beta distributions result:

Expert 1

	Beta Parameters x_0	Beta Parameters n_0	Mean
$E\|T$	0.17	0.83	0.21
$E\|P$	0.18	4.32	0.04
$E\|F$	0.18	4.32	0.04

Expert 2

	Beta Parameters x_0	Beta Parameters n_0	Mean
$E\|T$	0.17	0.83	0.21
$E\|P$	0.18	4.32	0.04
$E\|F$	0.18	4.32	0.04
T, P, F	0.11	0.22	0.50

Using Monte Carlo simulation of beta distributions for the above equation for $P(E)$, the analyst finds a distribution for the probability of E for each expert. For expert 1, $P(T)$, $P(P)$, and $P(F)$ are constant values (0.01), and the conditional probabilities of E given T, P, and F are the beta distributions listed in the above table. For expert 2, the analyst determines $P(E)$ as the sum and products of the beta distributions listed in the above table. Those empirical distributions have the following characteristics:

	Expert 1	Expert 2
Mean	0.0028	0.045
Standard deviation	0.0031	0.094
1st percentile	0.00001	0.00016
5th percentile	0.00006	0.00080
10th percentile	0.00013	0.0016
20th percentile	0.00025	0.0032
25th percentile	0.00034	0.0040
50th percentile	0.0016	0.0080
75th percentile	0.0042	0.043
80th percentile	0.0053	0.064
90th percentile	0.0082	0.14
95th percentile	0.0097	0.24
99th percentile	0.011	0.47

The empirical aggregation program in Appendix C combines the two experts' distributions. Equal weights pool the two distributions and give the following results:

Mean	0.030
Standard deviation	0.060
1st percentile	0.00037
5th percentile	0.0010
10th percentile	0.0017
20th percentile	0.0028
30th percentile	0.0038
40th percentile	0.0049
50th percentile	0.0073
60th percentile	0.013
70th percentile	0.020
80th percentile	0.036
90th percentile	0.078
95th percentile	0.14
99th percentile	0.33

■

As indicated in example 16.11, the decomposition, aggregation problem is a multistep, complex procedure. First, the experts must provide much information. That information is then formulated into distributions. This step requires some assumptions. The next step requires combining the distributions for each expert. The final step forms an aggregation of all the experts' combined distributions. All these steps involve assumptions, information transfer, and quantification. These steps also involve all the problems associated with imposing new information not originally present and changing granularity.

Granularity can change several times in this series of steps, making the interpretation of the final results meaningless. The mean value and its standard deviation might be meaningful, as in example 16.11; however, this example is a relatively simple one. Most problems usually have more than three simple conditions and more than two experts. Recombining such decomposed information requires more and more assumptions and makes final inference more and more difficult.

In spite of these obstacles, information is gained that is useful for the inference process. For instance, note in this example that the experts agreed on the basic conditions affecting the event. This agreement indicates that they are solving the same problem in a similar way. They disagreed as to how these conditions affected the event on a more detailed level, however, which led to a disagreement in their estimates. Such numerical differences can be interpreted by considering the wide ranges of values as representing the true uncertainty in estimating the event.

Application Environments

The problem or application environment plays a crucial role in determining who makes the inferences: the experts, the decision maker, the analyst, or a combination of these. Choosing an aggregation method involves considering the application environment. Examples given below demonstrate the differences in methods for some application environments. The specific cases examined follow:

1. One expert and one decision maker (DM)
2. DM and several (n) experts
3. One analyst and n experts.

A DM usually participates in a project when he or she has some relevant information to the problem. The DM may be an expert of equal expertise, or he or she may have erroneous or dated information. The DM controls the influence that his or her information has over the aggregation process. However, a DM can change his or her views after seeing the information from the other experts. Because the DM has such influence, he or she does not function independently of the experts (Morris, 1977; Genest & Zidek, 1986).

On the other hand, the analyst should act as a neutral party—the analyst should be independent of the experts. He or she never imparts his or her own information or biases to any assumptions, definitions, or information transformations. This ideal is unrealistic; however, the steps and recommendations in this book are designed to minimize problem areas and approach the ideal.

Decision Maker and One Expert

In this situation, the DM has his or her own information (DM's prior) for solving a problem, but, in addition, he or she consults a chosen expert (usually one the DM feels has more information). That expert gives information (expert data) to the DM. The DM then combines the information, forming a posterior (Morris, 1977). The DM imparts information into the final posterior in several ways:

1. The DM's prior
2. The choice of the expert
3. The aggregation of the information
4. The inference.

As previously noted, Bayesian methods combine information from various sources to form the posterior (in this case the DM's prior and the expert's data) in such a way as to emphasize one source over another. The DM determines which source is emphasized in the aggregation (item 3 above). If the DM feels uncomfortable with his or her prior, he or she reduces its influence even to the point of using a noninformative prior that imparts little (but some) information into the aggregation process.

Another effect results from item 3. After examining the expert's information, the DM may revise his or her prior to either match or differ from the expert's information. This is also a part of the aggregation process because the DM decides on how his or her information should be combined with the expert's information. Example 16.12 illustrates the effects from this process.

EXAMPLE 16.12: ***Decision maker and one expert***

The decision maker has knowledge about an event. He or she has never seen or heard of a particular component failing in 10 plants in a combined 60 years of operations. The DM asks his or her favorite plant operator to estimate a failure rate for this component. The expert estimates a possibility of one failure in 60 operation-years or one failure in 120 operation-years. Thus, the expert gave a range of values from 1/60–1/120 failures/operation-years. Using a Bayesian context for this problem, the DM can aggregate in many different ways.

I. The DM considers his or her information as valid data and combines it with the expert's information, using the expert's information as his or her prior. The DM uses the binomial process to characterize his or her information (0 failures in 60 years) and uses the expert's prior information as a beta distribution with 40th and 60th percentiles of 0.008 and 0.017, respectively.

The DM finds that the mean of the expert's beta is 0.0205, with parameters $x_0 =$ 0.677 and $n_0 =$ 33.096. The posterior distribution resulting from the Bayesian combination of the DM's binomial and expert's beta prior is also a beta with mean equal to

$$x'/n' = (x_0 + x)/(n_0 + n) = (0.677 + 0)/(33.096 + 60) = 0.677/93.096 = 0.0073,$$

or a mean value of 1 failure in 137 operation-years and a variance equal to

$$(x_0 + x)[(n_0 + n) - (x_0+x)]/\{(n_0 + n)^2[(n_0 + n)+ 1]\}= 0.000077.$$

Chapter 17's *Bayesian Methods* gives a detailed discussion of these aggregation estimates.

II. The DM believes his or her information is incorrect and uses a noninformative prior to characterise this belief. The expert's information can be characterized as the data from a binomial process with 0.677 failures in 33.096 trials. With a noninformative prior, the resulting posterior distribution is a beta distribution with mean equal to

$$x'/n' = (x_0 + x)/(n_0 + n) = (0.677 + 0.5)/(33.096 + 1) = 1.177/34.096 = 0.035,$$

or a mean value of 1 failure in 29 operation-years and a variance equal to

$$(x_0 + x)([(n_0 + n) - (x_0 + x)]/\{(n_0 + n)^2[(n_0 + n) + 1]\} = 0.00095.$$

III. The DM believes the expert acts too cautiously and decides to use a noninformative prior for the expert. The result relies more on the DM's information than on the expert's information. The DM claimed 0 failures in 60 trials. With a noninformative prior on the expert, the resulting posterior is a beta with mean equal to

$$x'/n' = (x_0 + x)/(n_0 + n) = (0.0 + 0.5)/(60 + 1.0) = 0.5/61.0 = 0.0082,$$

or a mean value of 1 failure in 122 operation-years and a variance equal to

$$(x_0 + x)([(n_0 + n) - (x_0 + x)]/\{(n_0 + n)^2[(n_0 + n) + 1]\} = 0.00013.$$

IV. After observing the expert provide a range of values for an estimate, the DM realizes that he or she should also provide a range. The DM likes the expert's evaluation of 1 failure in 60 years and decides to use that for an upper bound. The DM now has a beta distribution with 40th and 60th percentiles as 0.0 and 0.017, respectively. Because the beta distribution has a minimum value of 0.0, the 40th percentile represents an extremely small value (relative to the other estimates) such as 0.0000000001 instead of exactly 0.0. By doing so, the expert's beta distribution and the DM's beta distribution are pooled into a final distribution with the following mean:

$$x'/n' = (x_0 + x)/(n_0 + n) = (0.021 + 0.677)/(0.062 + 33.096) = 0.698/33.158 = 0.021,$$

or a mean value of 1 failure in 48 operation-years and a variance equal to

$$(x_0 + x)([(n_0 + n) - (x_0 + x)]/\{(n_0 + n)^2[(n_0 + n) + 1]\} = 0.00060.$$

In these four cases, the DM uses the expert's information in different ways. The different ways produce different final mean and variance estimates.

DM binomial and expert beta prior	0.0073	0.00008
DM noninformative and expert beta prior	0.035	0.00095
DM binomial and expert noninformative	0.0082	0.00013
DM beta and expert beta	0.021	0.00060

Note that the lower estimates occur when the DM uses a single estimate. Higher estimates occur when the DM's information is formulated into a distribution, either noninformative or beta. In this case, the DM's choice on how to handle his or her own data drives the final results.

■

The DM should play an important role in the inference process regardless of how many experts provide information. In example 16.12, the illustrations concentrated on the answers only. The DM has access and needs to use all the ancillary information about the expert gathered at the elicitation.

1. Is the expert solving the correct problem?
2. Are the expert's assumptions, cues, definitions, and problem-solving methods reasonable and in agreement with the DM?
3. Can the DM spot any relevant effects that this ancillary information might have on the expert's given answer?

Because the DM is also a knowledgeable party, he or she can answer these questions. This simple exercise ensures that the DM will monitor conditionality. Using what the DM has learned by answering these questions, he or she can make any adjustments or combine his or her information with the expert's in different ways.

Decision Maker and *n* Experts

The aggregation becomes more complicated when a DM combines his or her information with that of several other experts. The DM must immediately make a choice.

Choice 1: The DM can decide to aggregate all the experts into a single distribution or estimate (accompanied by a variance or uncertainty estimate) and then combine that result with his or her information. The following equation represents this choice, where the DM's distribution is f_0 and the combined distribution of the experts is f_c:

$$f_{final}(x) = K \bullet f_0(x) \bullet f_c(x) \qquad \text{(Morris, 1977)},$$

where K is a normalizing constant necessary to preserve the property that the area under a pdf is equal to 1.0.

Choice 2: The DM can decide to aggregate his or her information with the experts as though he or she was just another expert. The following equation represents this choice where the expert distributions are f_i for i=1, n:

$$f_{final}(x) = K \bullet f_0(x) \bullet f_1(x) \bullet \ldots \bullet f_n(x) \qquad \text{(Morris, 1977)},$$

or

$$f_{final}(x) = K \bullet w_0 \bullet f_0(x) + \sum_{i=1}^{n} w_i \, f_i(x) \qquad \text{(Winkler, 1968)} .$$

The second choice applies the aggregation techniques provided at the beginning of the chapter. However, the first case requires some modifications of these techniques. Specifically, the process is to first determine a composite distribution of the experts and then combine that distribution with the DM.

The DM's behavior is not necessarily the same in the one expert case versus n experts. The DM will more likely change his or her own views (information) based upon information from several experts than information from one expert. The exception to this would be if the DM is extremely dogmatic in personality. In this case, the DM will not reduce the influence of his or her information or change his or her information in view of the other expert or experts.

The methods for applying the above combinations are modifications of the methods already mentioned in the *Aggregation Distributions* section. Examples 16.13 through 16.15 illustrate these methods. The examples include using Bayesian methods (example 16.13), assumed distributions such as the normal (example 16.14), and empirical distributions with simulation (example 16.15). Weight determinations for combining distributions use the same methods as discussed in the weight determination section. These determinations also suffer from the same problems mentioned in that section, and the case for using equal weights is also applicable (Seaver, 1978). Example 16.15 also illustrates weight determinations and results.

EXAMPLE 16.13: *Decision maker and n experts: Bayesian aggregation*

A DM obtains the information elicited from the six experts in example 16.8. The information was formulated into six beta distributions, one for each expert. The parameters of these distributions follow (Winkler, 1968):

	x_0	n_0
$f_1(x)$	0.82	242.59
$f_2(x)$	0.19	11.04
$f_3(x)$	0.18	1.52
$f_4(x)$	0.41	41.80
$f_5(x)$	0.41	8.64
$f_6(x)$	0.65	15.30

The combination distribution for the six betas is also a beta with parameters equal to x' and n', where

$$x' = \sum_{i=1}^{n} w_i x_{0i}$$

and

$$n' = \sum_{i=1}^{n} w_i n_{0i} \ .$$

The DM reviews the ancillary information and analysis of this data and agrees with the analyst that no special conditions or circumstances make any expert's estimate different from any other. All experts use reasonable assumptions, definitions, cues, and problem-solving processes to solve the same problem. Both the DM and analyst agree that unequal weights are not appropriate. With equal weights, the values for x' and n' are

$$x' = 2.66$$

and

$$n' = 320.89.$$

The DM then combines his or her information with the other experts. The DM estimates the occurrence of the event as 1 in 100 trials. Because the DM agrees with the experts' problem-solving methods, he or she simply adds their estimate to the others, making a posterior beta with the following parameters:

$$x'' = x' + x_{DM} = 2.66 + 1 = 3.66,$$

and

$$n'' = n' + n_{DM} = 320.89 + 100 = 420.89.$$

Therefore, the aggregation result is a *beta* distribution with a mean and variance of

$$3.66/420.89 = 0.0087$$

and

$$[3.66(420.89\text{-}3.66)]/[420.89^2(421.89)] = 0.000020.$$

■

EXAMPLE 16.14: ***Decision maker and n experts: normal aggregation***

Using the three experts from example 16.9, the DM wants to combine his or her estimates of a mean value of 65 and a variance of 25 into the resulting posterior normal distribution formed from the experts. The experts' distribution, $f_c(x)$, is a normal with mean m_{post} = 62.02 and variance σ^2_{post} = 22.83 (Winkler, 1981). Adding the DM's normal distribution as the prior results in the following *normal* posterior with mean and variance parameters

$$m'_{post} = (m_{DM}/\sigma^2_{DM} + m_{post}/\sigma^2_{post})/(1/\sigma^2_{DM} + 1/\sigma^2_{post})$$

$$= (65/25.00 + 62.02/22.83)/(1/25 + 1/22.83) = 63.44$$

and

$$\sigma^{2\prime}_{post} = 1/(1/\sigma^2_{DM} + 1/\sigma^2_{post}) = (1/25 + 1/22.83) = 11.93.$$

■

EXAMPLE 16.15: ***Decision maker and n experts: empirical aggregation with Saaty-based weights***

The two experts in example 16.10 give a DM their empirical distribution functions. However, the analysts inform the DM that certain conditions were highly significant in the experts' determining their answers. These conditions were a set of five cues used by the experts. The cues came from a set of descriptions listed in the problem statement that the experts used to solve the problem.

To use these conditions in formulating weights for the experts, the DM arranged the cues into a hierarchical structure. The DM then evaluated the usage of the cues for each using Saaty's pairwise comparison method. This method allowed the DM to formulate a set of relative weights for the two experts based on their usage of the cues.

First, the DM evaluates the importance of the cues (C1–C5) relative to the problem being solved. His or her pairwise comparisons of the cues follow:

C1 vs C2	w
C1 vs C3	s
C1 vs C4	b
C1 vs C5	b
C2 vs C3	s
C2 vs C4	b
C2 vs C5	b
C3 vs C4	b
C3 vs C5	b
C4 vs C5	s

The b = important (Saaty weight = 2.72, the natural log base, e), s = neutral (Saaty weight = 1.00), and w = detrimental (Saaty weight = 0.37, 1/e). The resulting relative weights for C1 through C5 are

(0.22, 0.33, 0.26, 0.09, 0.10).

The DM then evaluates the experts, given each of the five cues:

Expert 1 vs 2, given C1	b
Expert 1 vs 2, given C2	b
Expert 1 vs 2, given C3	b
Expert 1 vs 2, given C4	w
Expert 1 vs 2, given C5	w

The following relative weights result for the experts, given each cue:

(0.73, 0.27) for C1
(0.73, 0.27) for C2
(0.73, 0.27) for C3
(0.27, 0.73) for C4
(0.27, 0.73) for C5

To calculate the final relative weights for the two experts, multiply each cue weight by the expert weight for that cue, then sum these products over all cues:

$$\text{Expert 1 weight} = 0.73(0.22) + 0.73(0.33) + 0.73(0.26) + 0.27(0.09) + 0.27(0.10) = 0.64,$$

and

$$\text{Expert 2 weight} = 0.27(0.22) + 0.27(0.33) + 0.27(0.26) + 0.73(0.09) + 0.73(0.10) = 0.36.$$

Therefore, the relative weights of the experts are 0.64 and 0.36. These are relative weights and should not be taken at their numerical values. (This is a granularity issue.) From this analysis the DM can see how the two experts rate relative to each other regarding the important cues in the problem solving. The DM can then assign numerical values based on this relative assessment. The DM decides to use weights of 0.7 and 0.3 for the two experts, indicating that he or she feels that the relative weights closely represent true weights for the experts.

The DM now wishes to combine the empirical distribution functions of the two experts using the following aggregation function:

$$f_c(x) = 0.7f_1(x) + 0.3f_2(x).$$

The resulting distribution characteristics for $f_c(x)$ follow:

Median	0.25
30th	0.20
70th	0.30
5th	0.05
95th	0.45

The DM also wishes to combine his or her information (empirical distribution function) with the experts' information. The DM uses equal weights for his or her distribution, f_0 and f_c:

$$f(x) = 0.5f_0(x) + 0.5f_c(x) = 0.5f_0(x) + 0.35f_1(x) + 0.15f_2(x).$$

Using the empirical code in Appendix C, the following aggregation distribution (f) results:

Percentile Level	Value
1st	0.08
5th	0.12
10th	0.14
20th	0.18
30th	0.20
40th	0.22
50th	0.25
60th	0.27
70th	0.29
80th	0.32
90th	0.37
95th	0.43
99th	0.58

Mean = 0.26
Variance = 0.0097
Standard deviation = 0.098

This aggregation is not very different from the original two experts (i.e., equally weighted) in example 16.10. One reason for the similarity is that the DM gave estimates similar to both experts. Another reason is that even though different weights were applied, the weights did not have a significant impact.

■

Analyst and *n* Experts

This section intends to help the analysts understand their role in a project, especially if no DM is available. In the application environments involving a DM, the text and examples mentioned that the analyst supplies the DM with vital information and results in addition to the answers from the experts. The analyst must also help the DM synthesize and assimilate this information. To do this, the analyst should make the DM aware of the concepts of granularity, conditionality, and quantification; in addition, the analyst also should guide the DM in the aggregation process.

The DM may act like just another expert and rely on the analyst to take the DM's information, do the analyses, and report back to the DM. In some cases, the DM is not an expert at all. In this case, the analyst is faced with a similar situation: do the analysis without the DM's added information and report the findings.

In either case, the analyst alone is responsible for the information (data). The analyst does his or her job and reports his or her findings.

This application environment has advantages. First, the analyst is the only source of influence on the information elicited. He or she is familiar with the cautions and recommendations designed to minimize his or her influence on the data. Second, the analyst either elicited the data or worked closely with the elicitor. Therefore, the analyst is familiar with the data and helps with the development of the elicitation. Third, the analyst functions as an objective bystander, showing no concern over implications from the results and conclusions of the study. The analyst can be free to *let the data speak* and to make inferences accordingly.

The analyst may find it extremely difficult in the reverse environment, in which the DM takes the analyzed results and changes them without the benefit of the analyst or the guidelines presented in this book. Careful analysis of the information can be quickly destroyed in such a case.

Part III focuses on the analyst's viewpoint. Most of the examples given in this book reflect the analyst and *n* experts environment. However, we also recommend that the analyst play an integral part in the elicitation. Facilitators or moderators of the elicitation and the analyzers of the data must work together. If they do not, then definitions, assumptions, problem solving, and conditions can change during the study. In addition, granularity can change without notice, making interpretations meaningless. Common sense and simple consistency in designing and implementing the study from start to finish are the keys to success.

Aggregation and Uncertainty Analysis

Chapter 17 addresses uncertainty analysis in more detail. However, many of the problems associated with aggregation are related to uncertainty characterizations. In fact, many of the techniques are useful for both, such as simulation methods, Bayesian methods, tree diagrams for decomposition, and the use of distributions rather than single estimates.

Both this chapter and chapter 17 use the same methods to characterize uncertainties and to aggregate estimates and distributions. The difference is one of interpretation. This chapter does not specifically label the ranges and distributions provided by the experts as characterizing uncertainties, although they did represent uncertainty. The primary goal was to aggregate all the information into either a convenient estimate with a variance or into a distribution. The interpretation of those final estimates and distributions incorporate and reflect the uncertainties in the information given by the experts and also the uncertainties among the experts. Chapter 17 emphasizes identifying and characterizing the uncertainty.

17

Characterizing Uncertainties

Many methods can help characterize and handle uncertainties, even in the restrictive environment of expert judgment applications. This chapter examines some of the easier ways. This book's concept and definition of uncertainty cover a very broad sense, including the four basic sources of uncertainty that prevail throughout the data and the analysis.

This chapter suggests the following steps to control, or at least understand, this total uncertainty characterization. First, the section *Obtaining Uncertainty Measures* discusses which uncertainty measures the answer data require. Second, the section *Modeling Uncertainties* discusses how to model uncertainties in the data. This chapter concludes in *Comparison of the Methods* with a comparison of various methods for handling uncertainties. Chapter 18 discusses the relationship between uncertainty and the inference process.

Living with Uncertainties

Different kinds of uncertainties become important in any sampling or experimental data-gathering process. Some uncertainties can be controlled and, therefore, reduced to an acceptable noise level of influence simply by taking larger samples or by doing careful experimental design and measurements. Other uncertainties cannot be controlled or reduced by any practical means. We must learn to accept these uncertainties. The most we can do is understand their importance and effects.

Uncertainties can stem from different sources, definitions, sampling errors, nonsampling errors (e.g., missing data), and scientific or modeling techniques (Stoto, 1988). All four sources create uncertainties in expert judgment applications. Careful elicitation can help reduce the uncertainty by following the procedures in Part II. Large sample sizes can reduce sampling error uncertainty; however, this method may not always be practical in expert judgment applications. No simple or practical means can reduce nonsampling uncertainty. Proper experimental design (where possible) and cross validation in the analyses can reduce modeling uncertainty.

Obtaining Uncertainty Measures

Analysts can obtain uncertainty values for the experts' answers by either directly eliciting uncertainties from the experts or by indirectly inferring uncertainties from the post-elicitation data. In either case, the expert or the analyst may have to make additional assumptions or provide estimates that stretch the limits of their current knowledge. However, the purposes of estimating uncertainties are to represent the possible inaccurate estimation by the experts of the variables of interest and to increase the chance of estimating (covering) the true answer by allowing for a range of possible values rather than relying on a single value.

Using Elicitation

On the other hand, asking experts to estimate uncertainty measures for an answer is like asking them to estimate a variance or a distribution of values. As noted in chapter 7, these estimations are difficult and may be highly inaccurate for most experts. On the other hand, asking for a simple range of possible values from an expert is not much different from asking for the original answer. Eliciting a range of values requires the same care as eliciting a single answer. One advantage of eliciting a range of values is that many experts are comfortable with providing uncertainties in this form because they realize the existence and importance of uncertainties. In fact, many experts prefer to give a range of possible values because they feel uncomfortable with the pinpoint accuracy implied by a single value estimate.

Chapter 7 offered several dispersion measures including error bars, variances, percentiles, and ranges. All these can characterize uncertainties in the answers and can be elicited from the experts along with the answers. The advantages and disadvantages for each follow.

Error Bars

Most engineers use the term *error bars* to connote some measure of uncertainty; however, they do not agree on how much uncertainty is characterized by error bars. Therefore, establish some sort of definition before eliciting error bars from experts. Most definitions will overlap with the other uncertainty measures. For example, error bars could be plus or minus one standard deviation from the best estimate (a standard deviation definition), or error bars could be the middle 90% (percentiles definition) of the distribution.

The philosophy of this book is to keep restrictions on the expert to a minimum. By doing so, the expert would be asked to provide error bar values and then provide his or her definition of what those values represent. We recommend this use for error bars.

Variances or Standard Deviations

Most engineers have heard and used the terms variance or standard deviation, but they may not have a good understanding of them. We also do not recommend that the experts be asked if they feel comfortable with these concepts. In general, most people do not like to admit a lack of knowledge or understanding of any concept. The experts can be trained in these concepts during the elicitation. However, studies indicate (Martz, Bryson & Waller, 1985) that even experts with expertise on these concepts fail to accurately estimate variances. In general, variances are underestimated in value, sometimes by a factor of two or more. Therefore, using variances and standard deviations will result in large underestimates of uncertainty.

Because many experts are not familiar with these concepts, we cannot recommend their use as uncertainty measures.

Percentiles

Percentile estimation involves a probability distribution of values. The 95th percentile is the value such that 5% of the distribution is larger and 95% is smaller than that value. As with variances, studies have demonstrated that even statisticians have difficulty in estimating percentiles. People tend to underestimate uncertainty and thus tend to incorrectly estimate percentiles. When asked to estimate 95th percentile values, people only estimate about the 60-70th percentile values, and 5th percentile values really only represent about the 30-40th percentile values. Another problem is inherent in this process. Even if the expert feels comfortable with the concept of a distribution, he or she will tend to think in terms of a symmetric, bell-shaped distribution. Such an assumed distribution may be totally inappropriate for the problem. The result is a distortion of the values and percentiles that the expert is attempting to estimate.

Again, because of the difficulty in defining and using the concept of percentiles, we cannot recommend their use as uncertainty measures.

Ranges

Many experts prefer to give a range of possible values instead of a single point estimate. This preference reflects their uncertainty in providing a single value. Many experts will give a range of values whether a range is elicited or not. Interpreting the given range presents a problem because experts usually are unwilling (or unable) to provide a definition of what the range represents. These definitions might also involve other uncertainty characterizations (e.g., variance) that should be avoided.

We recommend that definitions of ranges not be provided in terms of the other uncertainty measures. We also recommend that some rationale be gathered concerning what the expert has in mind about the ranges: Are they equally possible values? Do they represent extreme values? The analyst will find it difficult to use ranges that possess no meaning; however, the next section demonstrates how undefined ranges can be a source of information for analysis.

Using Post-Elicited Data

This section provides some suggestions on obtaining uncertainty measures about the data and on applying such information. Uncertainty characterizations can come from several sources: (1) directly from the experts, (2) indirectly from the experts, (3) assumed by the analyst from the data, or (4) any combination of these three. In any of these four sources, the experts, the analyst, or both make assumptions. Therefore, view the following suggestions with extreme caution because of these assumptions. If possible, try and compare different interpretations of the uncertainties. The section *Modeling Uncertainties* indicates how these uncertainty measures can be analyzed. The last section of this chapter provides a final comparison of the methods.

1. As mentioned above, many experts will volunteer ranges of values when answering a question. Make sure to query them about the meaning or interpretation that they attach to these ranges. Otherwise, the analyst is forced to assume some interpretation such as the 40th and 60th percentile values. Volunteered range values can be used in many ways: as uncertainty analyses calculations as described below, and as repeated measures for determining variations or parameters for assumed distributions, and as repeated measures for supplementing a sparse or small data set. For example, if 5 experts give 5 best guesses and 5 minimum values with 5 maximum values, then the data set has a total of 15 values.
2. Experts may also supply ranges indirectly. This can be done in several ways.

   First, the expert may give his best estimate and, later in the session, revise this estimate or give another possible estimate. The rationale for changes in values and assumptions (if changed) can be noted and recorded. If an expert changes his or her estimate, it is vital to find out why. In some cases, he or she may be updating his or her thought processes; in other cases, he or she may be changing the estimate because the problem has changed. If he or she changes the problem, then the new estimate cannot be used as a range value.

   Second, the expert may have recorded his or her response in such a way as to indicate a range of values. For example, if a continuous number scale is the response mode, an expert may use a wide mark or smear his or her response along the line, indicating a spread of values. This may only indicate a narrow range, but it is a useful range nonetheless. This kind of volunteered range value is best considered as a very narrow uncertainty measure; however, the analyst can use it as repeated measurements to increase the sample size.
3. The analyst can make many assumptions about either the direct or indirect ranges volunteered by the expert. The analyst can assume these ranges represent percentile values, fractions of standard deviations, or multiples of standard deviations. A common rule of thumb (Kahneman, Slovic & Tversky, 1982) is to take the experts' uncertainty range and double or quadruple it to make a 90% coverage interval (the difference between the 95th and 5th percentiles). In determining the validity of such a rule, the analyst can compare and study the effects of different uncertainty measures by using simulation.

   Analysts can also use uncertainty measures to increase the sample size by using the direct or indirect ranges as additional estimates. This method is especially useful for applications which have only a few available experts.

Again, a simulation technique (e.g., the bootstrap) is useful for using the range values to increase sample size.

The analyst can also make assumptions when the experts do not directly supply ranges. Such possible assumptions come from examining the variation in the given answers. For example, induce a range of values by doubling the variance of the original set of single estimates and assume a distributional form for the estimates with this doubled variance and original mean. Doubling the original range (maximum value–minimum value) of the data can also produce a similar uncertainty measure. To assist the analyst in making such determinations, the experts may provide some verbal clues about the uncertainties in their estimates. The analyst must transform any such qualitative statements into quantitative values of uncertainties. Therefore, any and all such information the experts provide should be viewed as part of the uncertainty characterization.

Modeling Uncertainties

After using the above methods to determine some uncertainty measures for the experts' estimates, uncertainties can be modeled by different methods. A few more commonly used techniques are described in this section, and a comparison of these techniques is given at the end of this section for the application described in the examples.

Bayesian Methods

Bayesian methods use probability distributions to characterize uncertainties. Some cases represent uncertainty with a single prior distribution. Other cases represent uncertainty by multiple priors, each representing different sources of uncertainty. These two cases follow.

Using One Prior

Many risk and reliability problems use expert judgment to supplement existing (but usually small amounts of) data. The expert judgment information formulates a prior distribution that is combined with the sample data and its distribution by applying Bayes theorem to form a posterior distribution that reflects a composite of all the available information. In this type of application, the expert judgment information serves two purposes: it augments the data, and it serves to characterize the uncertainty in the data.

The major disadvantage of this approach is that probability distribution forms are needed for both the expert judgment data (as the prior) and for the existing data. Therefore, the process of combining these distributions can be mathematically difficult as well. Some difficult cases require numerical analysis techniques to find a solution. Other difficult cases use simulation to find a solution. To avoid mathematical difficulties in calculating a posterior result, many analysts assume commonly used distributions that have convenient mathematical properties that provide easy forms for the posterior distribution. One such commonly used combination of the prior information with the data is the binomial-beta combination. In this case, the existing data are assumed to follow a binomial process—x failures occur in n independent trials, with each trial having a probability of failure equal to p. The prior information on p (usually from the experts) also takes the form of x_0 failures in n_0 trials. However, the latter case's information forms a beta prior distribution. The beta is convenient for three reasons: the beta distribution can have many different shapes, the beta ranges from 0 to 1 just as p does, and the beta parameters are interpreted as x_0 failures in n_0 trials. The resulting posterior of this binomial-beta combination is also a beta distribution. As shown in chapter 16, the posterior's mean and variance are easily calculated functions of n, x, n_0, and x_0 values as follows:

$$\textit{posterior mean} = \frac{x + x_0}{n + n_0},$$

and

$$\textit{posterior variance} = \frac{(x + x_0)(n + n_0 - x - x_0)}{(n + n_0)^2 (n + n_0 + 1)}.$$

Two special cases of the beta prior also combine easily with the binomial process and are commonly used in situations in which little or no prior information is available. These are the uniform distribution on the (0,1) interval and the noninformative prior $k(p - p^2)^{-1/2}$, where k is any constant. By choosing k in terms of the gamma functions (Martz & Waller, 1982), this noninformative prior becomes a beta distribution, with $x_0 = 0.5$ and $n_0 = 1.0$. The posterior distribution is then also a beta distribution function with the above mean and variance formulas. The uniform distribution is a special case of the beta distribution, with parameters $x_0 = 1.0$ and $n_0 = 2.0$. Again, the resulting posterior is a beta distribution with mean and variance formulas given above. Thus, for either the uniform or noninformative priors the above formulas can be used to determine the posterior mean and variance. One word of caution when using either of these priors: the uniform prior is used when no prior information is available and is sometimes called the ignorance prior. This prior spreads the information evenly across the entire range of values, from 0.0 to 1.0. It has almost a negligible impact on the posterior. On the other hand, the noninformative prior is really a misnomer. This prior is informative and does have an impact on the posterior. Its name should be the prior of little information.

Choosing the distributional form and corresponding parameters for the prior distribution can make a substantial difference in the posterior. Using the binomial-beta combination to form the posterior, example 17.1 shows the influence that the prior parameters can have on the final results.

EXAMPLE 17.1: ***Using Bayesian methods for uncertainty—forming a single prior***

Ten experts provide estimates for the probability of failure per year of a subsystem in a reactor as follows:

Expert	Estimate of Failure Rate
1	0.00250
2	0.00100
3	0.05000
4	0.00500
5	0.01000
6	0.02500
7	0.00100
8	0.00250
9	0.00010
10	0.00005

This subsystem has been operational for 12 years without any failures. This data follows a binomial process, with $x = 0$ failures in $n = 12$ years. The 10 experts' estimates can form a prior distribution that is combined with this binomial data. The major advantage in this case is that the data alone do not contain enough information to form a failure-rate estimate. Just using the data would give a failure-rate value of 0/12 = 0. Combining the data with the experts' information yields a more reasonable estimate.

The variation in the 10 experts' estimates represents the uncertainty in the value of the failure rate. The distribution formed out of these 10 values represents the uncertainty. The mean of the 10 values is 0.010; the standard deviation is 0.016. A beta distribution with that mean and standard deviation has parameters $x_0 = 0.40$ and $n_0 = 39.5$. These parameters represent an average failure rate of 0.4 failures in 39.5 years or about 1 failure in 98 years. A beta distribution with the above parameters x_0 and n_0 has the following characteristics:

Mean	0.0102
Variance	0.00025
Minimum	0.00000023
5th percentile	0.000012
50th percentile	0.0038
95th percentile	0.042
Maximum	0.074

Combining the expert information (beta prior distribution) with the data (binomial process) gives the mean and variance of the posterior beta distribution as

$$\frac{0.4+0}{39.5+12} = 0.0078$$

and

$$\frac{(0.4+0)[(39.5+12)-(0.4+0)]}{(39.5+12)^2[(39.5+12)+1]} = 0.00015.$$

Examining the changes in the data demonstrates the influence of the data and the prior information on the final estimate. Suppose 1 failure occurred in 12 years, then the final mean estimate becomes 0.027 with a variance of 0.00050. In this case, the data influences the increase in the mean estimate. Suppose the data was 0 failures in 144 years. The final estimates of the mean and variance are 0.0076 and 0.000041, respectively. In this case, the final mean is more influenced by the experts.

Of course, changes in the experts' estimates affect the posterior in a similar fashion. The purpose of this exercise is to emphasize that both the prior and the data are influential and that care must be taken to represent each appropriately.

■

Assuming distribution forms and determining posteriors can be arbitrary and difficult. Therefore, we recommend that Bayesian methods be used only in the simple binomial-beta cases such as in example 17.1. Consult a statistician for the other cases.

Using Multiple Priors

To characterize the uncertainty in each expert's estimate (answer), Bayesian methods establish a prior distribution for each expert. Each expert provides an estimate or best guess and a corresponding range of values for that estimate. These ranges represent the uncertainties that the experts have about the accuracy of their single-point estimates. Using ranges and estimates can help form distributions representing the estimates and their uncertainties for each expert. Example 17.2 discusses some of the ways of forming these distributions and shows one method in detail.

EXAMPLE 17.2: ***Using Bayesian methods for uncertainty—forming multiple priors***

The 10 experts in example 17.1 provided uncertainty ranges with their estimates as follows:

Expert	Estimate of Failure	Rate Ranges
1	0.00250	0.00100 - 0.0040
2	0.00100	0.00010 - 0.0100
3	0.05000	0.00100 - 0.1000
4	0.00500	0.00100 - 0.0100
5	0.01000	0.00500 - 0.0500
6	0.02500	0.01000 - 0.0500
7	0.00100	0.00500 - 0.0025
8	0.00250	0.00100 - 0.0050
9	0.00010	0.00010 - 0.0100
10	0.00005	0.00005 - 0.0005

The analyst can use this information in several ways for formulating distributions for each expert. The first decision involves determining what type of distribution to use. For convenience, the beta distribution is chosen. Several ways to form the beta distribution follow:

1. The single estimates represent the mean of the beta. The lower range values represent specified percentiles. Expert 1's resulting beta distribution would have parameters x_0 = 1.90 and n_0 = 761.9, if his or her lower range was the 20th percentile. This beta distribution has a mean of 0.0025 and a variance of 0.0000033.
2. The single estimates represent the mean of the beta. The upper range values represent specified percentiles. Expert 1's resulting beta distribution would have parameters x_0 = 1.08 and n_0 = 430.8, if his or her upper range was the 80th percentile. This beta distribution has a mean of 0.0025 and a variance of 0.0000058.
3. The single estimates represent the median of the beta. The lower range values represent specified percentiles. Expert 1's resulting beta distribution would have parameters x_0 = 0.64 and n_0 = 142.5, if his or her lower range was the 30th percentile. This beta distribution has a mean of 0.0045 and a variance of 0.000031.
4. The single estimates represent the median of the beta. The upper range values represent specified percentiles. Expert 1's resulting beta distribution would have parameters x_0 = 1.32 and n_0 = 401.8, if his or her upper range was the 70th percentile. This beta distribution has a mean of 0.0033 and a variance of 0.0000081.
5. The range values represent specified percentiles. The single estimate is not used. Expert 1's resulting beta distribution would have parameters x_0 = 0.31 and n_0 = 38.6, if his or her lower range was the 40th percentile and his or her upper range was the 60th percentile. This beta distribution has a mean of 0.0080 and a variance of 0.00020.
6. The range values represent specified percentiles such as the 40th and 60th percentiles. These percentiles form betas, acting as priors, and combine with

the information in the single estimate where the single estimates represent 1 in p failures such that $x = 1$ and $n = 1/p$. Expert 1's range values form a beta prior with parameters $x_0 = 0.31$ and $n_0 = 38.64$. If the single estimate represents a binomial process with $x = 1$ and $n = 400$, then a resulting beta for expert 1 has parameters $x_0 + x = 1.31$ and $n_0 + n = 438.64$. The resulting beta has a mean of 0.0030 and a variance of 0.00000014.

As seen in the example of using expert 1, the beta parameters and the variances can greatly change, depending on the interpretation of the uncertainty range values. However, the means remain fairly similar to the original single estimated value (except in item 5, where the mean is not used).

Using the method in item 5, the lower ranges are the 40th percentiles and the upper ranges are the 60th percentiles. With these percentiles, the following beta parameters emerge for each expert:

	Beta Distribution Parameters	
Expert	x_0	n_0
1	0.31	38.64
2	0.09	0.60
3	0.09	0.30
4	0.18	4.32
5	0.18	1.25
6	0.26	2.60
7	0.26	42.76
8	0.26	21.64
9	0.09	0.60
10	0.18	76.19

Distribution aggregation methods from chapter 16 (e.g., such as using Monte Carlo simulation) can combine these 10 beta distributions. The resulting combination distribution, now used as a prior, can combine with the data or with a decision maker's distribution to form a posterior. The next section on Monte Carlo simulation aggregates these distributions using simulation in example 17.3.

■

Simulation Methods

Chapter 11 introduces the bootstrap and Monte Carlo simulation techniques. Chapter 14 uses these techniques for exploring correlation among experts while chapter 16 applies these techniques for aggregation of expert estimates. Both techniques effectively characterize uncertainties.

Monte Carlo Simulation

One of the easiest and most successful methods (Martz *et al.*, 1983) of determining the effects of uncertainties through a model is to use Monte Carlo simulation.

Simulation can combine various types of distribution functions without relying on difficult or complex mathematical formulations of the combinations. Chapter 16 showed the mathematical difficulties encountered in aggregating expert estimates. Similar problems can also arise in characterizing uncertainties because distributions commonly represent them. One way to obtain an overall effect of all the uncertainties is to combine these uncertainty distributions with original estimates or with other uncertainty distributions.

Example 17.3 illustrates how to combine uncertainty characterizations for the 10 experts' estimates from example 17.2 to form a single distribution that represents the uncertainties in all 10 distributions. In example 17.2, each expert's range was used to formulate a beta distribution as the uncertainty distribution for his or her estimate. Example 17.2 gave several methods of formulating such beta distributions using the ranges and the original estimates. Example 17.3 illustrates the effect of choosing different distributional forms for the uncertainty distribution. First, the ranges establish the upper and lower limits of a uniform distribution. Second, the ranges also determine the parameters of a beta distribution. In both cases the 10 distributions are then aggregated to determine the distribution of the median (50th percentile value) of the 10 experts' estimates using Monte Carlo simulation. To initiate the simulation, randomly choose one value from each of the 10 expert distributions. Repeat this sampling process 1000 times, forming 1000 different samples. For each sample, calculate the median of the 10 values. The end result yields a distribution of 1000 medians. The variance, percentiles, and mean of this final distribution provide the estimates of the variance, percentiles, and mean for the median of the 10 experts.

EXAMPLE 17.3: ***Uncertainty characterization using Monte Carlo simulation***

Using the 10 experts' ranges given in example 17.2, the analyst determines individual uncertainty distributions for each expert. Combining these distributions forms a distribution for an overall estimate of the 10 experts. According to chapter 16 on aggregating expert judgment, the median of the experts is a commonly used aggregation estimator. Monte Carlo simulation allows the analyst to determine the distribution of the median.

This example uses the experts' ranges to form uncertainty distributions in two different ways: the lower and upper range values form the 40th and 60th percentiles of uniform distributions and the lower and upper range values form the 40th and 60th percentile values for beta distributions, as done in example 17.2. Beta distributions can have a variety of shapes including asymmetric shapes.

The analyst performs two simulations. In each simulation, the sampling process is the same: each sample is formed by randomly selecting a value from each expert's distribution. The 1000 median values calculated from the 1000 samples of size 10 form a final distribution for the median of the 10 experts. The first simulation uses uniform distributions to represent the experts' uncertainties in their estimates while the second uses beta distributions. The two resulting median distributions have the following characteristics:

Uniform Uncertainty Distributions for the Experts

Mean	0.0047
Variance	0.0000072
Minimum	0.0000013
5th percentile	0.00030
50th percentile	0.0038
95th percentile	0.012
Maximum	0.026

Beta Uncertainty Distributions for the Experts

Mean	0.0091
Variance	0.00024
Minimum	0.00000033
5th percentile	0.00038
50th percentile	0.0041
95th percentile	0.034
Maximum	0.14

The variance for the medians is much larger for the beta uncertainty distributions than for the uniforms. The maximum for the beta is a little larger, while the minimum for the beta is much larger. The different shapes of the beta and the uniform account for these differences. However, both medians are the same, and the means are also quite close. Thus, the choice (beta or uniform) of the uncertainty distribution form used for the experts does not affect the central measures of the final distributions for the median. The greatest effect from distributional choice is in the variances and the tail shapes of the resulting distributions.

■

Examples 17.2 and 17.3 identified two problems in translating the experts' range values into uncertainty distributions. The first problem (example 17.3) involves deciding on a form for the distributions. The second problem (example 17.2) involves deciding how to use the ranges to form the parameters of the chosen distributions.

Example 17.3 showed the effect of the first problem. The choice of the uncertainty distributions (beta or uniform) affected the variances and tails of the final distributions for the median. One way of solving this problem is to run several simulations, each using a different but reasonable distribution choice. Results from different choices should be consistent with each other, or inconsistencies should be resolved based on predetermined assumptions. For example, in example 17.3 the wider variance in the beta case is a result of the interpretation made about the experts' range values (the second problem). If the ranges had been used to represent the 5th and 95th percentiles, then the resulting variance from the beta case would have corresponded better to the uniform case.

Many types of logical choices for distributions can be tried and compared for various types of estimates. For example, the uniform and beta distributions are logical choices for probability estimates. The normal distribution is also commonly used; however, take care not to use a normal distribution that gives values for probabilities which are negative or greater than 1.0. The normal, lognormal, and gamma distributions are often used to represent estimates of physical quantities, such as temperatures or failure rates.

Another way to solve the first problem is to do the simulation using a technique that does not require a distributional assumption (e.g., the bootstrap). The bootstrap technique uses the data itself to form an empirical distribution for the simulation. Details on this technique follow.

Understanding that experts underestimate uncertainty can prove useful when solving the second problem. Because experts underestimate uncertainty, it is recommended that either the range be doubled for use as tail percentile values (e.g., 5th and 95th) or that the range values represent inner percentiles, such as the 30th to 40th and 60th to 70th percentiles. As in example 17.2, analysts should try different interpretations. The different interpretations will usually tend to affect the variances of the uncertainty distributions rather than the centers of the distributions. The basic idea is to represent the uncertainties in the quantity under estimation. If the experts feel that the uncertainty is great, then larger variances are to be expected. As a result, these variances may be larger than the experts expect.

Bootstrap Simulation

Sampling and resampling from the original data set can form distributions. This sampling and resampling process, referred to as sampling with replacement, forms the basis of bootstrap simulation (see chapter 11). To form a single sample of size n, randomly choose n values from the original data set. Values can be chosen once. Then form N such samples ($N = 1000$). As in the Monte Carlo simulation, a calculation or model is formed for each sample. The N results of this calculation or model are collected to form a final distribution.

The data require no distributional assumption for bootstrap sampling. However, the sampling/resampling procedure produces a final distribution with a small variance. In other words, the variation of the final distribution is limited to the variation of the original data set.

To overcome the restricted variance problem, induce more variation (more uncertainty) into the original data set. The ranges of values that represent the uncertainties in the values serve to expand the variation of the original data. Therefore, if the ranges are included in the original data set, the results from a bootstrap simulation will have a wider variation than the original sample without ranges. Example 17.4 illustrates the difference in the bootstrap final results when calculating the median of the 10 experts' estimates for cases *without* the range values and *with* the range values.

EXAMPLE 17.4: ***Uncertainty characterization using the bootstrap***

Using the data from the 10 experts from example 17.1, a bootstrap sampling procedure of those 10 estimates represents the uncertainty in the estimates. Samples are formed in a similar manner to the Monte Carlo simulation, except random selections are taken from the original data set and not from some specified distribution. For each sample, the random selection of a single datum is done either once, more than once, or not at all.

Forming 1000 such random samples from the original 10 estimates and calculating the median value of each sample forms a distribution of 1000 medians with the following characteristics:

Mean	0.0044
Variance	0.0000085
Minimum	0.0010
5th percentile	0.0018
50th percentile	0.0037
95th percentile	0.010
Maximum	0.027

The central values (mean and median) are similar to those in example 17.3. In this case, however, the variance lies between the two results in 17.3. Note that the example's smaller variance probably reflects the tendency of the experts to underestimate uncertainty. The bootstrap simulation also produces a small variance—a variance limited by the variation in the original data set. Therefore, the resulting variance of the median may also be too small to adequately represent the uncertainty.

To solve this underestimation of uncertainty, analysts can expand the variance of the original data set by including the range values given by the experts. One way of doing this is to supplement the data set by adding these upper and lower range values as if they were additional estimates. As a result, the original data set is increased to 30 values. However, when performing the bootstrap simulation, we recommend a sample size of 10 so that false benefits from an increased sample size are not induced into the simulation.

Results for the bootstrapped median with range values follow:

Mean	0.00083
Variance	0.00000089
Minimum	0.0000010
5th percentile	0.0000010
50th percentile	0.00055
95th percentile	0.0030
Maximum	0.0063

The final distribution for the median is stretched over a wider range of values than the final distribution without range values. However, because the range values are biased toward the lower values, this final distribution shifts in that direction. This shift may not be a desirable result. To avoid this shift, add range values that are symmetric about the original estimates.

■

In conclusion, both simulation techniques (Monte Carlo and bootstrap) have their advantages and disadvantages. Decision makers and analysts must exercise care in deciding which advantage is most desirable and which disadvantage can prove most harmful. In the case of uncertainty characterization, the Monte Carlo distribution assumption disadvantage is less harmful than the bootstrap restricted variance disadvantage because experts tend to underestimate uncertainties and more diversity is generally needed to adequately represent the true uncertainty. This is a conservative approach to uncertainty. However, unless more information about the size and effect of uncertainties surfaces, the risk and reliability communities will continue to use the conservative approach.

Decision Analytic Methods

The decision analysis community has adopted and developed many ways of characterizing uncertainties, including the Bayesian and simulation methods mentioned above (Booker & Bryson, 1985). Many of these techniques rely on strict model formulations. The following example shows how to implement one decision analytic technique: the maximum entropy technique.

The idea behind maximum entropy is to formulate a distribution for the data such that the distribution maximizes the uncertainty in the data. Determining this distribution requires several values; the distribution also requires a choice for the prior distribution on the variable of interest (e.g., probability of an event, x). At least two percentile values are needed, plus the absolute maximum and minimum values of x. The uniform and the noninformative prior are two commonly and easily used prior distributions (Cook & Unwin, 1986).

Using a uniform prior on x from the absolute minimum value a to the absolute maximum value b and two percentiles x_l and x_u, the maximum entropy distribution for x is

$$f(x) = \begin{cases} L(x_l - a) & a \le x < x_l \\ (U - L)/(x_u - x_l) & x_l \le x < x_u \\ (1 - U)/(b - x_u) & x_u < x \le b, \end{cases}$$

where U and L are the percentage values (e.g., 0.95 and 0.05) for the x_l and x_u percentiles.

Using a noninformative prior on p with two percentiles (a and b), the maximum entropy distribution for p is

$$f[\log(x)] = \begin{cases} L/\log(x_l/a) & \log(a) \le \log(x) < \log(x_l) \\ (U - L)/\log(x_u/x_l) & \log(x_l) \le \log(x) < \log(x_u) \\ (1 - U)/\log(b/x_u) & \log(x_u) < \log(x) \le \log(b), \end{cases}$$

where log is the base 10 logarithm.

These two distributions will be shaped as three blocks or steps. If other percentiles are easily estimated, the above formulas can be expanded and the distribution will have more steps:

$$f(x) = \begin{cases} L_1/(x_1 - a) & a \le x < x_1 \\ (L_2 - L_1)/(x_2 - x_1) & x_1 \le x < x_2 \\ (L_3 - L_2)/(x_3 - x_2) & x_2 < x \le x_3 \\ \cdot & \\ \cdot & \\ \cdot & \\ (1 - L_n)/(b - x_n) & x_n < x \le b, \end{cases}$$

for n percentile estimates ($x_1, x_2, \ldots, x_n$).

Example 17.5 illustrates the use of these formulas for the 10 experts' estimates from example 17.1. Prior distributions other than the uniform and noninformative can be used; however, forming the maximum entropy distribution for those can be mathematically difficult. For such complex cases, use simulation to find a solution.

EXAMPLE 17.5: ***Forming a maximum entropy distribution***

For the 10 experts in the previous examples, the smallest range value is 0.00005, and the largest is 0.1. These numbers for the 5th and 95th percentile values in the maximum entropy formulas form the uniform prior distribution. The problem also needs the values for the absolute minimum and maximum. These values are listed below with the percentiles:

Absolute minimum value, a =	0.000001
Absolute maximum value, b =	0.20
Number of specified percentiles =	2
Percentile levels, U =	0.95 and L = 0.05
Percentile values, x_u =	0.10 and x_l = 0.00005.

The following is the distribution of the failure rate, $p(x)$:

$$f(x) = \begin{cases} 1020.41 & 0.000001 \le x < 0.00005 \\ 9.00 & 0.00005 \le x \le 0.10 \\ 0.50 & 0.10 < x \le 0.20. \end{cases}$$

This distribution has the following characteristics:

Mean	0.00056
Variance	0.000041
Minimum	0.0000010
5th percentile	0.000050
50th percentile	0.025
95th percentile	0.10
Maximum	0.20.

■

Comparison of the Methods

This chapter presents methods for characterizing uncertainty that are easy to implement and require a minimal amount of assumptions. Example 17.6 shows how these methods for characterizing uncertainty compare by using the 10 expert estimates and their range values.

EXAMPLE 17.6: ***Comparison of uncertainty characterizations***

This graph plots the maximum and minimum values for the final distribution at the ends of the straight line. The narrow boxes connect the 5th and 95th percentile values for the median. The point inside the box is the 50th percentile (or median). The final distributions for the bootstrap and Monte Carlo plots are the distributions for the median. The final distributions for the experts' estimates, Bayesian single prior, and maximum entropy plots are the distributions for the failure rate (not the median of the failure rates).

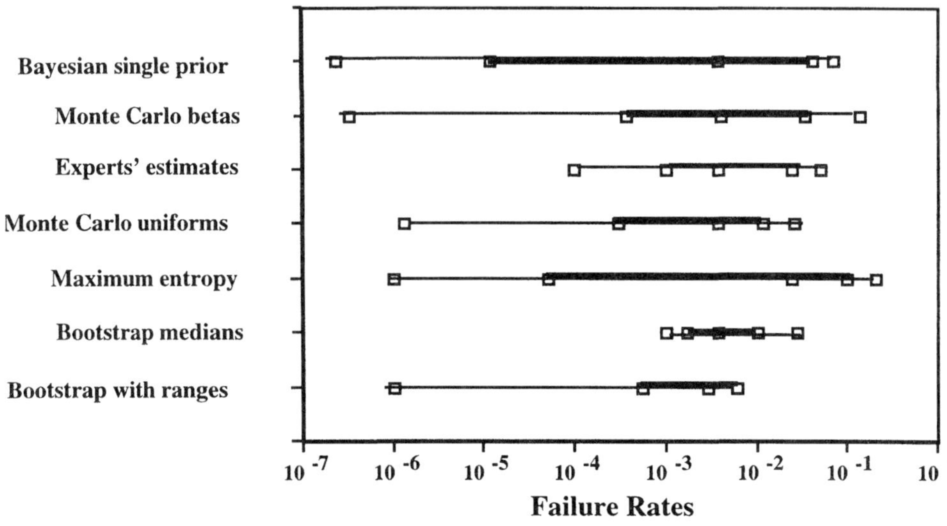

■

The graph indicates some interesting results. First, the bootstrap without ranges is very narrow, indicating little variation in the values. This variation is even smaller than the raw data, which is not unexpected. Small variations are indicative of the bootstrap (Efron, 1979), and the variation of the median is expected to be smaller than the variation in the raw data. The bootstrap for the median with ranges has a wider spread than the bootstrap from just the raw data because of an increased variability of the data set from inclusion of the range values. Second, most plots indicate a skew of the values with more of the distributions shifted to the right or to higher failure rates. Such a shift is not surprising because the raw data and range values shift to the higher failure rate values. Third, the Monte Carlo with beta priors, the maximum entropy, and the single beta prior indicate the most spread and are very similar to each other. The maximum entropy and single prior also have similar variability in the center (the boxes). The Monte Carlo betas have a narrower center because plot represents the distribution of the median and not the raw estimates. Finally, the graph shows an increased variability over the raw estimates indicated in the plots of the methods that use ranges to represent uncertainty.

The different assumptions made about the distributional forms (beta, uniform, and empirical) and the interpretations of the range values (where used) yielded the differences in the methods. Do not expect complete consistency. However, inconsistencies should be explained. In this comparison, the plots for all methods are quite similar, with the exception of the raw data and the bootstrap median of the raw data. These plots should show very little variation because the range values do not represent uncertainties in the raw estimates. As previous chapters have advised, analysts should try and compare several different uncertainty characterizations. However, if time constraints do not allow such experimentation, we recommend using the Monte Carlo with uniform priors (Martz *et al.*, 1983).

18

Making Inferences

The purpose of eliciting and analyzing expert judgment is to use all the available information in cases where data does not exist or where data is sparse. The analyst uses the experts' information to draw conclusions. This process is called making inferences. This chapter describes two types of inferences: general and statistical. This chapter discusses the inferences of both types that are possible, accounting for the design features and analysis methods used in the problem. This chapter also presents some comments about inference relating to expert judgment applications.

What Inferences Can Be Made

Statistical inferences are based on statistical principles of sampling. For example, analysts use results from a sample that is representative of a larger population to make statements regarding that population. In expert judgment applications, however, such extended inference is usually not possible. For one reason, the information from the experts is not a random sample (or representative sample) of the true state of knowledge. In most cases, the analyst may not even be able to form a random sample of the experts for the elicitation. Therefore, making inferences from expert information about the true state of the universe is not recommended.

In most expert judgment applications, the experts' knowledge represents the state of the existing or available knowledge. In that sense, general inferences can be made as follows: the results from the experts' information can be used to draw conclusions about the existing or available knowledge base which may or may not represent the true state of nature. In other words, the inferences that can be made are not necessarily relevant to truth; nor are they statistically based inferences. Example 18.1 illustrates this distinction.

EXAMPLE 18.1: ***Expert judgment inference versus statistical inference***

When a sample is properly drawn (usually randomly drawn) from an underlying population, the estimates from that sample can be used to draw conclusions about the parameters of the population. The analyst infers the characteristics of the population from the characteristics of the sample. This is statistical inference. The probability that a statistical inference is incorrect accompanies the inference statement. For example, the parameter of interest P is the probability of an earthquake scaled at 8.0 or greater at a particular location. A true value exists for this probability. In an ideal situation, the analyst draws a random sample of times n throughout the planet's history and future at this location. He or she counts the number of instances that such an earthquake occurred (x) and estimates the true probability of such a quake (p) by x/n. The true value of P is inferred by p.

Such an example is ridiculous in reality. The information base required is not available. Some information about earthquakes and the necessary potential conditions are available to the expert. Only the expert's knowledge and expertise are available for estimating the probability at this location. Suppose that existing historical data show no record of an earthquake. Such information does not mean that one never occurred. The expert is still the primary source of all information. The expert is carefully interviewed using the techniques in this book. The information is recorded. His or her final estimate (p) is given under sets of plausible conditions based upon what is known.

The expert's estimate (p') is not the same as the mythical p from a properly drawn random sample. The true value of P cannot be statistically inferred from p', whereas p provides a good estimate of P.

What is p'? The value p' is the only existing information about P and may be the best information that will ever be available in the history of man. It is subject to bias, change, and misinterpretation. It is not a statistical estimate of the parameter P. It may be considered a single datum from a sample of possible estimates. As such, statistical inferences are not possible from a sample size of 1. However, p' is relevant, useful information.

How is p' useful if statistical inference is not possible? The expert's estimate represents the state of knowledge. It represents previously unknown and undocumented information. Analysts can interpret the expert's information in a general inference manner in conjunction with any caveats from conditions and assumptions the expert made.

■

Analysts who are accustomed to drawing statistically based inferences about a population (i.e., the truth) from a statistical sample (i.e., the data) may find this limited ability to infer bothersome. This limited ability to infer also leads many analysts and many experts to believe that expert information is not valid data and cannot be used. One of the misconceptions listed in chapter 2 deals with the issue of how to interpret expert judgment as valid data. That section effectively claimed that information from experts (data) was like any other data in that it must be carefully gathered, analyzed, and interpreted. Example 18.2 illustrates how an expert judgment application is identical to an experiment regarding the treatment of the data.

EXAMPLE 18.2: ***Expert judgment data versus experimental data***

A chemist determines the composition of a chemical mixture. He or she measures the easily identifiable elements and compounds first to determine composition. To do this, the chemist uses calibration standards, instruments, and test procedures. He or she then tackles the more difficult remaining items. The chemist eliminates some and suspects some as being present based on his or her own experience of how elements and compounds could mix together. He or she completes the analysis and presents the results.

Most scientists, analysts, and laymen would consider the chemist's results good experimental data. They would even make statistical inferences based upon the results. Is this a good idea? What are the problems with the above experimental data that affect the inference process? Answering these questions reveals that this experimental data have no advantages over or better quality than expert information. Here is why.

First, the chemist used judgment and expertise to make decisions about which tests he or she would run. He or she went so far as to decide, using experience, which items were likely to be there or not. These conditions and assumptions that the chemist makes parallel the conditions and assumptions any expert makes in solving a problem.

Second, the instruments used have measurement problems. Calibration, standards, and operator errors are commonplace sources of bias and lack of precision. The data recorded from these do not represent the ideal samples required to make a statistically based inference. These biases and lack of representativeness of the data parallel the same concepts discussed in this book in expert data elicitation (see particularly *Pitfalls* in chapter 2).

Why then is the chemist's experimental data more acceptable than data from an expert's answer to a question? From this experimental description, the inferences possible for the chemist's data are no better than those for expert data. From a statistical perspective, neither data can accurately develop a statistically based inference. However, both reflect the best state of knowledge available.

■

The above illustrations indicate the difficulty in distinguishing statistical inference from general inference and the different ways in which data are interpreted. The distinction is particularly important in expert judgment applications. In these applications the interpretations of the results should assume general inference, not statistical inference. However, expert judgment data can still provide valuable information and valid conclusions. In addition, expert judgment data can improve the process of making general inferences.

Improving the Inference Process

The analyst should do everything possible to accurately elicit and analyze the information from the experts, thereby getting the best existing knowledge. Because inference stems from the interpretation of the results, careful interpretation becomes crucial. Analysts should practice care in two crucial areas: the design of the elicitation and the analyses. Improvements in the inference process come from both these areas.

Design-Based Improvements

Proper experimental design always provides the key to obtaining the most and best information for making inferences. This statement is true for both experimental and expert judgment applications. We will refer to the ways of accomplishing design-based improvements by chapter numbers for this discussion. Basically, proper design includes coordinating elicitation with analysis methods by structuring the questions and response modes, monitoring granularity, recording all information from the experts (conditions), and performing quantification.

Synergism of Elicitation and Analysis

The interrelationship between Parts II and III of this book may not appear too obvious. However, the elicitation (Part II) focuses on the experts, while the analysis (Part III) focuses on the elicitation.

Many different methods can effectively analyze expert answer data (usually quantitative in structure). The methods in Part III are chosen specifically to match with the elicitation methods in Part II. The authors' choice of methods in this book is based on personal experiences while working with elicitation and analysis.

Too often the elicitation and analysis remain disconnected until the elicitation is completed. As a result, the process of inference becomes pure guesswork. The analyst may be unfamiliar with how the data are gathered or with what forms the data will take. Moreover, the analyst may have his or her own methods which may require assumptions that are not appropriate for the data or for the way in which it was gathered. This situation can result in poor inferences. Avoid this approach.

If the analyst is not the data gatherer, he or she should at least work with the person gathering the data at all stages of the project, from designing the questions (chapters 4 and 5), to selecting the elicitation components and tailoring the elicitation (chapters 6 and 7), to pilot testing and final elicitation (chapters 9 and 10). The decisions the data gatherer makes concerning the methods and components of the elicitation provide the analyst with important information on the problem. In addition, the analyst needs to know what form the gathered information will take for the analysis. The analyst can help during the elicitation by watching for some of the problems involving granularity, quantification, and conditionality.

Granularity

The analysis part of expert judgment applications emphasized monitoring the level of detail reflected in the information at each analysis step. Granularity can change within or between the analysis steps without notice. Granularity can also change during the elicitation. Proper recording of the elicitation to monitor such changes provides a strong defense against granularity changes.

Analysts make inferences at the most general granularity for all the steps. Therefore, it may not be possible to improve inferences made by monitoring granularity; however, inferences may be erroneous if proper attention to changing granularity is denied.

The data gathering and data base formation steps are likely to produce a data base (chapters 12 and 13) containing variables (information) with different granularities. The qualitative information can be thought of as more general than quantitative information for analysis purposes. The quantitative variables can be categories or ranks; categories are more general than ranks and both are more general than variables measured on a continuous numerical scale. The model formation step (chapter 15) can model variables of differing granularities. Example 18.3 illustrates how the granularity should be monitored and what inferences can be made as a result of different granularities.

EXAMPLE 18.3: ***Inference and granularity***

In an expert judgment application, six experts were asked to provide estimates of how well experimental results matched results from a simulation code. The data base included variables on problem-solving features. Three variables indicated assumptions used, two indicated cues used, and one related to a definition. The data follows.

Expert	A_1	A_2	C_1	C_2	D_1	*ANSWER*
1	1	13	2	N	B	0.25
2	1	20	2	Y	A	0.10
3	2	33	1	Y	C	1.00
4	3	27	2	Y	A	0.50
5	3	25	1	N	B	0.45
6	2	30	2	Y	B	0.95

The variables A_1 and D_1 are category variables in which the numbers 1, 2, and 3 and the letters A, B, and C refer to qualitative descriptions. Variable C_1 is a rank variable where the values 1 and 2 mean that 2 is twice as important as 1. Variable C_2 is also categorical where Y is *yes* and N is *no,* indicating whether or not a certain cue was used. Variable A_2 is a numerical variable indicating the values assumed for a parameter important to the problem. The answers are on a continuous linear scale that describes degree of agreement from 0.0 to 1.0 in value.

Regression analysis (GLM) produces a model with A_2 as the single best predictor variable for *ANSWER*. If these 6 experts were a representative sample of all experts, then a statistical inference about A_2 would be possible. However, no such claim can be made, and only the more limited inference is possible: A_2 may be an important variable for determining answers for these 6 experts. In addition, the variables D_1, C_1, and C_2 are significant variables in this regression. Their ordered list indicates their degree of importance as predictor variables, with D_1 being the best.

This set of 4 predictors from a regression analysis represents a mixture of granularities. With such a mixture and the lack of a statistically valid sample, only a limited inference is possible. In this case, the information gained is merely that these 4 variables are possible conditions for determining the answers. Do not use these variables as predictors for other answers, as would be the inference made with significant variables from a regression analysis. As possible conditional variables, they can be considered for use in aggregation; they can also be used to answer the important question: Are all the experts solving the same problem?

■

Example 18.3 illustrated granularity problems with modeling. The aggregation step (chapter 16) has similar problems. However, aggregation involves combining final answers or answer distributions using some weighting scheme. In most cases, these answers are already at the same granularity. The same granularity might not hold if different aggregation methods were used on the same data set. Therefore, analysts must still monitor granularity.

In characterizing uncertainties (chapter 17), granularity can become a severe problem because the uncertainties may represent one level and the answers represent another. In addition, the uncertainties from different experts may represent different granularities, as illustrated in example 18.4.

EXAMPLE 18.4: ***Granularity and uncertainty characterization***

Suppose that the six experts in example 18.3 provided the answers given above and provided uncertainty estimates as follows:

Expert	Answer	Uncertainties	Expert's Definition
1	0.25	(0.10, 0.50)	Range
2	0.10	(0.00, 0.20)	5th and 95th percentiles
3	1.00	(0.90, 1.00)	Range
4	0.50	(0.25, 0.75)	5th and 95th percentiles
5	0.45	(0.00, 1.00)	Absolute min. and max.
6	0.95	(0.90, 1.00)	Range

These uncertainty estimates do not have matching definitions (or matching granularities). The percentile definitions from experts 2 and 4 will not represent a 90% coverage interval. The definitions might represent more of a 30% to 40% coverage interval. The ranges from experts 1, 3, and 6 have no obvious interpretation. The values from expert 5 appear useless. However, if expert 5 claims that his or her range really represents his or her assessment of the true uncertainty, then those values are valid uncertainty characterizations even though they cover the entire possible range of answers.

All experts need a common but more general definition for the uncertainty values. One solution involves assuming uniform distributions for each expert using the ranges provided as the upper and lower limits for the uniform distributions. Monte Carlo sampling from these uniform distributions provides a useful uncertainty analysis method in this case. However, the results of such an analysis would reinforce what is already evident from the ranges of values provided by the experts. The uncertainty is large enough to cover the entire possible range of values for the answer. In such a case, a large uncertainty overshadows the granularity problem.

■

The section *Inferences with Modeling, Aggregation, and Uncertainties* discusses how these analyses are important for making inferences in ways other than those regarding granularity.

Quantification

The broad definition of quantification in this book actually includes granularity. Quantification is defined as transformations from one type of data to another, rather than using the traditional definition of transforming qualitative information to numbers. Qualitative information is considered more general than quantitative; ranks (or integers) more general than (continuous) numbers; categories more general than ranks; and descriptions more general than categories. The granularity/quantification connection affects the inference process.

In cases where data are sparse or missing for a set of variables, analysts sum or collapse categories, ranks, or numerical values to form new variables or new categories for a variable. Summing or collapsing variables also changes granularity from specific to general. The inferences must be made in terms of the new, more general variables and not in terms of the original information. Example 18.5 provides an example of changing granularity by using the Saaty method of quantification based on numerical weights from qualitative information.

EXAMPLE 18.5: ***Granularity and quantification using Saaty's method***

An expert compares and evaluates the likelihoods of seven different events. By using Saaty's pairwise comparison method, the expert specifies the relative evaluations (a general granularity) using the Saaty scale as follows:

1 versus 2 = 0.50	2 versus 3 = 0.50
1 versus 3 = 0.33	2 versus 4 = 3.00
1 versus 4 = 1.00	2 versus 5 = 4.00
1 versus 5 = 2.00	2 versus 6 = 6.00
1 versus 6 = 3.00	2 versus 7 = 5.00
1 versus 7 = 2.00	
3 versus 4 = 2.00	4 versus 5 = 2.00
3 versus 5 = 3.00	4 versus 6 = 3.00
3 versus 6 = 3.00	4 versus 7 = 2.00
3 versus 7 = 3.00	
	5 versus 6 = 2.00
6 versus 7 = 0.50	5 versus 7 = 1.00

The resulting relative weights are

(0.13, 0.27, 0.28, 0.13, 0.07, 0.05, 0.07).

Although the weights have a numerical form, the analyst must make the interpretation on a relative, qualitative basis. For example, events 5 and 7 are *not* half as likely as events 1 and 4, nor are they one-fourth as likely as events 2 and 3. The only interpretation possible for the 7 events is that the events judged most likely are 2 and 3, and the events judged least likely are 5, 6, and 7.

■

Conditionality

The concept of conditionality refers to the fact that the answers the experts provide are conditioned on many aspects of the problem, the elicitation, and the experts themselves. Chapters 12 and 13 discuss analysis techniques that emphasize how to search for these conditions. These conditions play a crucial role in interpreting the answers.

The major focus for investigating conditions is to guarantee that all the experts are to solve the same problem and that the problem being solved is indeed the one being asked. If conditions relating to problem-solving features (e.g., definitions, assumptions, and heuristics) influence the answers given, then analysts must examine these features to determine if they somehow change the problem.

We also recommend gathering background information on the experts as possible sources for influential conditions. Many analysts and researchers have speculated for a decade that experts' backgrounds are a source of correlation among experts (Baecher, 1979); however, recent studies (Booker & Meyer, 1988*a*; Meyer & Booker, 1987*b*) have indicated that evidence for influential background features is lacking.

Design features of both the elicitation and the environment contain conditions. To minimize biases and other conditions that can affect the answers given by the experts, analysts and data gatherers must carefully formulate questions (chapters 4 and 5) and implement the elicitation (chapters 8 and 10). Other uncontrollable conditions, such as the expert's mood, an uncomfortable room, or a disturbing recent event, should also be noted and documented as possible important conditions.

Recording and monitoring conditions helps determine if such conditions are important. The analyst and data gatherer must never be responsible for inducing any additional conditions by the methods they use to analyze and elicit the data. This warning includes the inference process. The analyst can impose his or her own views to the extent that he or she interprets the data in the way that he or she desires. Conditions that really might not be important can be used as excuses to disregard data that do not fit preconceived ideas. Even conditions that might be important can be falsely used for this purpose.

Placing too much importance on conditions can also result in erroneously made inferences. While analysts must learn not to ignore conditions, they must also avoid using them as excuses, cover-ups, or justifications. Conditions that play a significant or important role may only be masks or indirect effects for some other effect that cannot be monitored. Important conditions must also be relevant to the problem. Only then do analysts make interpretations with conditions given as caveats or qualifiers.

Three concepts reflect the basic philosophy regarding possible effects from conditions. First, proper elicitation and analysis reduces any effects induced by the data gatherer and analyst. Second, the data gatherer and analyst should control any controllable conditions; they should also record information on *any* observable conditions. Third, the analyst should state significant and relevant conditions as part of the conclusions and as reflecting the inferences made.

Analysis-Based Improvements

The methods chosen in the analysis section reflect a conservative, cautious interpretation of the results. This caution is borrowed from the reactor design and probabilistic risk analysis communities. Redundancy and cross validation play a major role in the analysis of expert judgment data. The philosophy encourages the use of more than one technique to determine the results. Therefore, the consistency of results is validated. A simple way to cross validate is to use simulation techniques. Simulation has the added advantage of not relying on assumed distributions or on methods based on those assumptions.

Cross Validation and Redundancy

Many methods can effectively analyze a data set. Many of the methods presented in Part III of this book have similar characteristics and produce similar types of results that together can validate each other or provide redundancy of analyses.

Several examples in Part III showed how different techniques validated results. The steps for detecting correlation in chapter 14 included redundant techniques.

A regression analysis (GLM) provides information on how condition variables affect the answers. Correlation also determines variable relationships. If two variables are significantly correlated, they will usually be significant terms in a regression model.

The various multivariate techniques are also interrelated. If a conditional (ancillary) variable functions as a discriminator, then it also functions effectively as a predictor in a regression model. If cluster analysis reveals definite clusterings, the analyst should search for a discriminating variable that is responsible for the clusterings. Factor analysis reveals, as can cluster analysis and correlation analysis, how the different variables interrelated.

Therefore, the analyst can run different techniques and determine if the results are consistent. If the results are consistent, that lends strength to the conclusions. If the results are inconsistent, then either the techniques were applied incorrectly (e.g., violation of the required assumptions) or the interpretations were interpreted incorrectly (e.g., a variable is not really important in determining the answers). Example 18.6 illustrates the use of redundant techniques.

EXAMPLE 18.6: ***Inference using redundant techniques***

Ten experts answered five questions (A_1 through A_5). The analyst constructed a general problem-solving variable (P_1 through P_5) for each question by totaling all the problem-solving features for that question. The analyst found that each problem-solving variable was significant in the regression analyses done for the answers. Individual problem-solving features and all other ancillary information (variables) were not significant variables in the regression models.

Answer variables A_1, A_3, and A_4 have a trimodal structure. The three modes formed three clusters. Variables P_1, P_3, and P_4 were found to be significant discriminating variables for their respective answer variables; they successfully predicted the three clusters for each answer variable.

Answer variables A_2 and A_5 had no identifiable structure. A cluster analysis of all the variables for questions 2 and 5 indicated that variables A_2 and P_2 formed a cluster; variables A_5 and P_5 also formed a cluster.

Answer variable A_1 has a unimodal structure. A significant correlation coefficient between A_1 and P_1 equals 0.87.

Therefore, additional evidence supports the claim that the P_i variables might be important for determining the A_i variables.

Having supportive evidence for the P_i variables as conditions for the answers, the next step involves specifying this result in the statement of conclusions. The answers the experts provide are conditioned on their general problem-solving processes (not specific problem-solving features). For aggregation, examine the values of the P_i variables for weight determinations. For inference, use the P_i

variables to determine whether or not all the experts solved the same (the given) problem.

Suppose the additional evidence was not very clear. If the A_2, P_2 and A_5, P_5 variables did not cluster, then the conditionality would not necessarily apply for these questions. Analysts would have to attempt other techniques. If none of the other techniques provided evidence of variable relationships between the A variables and the P variables, then the conditionality would be suspect. These potential conditional variables would have little or no importance for aggregation and interpretive purposes.

■

Because the inference process itself is weak in expert judgment applications, strong supportive evidence must exist for conditionality. Redundancy or cross validation helps determine the degree of strength.

Simulation

Simulation can provide cross validation and redundancy checks. Using simulation to improve the inference capability in expert judgment applications has three advantages: sample sizes are typically small (less than 10 experts); distributional forms for the answers do not follow convenient forms (such as the normal) and are usually multimodal or distribution mixtures; and estimations for variances, percentiles, or central measures (such as the median) are desired for inference but are difficult to obtain without specified distributional forms. At the very least, simulation allows the analyst the freedom to explore and check effects in an empirical (data-based) manner.

Making statistical inferences with small sample sizes can prove a most difficult task. Most statistical techniques rely on asymptotic or theoretic results that require larger sample sizes; such sizes are impractical in expert judgment applications for several reasons. For example, the elicitation would be too time consuming and expensive, or only a few individuals are experts in that particular field.

Making statistical inferences without specified distributional forms may prove equally difficult. Many statistical techniques require distributional forms (e.g., normality). This is true of many of the multivariate analysis techniques (e.g., discriminant analysis). Other techniques (e.g., regression) have less strict distributional requirements, requiring only that the residual or model errors be normal rather than the data itself.

We emphasize that variance estimates or interval estimates (e.g., percentile values) accompany all estimators (e.g., the mean or median). Providing such estimates is part of the inference process and part of establishing the uncertainty. A single-valued estimator does not provide any information about the variability or uncertainty surrounding it. A single-valued estimator implies a precision in the results that is not present. A variance or interval estimator conveys the appropriate state of uncertainty and variability. Estimating variances of estimators like the median is difficult without distributional forms. In some cases, estimating variances of estimators is difficult even with distributional forms because the formulas are not tractable.

Simulation can provide solutions to the difficulties from small sample sizes and required assumptions. Simulation allows the analyst to make the most of a small sample size using such techniques as the bootstrap. Reliance upon asymptotic or theoretic results is not necessary, and simulation provides a way of obtaining estimates for variances and other quantities without theory, distributional assumptions, or difficult calculations.

Inferences with Modeling, Aggregation, and Uncertainties

The discussion of granularity above introduced the relationship between the inference process and the results obtained from modeling (chapter 15), aggregation (chapter 16), and uncertainty (chapter 17). These three steps in Part III relate to the inference process in other ways.

Analysts used the models in chapter 15 to identify relationships between the conditional (ancillary) variables and the answer variables. GLMs represent the major category of these models, which are the backbone of statistical modeling techniques. Other conditional models were also suggested that did not have a strong statistical basis. The purpose of both types of models does *not* involve identifying functional relationships among the variables to be used for prediction purposes; the purpose does not include specifying model functions (equations); but the purpose is to provide clues about variable relationships that could be verified by using other procedures (or results from other procedures could be verified using these models). The reason for such a limited purpose is the probable lack of model assumption compliance.

The purpose of aggregation is to formulate the final results as a combined estimate with a variance or as a distribution with central values (mean or median) and dispersion measures (variance or 5th and 95th percentiles). The analyst interprets these final estimates or distributions as the cumulative knowledge for the parameter of interest (the answer to the question). This interpretation is part of the inference process. These results do not have a statistical interpretation that relates the estimates to the true value. The dispersion estimates do have a valid interpretation relating to the uncertainty in the state of existing knowledge. Therefore, conclusions made using the uncertainty characterization reflect this state of knowledge. Aggregation provides a convenient summary or combination of all the available information.

As mentioned in the simulation section, uncertainties play an integral part in the inference process. Representing or characterizing uncertainties acknowledges their existence and estimates their effects. Drawing conclusions without accounting for uncertainties makes the information appear more precise than is true (bad inference). As emphasized in chapter 17, uncertainties play an important part in every experiment or application and cannot be ignored, especially when the time comes to interpret the results.

Final Comments

With only weak inferences possible, a natural question becomes: Why take such care in gathering and analyzing the data? Part I of the book provided an answer: Expert judgment data is like any other data—it must be carefully gathered, analyzed, and interpreted. Unfortunately, careful interpretation in this case translates to limited inferences. To do otherwise, however, violates the true content of the information gathered. The analyst cannot get better information than exists.

We recommend these final suggestions pertaining to the philosophy of this book and to the inference process:

1. **Continue research efforts.** Resolving many of the problems in expert judgment elicitation and analysis requires more research. This book represents only the beginning of such intensive research. Better techniques and understanding of the expert information will lead to better inference.
2. **Follow procedures carefully.** All the steps and suggestions offered in this book may seem tedious and unnecessary, but to obtain good quality data takes time and effort. Appropriate procedures and techniques are required to yield even limited inferences.
3. **Provide caveats with results.** The results are accompanied by a list of caveats and conditions—interpretation must include these.
4. **Keep everything simple.** This book presents the methods with the data gatherers and analysts in mind. Many other techniques exist, and some of them are referenced. No evidence exists that the more complex methods omitted are better than the simple ones offered. In fact, many of the more complex methods do not perform as well as the simpler ones do.

In conclusion, the inference process may be disappointing in that the results and conclusions available do not directly extend to the underlying population; that is, statistical inference is not possible. However, the elicitation and analysis of expert judgment information has revealed information that was previously unknown. Because so many applications are data sparse, expert judgment data will continue to play an important role in the quest for new data.

Appendices

Appendix A—SAATY

```
      program saaty
c
c   Uses Saaty's vintage FORTRAN subroutine for calculating
c   the weights for a single level matrix. This code is used
c   with permission and is from Saaty's book, Decision Making
c   for Leaders (1982). For a more recent version of this
c   code that accommodates more than one level, free trial
c   software can be downloaded from Expert Choice at:
c   http://www.expertchoice.com
c
c   The input to this code can be done on the terminal or
c   by contructing input file SAATY.IN which has:
c     line 1: nfctr = number of experts (factors), maximum
c             =10.
c     line 2: contains (nfactr-1) pairwise comparisons in
c             f5.2 format, comparing item 1 with items 2-nfct
c             using scales **
c     line 3: contains (nfctr-2) pariwise comparisons in f5.2
c             format, comparing item 2 with items 3-nfactr.
c     etc.
c     line (nfctr-1): contains 1 pairwise comparison in f5.2
c          format, comparing the last two items [item nfctr
c          to item (nfctr-1)].
c
c    **Values for pairwise comparisons can be taken from
c    1)the Saaty integer scales: values = 2-9 where the first
c      of the pair is better or more likely than the second;
c      value = 1 where the pair is identical; values =
c      1/2-1/9 where the first of the pair is worse or less
c      likely than the second.
c    2)the pairwise comparisons can be qualitative, using a
c      triplet of choices (better, same, worse) rather than
c      the numerical scale, the values for this triplet are:
c      (2.72, 1.00, 0.37), respectively.
c
c
      dimension w(60)
      open (unit=11,file='saaty.in',status='old')
```

```
      open (unit=22,file='saaty.out',status='new')
      write(*,'("To use input file SAATY.IN, type 1 else
     *type 0: ",\)')
      read(*,*) infile
      if (infile.eq.1) then
        read(11,*) nfctr
        go to 61
      endif

      write (*,'("Enter the number of factors (up to 10):
     * ",\)')
      read  (*,*) nfctr

   61 continue
      call mtxin(w,nfctr,infile)
      stop
      end
c
c
c
      subroutine mtxin(w,nfctr,infile)
      dimension c(60,60),w(60),cw(60),w2(60),rct(10)
      data rct/.0,.0,.58,.90,1.12,1.24,1.32,1.41,1.45,1.49/
   16 continue
      if (infile.eq.0) write (*,9)
      nfctr1=nfctr-1
      do 10 i=1,nfctr1
    2 continue
      if(infile.eq.0)write (*,3) i
      i1=i+1
      if(infile.eq.0)read (*,4)  (c(i,j),j=i1,nfctr)
      if(infile.eq.1)read(11,4)  (c(i,j),j=i1,nfctr)
      if(infile.eq.0)write (*,5) i,  (c(i,j),j=i1,nfctr)
      if(infile.eq.1)write (22,5) i,  (c(i,j),j=i1,nfctr)
      if(infile.eq.0) write (*,1)
    4 format(10f5.2)
    5 format('Row',i3,' is:',10f5.2)
    1 format(' If not correct type 9, if correct hit
     *return.')
    3 format(/,'Enter row',i3,' (use f5.2 format): ')
    9 format(/,'The upper triangular part of the matrix:')
   18 format(f1.0)
      if(infile.eq.1) yn=4
      if(infile.eq.0)read (*,18) yn
      if(yn.gt. 6) go to 2
      do 10 j=i1,nfctr
      if(c(i,j).lt.0.) go to 6
      if(c(i,j).ge.0.) go to 8
    6 c(i,j)=-(1.0/c(i,j))
    8 c(j,i)=1.0/c(i,j)
   10 continue
```

```
      do 14 i=1,nfctr
   14 c(i,i)=2.
      ts=0.
      do 24 i=1,nfctr
      s=0.
      do 22 j=1,nfctr
   22 s=s+c(i,j)
      w2(i)=s
   24 ts=ts+s
      do 26 i=1,nfctr
   26 w2(i)=w2(i)/ts
      k=0
   27 ts=0.
      k=k+1
      do 30 i=1,nfctr
      s=0.
      do 28 j=1,nfctr
   28 s=s+c(i,j)*w2(j)
      w(i)=s
   30 ts=ts+s
      d=0.
      do 38 i=1,nfctr
      w(i)=w(i)/ts
   38 d=d+abs(w(i)-w2(i))
      if(k.gt.10000) go to 42
      if(d.lt.1.e-5) go to 42
      do 37 i=1,60
   37 w2(i)=w(i)
      go to 27
   42 continue
      if(infile.eq.0) write(*,47)
      if(infile.eq.1) write(22,47)
   47 format(/,'The final matrix is:')
      do 40 i=1,nfctr
      c(i,i)=1
      if(infile.eq.1) write (22,41) (c(i,j),j=1,nfctr)
   40 if(infile.eq.0) write (*,41) (c(i,j),j=1,nfctr)
   41 format(10f6.3)
      do 46 i=1,nfctr
      s=0.
      do 44 j=1,nfctr
   44 s=s+c(i,j)*w(j)
       cw(i)=s
   46 continue
      s=0.
      do 48 i=1,nfctr
   48 s=s+cw(i)/w(i)
      ymax=s/nfctr
      ci=0.
      cr=0.
      if(nfctr.le.1) go to 49
      ci=(ymax-nfctr)/(nfctr-1)
      if(nfctr.le.2) go to 49
      cr=ci/rct(nfctr)
   49 continue
      if(infile.eq.0) write (*,50) (w(i),i=1,nfctr)
      if(infile.eq.1) write (22,50) (w(i),i=1,nfctr)
```

```
50 format(/,'Normalized weights=',(10f6.3))
   if(infile.eq.0)write (*,52) ymax,ci,cr
   if(infile.eq.1)write (22,52) ymax,ci,cr
52 format(' principle eigenvalue, lmax =',f6.3,/,
  *'Consistency index (deviation of lmax from n)
  *=',f6.3, /,' consistency ratio (should be < .10)
  *=',f6.3)
   if(infile.eq.0) write (*,54)
54 format(/,'If you want to redo this matrix,',
  * 'type 9, else hit return:')
   if(infile.eq.0)read (*,18) yn
   if(yn.ge.5) go to 16
   return
   end
```

Appendix B—MCBETA

```
      program mcbeta
c
c   Monte Carlo uncertainty analysis code
c    for beta distributions.
c
c   Important!!!!!!!!!!!! user must change the function for
c    combining the set of experts, events, or sequence as:
c    FUNCTION EVAL(PR)
c    where pr is array of probabilities of primary events,
c    and the function is aggregation estimator for all the
c    betas.
c
c   Input is supplied using file MCBETA.IN, as unit 11,
c     see format below.
c   Output is sent to two files:
c     MCSTATS.OUT which contains the Monte Carlo results
c     BET.OUT which contains the fitted beta distributions
c     results.
c     MCSTATS.OUT is unit 22 and BET.OUT is unit 33
c
c   Betas are fit with two supplied estimates either as:
c    1) 2 percentile estimates and levels (e.g. 0.05 & 0.95)
c    2) 1 percentile estimate and level and a mean value.
c
      parameter (nbmax=500,klmax=100,ndmax=100,nhmax=10000,
     *nhp1=nhmax+1)
      dimension ppbig(10000)
      common/primary/pr(nbmax),param(klmax,ndmax,2)
      common/options/nb,nruns
      common/utility/dumy(nhp1)
      save /utility/, /options/, /primary/
      open (unit=22,file='mcstats.out',status='new')
      call input
      nout=0
      do 10 n=1,nruns
 5    call pgen
      prob=eval(pr)
      if (prob.ge.0..and.prob.lt.1.01) go to 8
      nout=nout+1
      write (22,1222) n,prob
      go to 5
 8    continue
      ppbig(n) = prob
 10   continue
      call finish (ppbig,nruns)
      write(22,1000) nout
 1000 format(//' Simulation generated',i5,
```

```
     *       ' values not in (0,1)')
 1222 format(' n=',i6,' prob=',e16.8)
      close (unit=22)
      stop
      end


      subroutine input
c
c    Read parameters from file 'MCBETA.IN'.
c
c      line     description
c
c      1        nb = number of primary events (le nbmax)
c               nr = number of monte carlo runs, max = 10000
c      2        iibet = 2 for 2 percentiles, = 1 for 1
c               percentile & mean
c      next nb lines free format
c               two percentiles for the beta case:
c               upper estimate, lower estimate,
c               upper level (.95), lower level (.05)
c               or
c               mean, percentile value, percentile level.
c
      parameter (nbmax=500,klmax=100,ndmax=100)
      common/primary/pr(nbmax),param(klmax,ndmax,2)
      common/options/nb,nruns
      save /options/,/primary/
      dimension t(2)
      open (unit=11,file='mcbeta.in',status='old')
      open (unit=33,file='bet.out',status='new')
      read (11,*) nb,nr
      read(11,*) iibet
      nruns=nr
      do 10 i=1,nb
c
c   Subroutines twoper & meanper find the beta parameters,
c   x0 and fn0
c
      if(iibet.eq.2) call twoper(x0,fn0)
      if(iibet.eq.1) call meanper(x0,fn0)
      t(1)=x0
      t(2)=fn0
       param(i,1,1)=t(1)
       param(i,1,2)=t(2)
 10   continue
      close (unit=11)
      close (unit=33)
      return
      end



      subroutine pgen
      parameter (nbmax=500,klmax=100,ndmax=100)
      common/primary/pr(nbmax),param(klmax,ndmax,2)
      common/options/nb,nruns
      common/utility/p(klmax,ndmax)
      save /primary/, /options/, /utility/
      do 100 n=1,nb
```

```
      prm1=param(n,1,1)
      prm2=param(n,1,2)-prm1
  82  if(prm1.gt.1.0) gx1=gt(prm1)
      if(prm1.eq.1.0) gx1=gs(1.)
      if(prm1.lt.1.0) call gl(prm1,gx1)
      if(prm2.gt.1.0) gx2=gt(prm2)
      if(prm2.eq.1.0) gx2=gs(1.)
      if(prm2.lt.1.0) call gl(prm2,gx2)
      p(n,1)=gx1/(gx1+gx2)
      if(p(n,1).gt.1.0) go to 82
 100  continue
      do 200 i=1,nb
  200 pr(i)=p(i,1)
      return
      end
c
c ***********************************************************
c
c   FUNCTION EVAL: must be changed for each new problem
c
c ***********************************************************
c
      function eval(x)
      dimension y(150),x(500)
      ymedian = (x(1) + x(2) +x(3))/3.0
      eval =  ymedian
      return
      end
c
c
c
      function gs(alp)
c
c  Routine for generating gamma variates with
c   shape parameter less than 1.  Adapted from:
c   Ahrens, J. H. and Dieter, U. (1974)
c   "Computer Methods for Sampling from Gamma,
c    Beta, Poisson, and Binomial Distributions,"
c   Computing, vol 12, p.223-246.
c
      data ex/2.718281828459045/
      bet=1.0
 1    u1=rnd(0.)
      b=(ex+alp)/ex
      p=b*u1
      if (p.gt.1.) go to 3
 2    x=exp(alog(p)/alp)
      u2=rnd(0.)
      if (u2.gt.exp(-x)) go to 1
      go to 10
 3    x=-alog((b-p)/alp)
      u3=rnd(0.)
      if (alog(u3).gt.(alp-1.)*alog(x)) go to 1
 10   gs=x*bet
      return
      end
c
c
c
      subroutine twoper(zpar,znpar)
```

```
c
c  This program computes the parameters of a beta
c  distribution given two percentiles of the distribution
c
c  The following values are to be read in from MCBETA.IN:
c     r2=upper percentile estimate
c     r3=lower percentile estimate
c     y=percentile level of r2 (e.g. .95)
c     w=percentile level of r3 (e.g. .05)
c
c  The following values are calculated as the beta
c  parameters:
c    zpar=the value of x0 in single precision
c    znpar=the value of n0 in single precision
c
      implicit real*8(a-h,o-y)
      common/per2sub/n,int,ifin
      save /per2sub/
      read(11,*)r2,r3,y,w
      a1 = .0001
      b1 = 1000.
      n = 0
      int=0
      ifin = 0
   30 call perint(an,bn,n)
      call betasub(an,bn,a1,b1,r2,r3,y,w,ppar,xnpar)
      if(ifin.ge.1)go to 20
      if(int.gt.2) go to 40
      if(int.ge.1) go to 30
   40 write(*,41)
   41 format(' parameter value is greater than 100000.' )
   20 continue
      zpar=ppar
      znpar=xnpar
      return
      end
c
c
c
      subroutine betasub(an,bn,a1,b1,r2,r3,y,w,ppar,xnpar)
      implicit real*8(a-h,o-y)
      common/per2sub/n,int,ifin
      save /per2sub/
c
c  Prints to output file for beta information, BET.OUT
c
      write(33,31)
   31 format(' RESULTS FOR THIS BETA DISTRIBUTION:',/,/)
      write(33,33) r2,y,r3,w
   33 format(' upper percentile= ',f12.6,' level=',f12.6,/,
     *' lower percentile=', f12.6,' level=',f12.6)
      x0=betpar(a1,b1,r2,an,y)
      ax= dbti(r3,an,x0)
      x1 = betpar(a1,b1,r2,bn,y)
      bx = dbti(r3,bn,x1)
   15 cn = (an + bn)/2.0
      x2 = betpar(a1,b1,r2,cn,y)
      cx = dbti(r3,cn,x2)
      if(cx-w) 50,60,60
   50 bn = cn
```

```
      go to 70
   60 an = cn
   70 dn = bn - an
      if(dn .lt. 10e-10) go to 80
      go to 15
   80 ppar = (an + bn)/2.
      qpar = x2
      prob = dbti(r2,ppar,qpar)
      diff = abs(prob-y) - .0001
      if(diff) 30,20,20
   30 prob = dbti(r3,ppar,qpar)
      diff = abs(prob-w) - .0001
      if(diff)35,20,20
   20 int = int+1
      go to 25
   35 xnpar = ppar + qpar
      bmean = ppar/xnpar
      ifin = ifin + 1
   91 format(' ',a4,/,a80)
      write(33,90) ppar,qpar,xnpar
   90 format(5x,'p = x0 = ',f12.6,5x,'q =  ',f12.6,5x,'  n0
     *= ',f12.6/)
      write(33,98) bmean
   98 format(2x,'the mean of the prior is  ',f12.8)
      var=(bmean*(1-bmean))/(xnpar+1.)
      std=var**.5
      write(33,85)var,std
   85 format(2x,'variance=',e14.6,'  std. dev.=',e14.6/)
      prob=dbti(r2,ppar,qpar)
      write(33,95) prob
   95 format(2x,'the upper percentile probability is
     *',f16.13)
      prob = dbti(r3,ppar,qpar)
      write(33,96) prob
   96 format(2x,'the lower percentile probability is
     *',f16.13,//)
   25 return
      end
c
c
c
      function betpar(a1,b1,r2,an,y)
      implicit real*8(a-h,o-y)
      a = a1
      b = b1
   10 c = (a + b)/2.0
      x = dbti(r2,an,c)
      if (x-y) 20,30,30
   30 b = c
      go to 40
   20 a = c
   40 d = b-a
      if (d.lt.10e-10) go to 50
      go to 10
   50 betpar = (a + b)/2.0
      return
      end
c
c
c
```

```
      subroutine perint(an,bn,n)
      n = n+1
      go to(10,20,30) n
   10 an=.0001
      bn=1000.
      go to 40
   20 an=1000.
      bn=10000.
      go to 40
   30 an=10000.
      bn=100000.
   40 return
      end
c
c
c

      subroutine meanper(z0,zn0)
c
c  This program computes the parameters of a beta
c   distribution given the mean and one percentile
c   of the distribution
c
c  The following values are read in from MCBETA.IN:
c       rmean = mean
c       perc = percentile
c       prob = percentile level.
c
c  The following values are calculated:
c       z0 = the value of x0 in single precision
c       zn0= the value of n0 in single precision
c
      implicit real*8(a-h,o-y)
      common/meansub/rmean,perc,prob,n,int
      save /meansub/
      read(11,*) rmean, perc, prob
      n = 0
      int = 0
   30 call parint(a,b,n)
      if(rmean.lt.perc)go to 10
      go to 20
   10 r1 = rmean
      r2 = perc
      y = prob
      go to 38
   20 r1 = 1. - rmean
      r2 = 1. - perc
      y = 1. - prob
   38 call betsub2(r1,r2,a,b,y,x0,fn0)
      if(int.eq.1) go to 35
      if(n.le.3)go to 30
      if(n.gt.3) go to 40
      go to 35
   40 write(*,42)
   42 format(' parameter value is greater than 100,000.' )
   35 continue
      z0=x0
      zn0=fn0
      return
      end
```

```
c
c
c
      subroutine betsub2(r1,r2,a,b,y,x0,fn0)
      implicit real*8(a-h,o-y)
      common/meansub/rmean,perc,prob,n,int
      save /meansub/
c
c   Prints beta information to BET.OUT
c
   92 format(' ',a4,/,a80)
      write(33,15) rmean,perc,prob
   15 format(2x,' the given mean is ',f8.5,//
     * '    the given percentile is   ',f8.5,//
     * '    the given percentile level   ',f8.5/)
      fn0=btpar(r1,r2,a,b,y)
      x0 = fn0*r1
      if(rmean.lt.perc)go to 30
      go to 40
   30 p = x0
      q = fn0 - x0
      go to 50
   40 q = x0
      p = fn0 - x0
      r2 = perc
      x0 = p
   50 prb = dbti(r2,p,q)
      diff = dabs(prb-prob) - .00001
      if(diff.ge.0.) go to 85
      int = int + 1
      write(33,60) fn0, x0
   60 format(2x,' n0 =   ',f12.6,'   x0 =   ', f12.6/)
      write(33,20) r2,prb
   20 format(/2x,' the probability at',f12.7,' is ',f19.17/)
      pm = dbti(rmean,p,q)
      write(33,70) rmean, pm
   70 format(2x,' the probability at the mean ',f7.5,'
     *is',f12.8/)
      pmean = p/(p+q)
      var = (pmean*(1.-pmean))/(p+q+1)
      std = var**.5
      write(33,80) pmean
   80 format(2x,' the mean of the beta prior is x0/n0 =
     *',f8.5)
      write(33,86) var,std
   86 format(2x,' variance=',e14.6,'  std. dev.=',e14.6)
   85 return
      end
c
c
c
      function btpar(r1,r2,a,b,y)
      implicit real*8(a-h,o-y)
      common/meansub/rmean,perc,prob,n,int
      save /meansub/
   10 c = (a + b)/2.0
      x=c*r1
      q=c-x
      p=dbti(r2,x,q)
      if(p-y)20,30,30
```

```
   30 b=c
      go to 40
   20 a=c
   40 d=b-a
      if (d .lt. 1.0e-10)go to 50
      go to 10
   50 btpar=(a+b)/2.0
      return
      end
c
c
c
      subroutine parint(a,b,n)
      implicit real*8(a-h,o-y)
      n = n+ 1
      go to(10,20,30) n
   10 a = .0001
      b = 1000.
      go to 40
   20 a = 1000.
      b = 10000.
      go to 40
   30 a = 10000.
      b = 100000.
   40 return
      end
c
c
c
      subroutine gl(alp,x)
c
c  Finds gamma values for alpha less than 1.0
c
    1 u1=rnd(0.)
      b=(2.718281828+alp)/2.718281828
      p=b*u1
      if(p.gt.1.0) go to 3
    2 x=exp(alog(p)/alp)
      u2=rnd(0.)
      if(u2.gt.exp(-x)) go to 1
      return
    3 x=-alog((b-p)/alp)
      u3=rnd(0.)
      if(alog(u3).gt.(alp-1.)*alog(x)) go to 1
      return
      end
      function gt (alp)
c
c   for alpha .gt. 0.5  Adapted from:
c   Cheng, R. C. H. and Feast, G. M. (1979)
c   "Some Simple Gamma Variate Generators,"
c   Applied Statistics, vol 28, p. 290-295.
c
      data aset/-1./
      if (aset .eq. alp) go to 1
      aset = alp
      a = alp - 0.5
      b = alp / a
      c = 2.0 / a
      d = c + 2.0
```

```
      s = sqrt(alp)
      h1 = (0.865 + 0.064/alp) /s
      h2 = (0.4343 - 0.105/s) / s
  1   u1 = rnd(0.)
      u = rnd(0.)
      if(u1.le.0. .or. u1.ge.1.)go to 1
      if(u.le.0.  .or.  u.ge.1.)go to 1
      u2 = u1 + h1*u - h2
      if (u2 .le. 0.) go to 1
      if (u2 .ge. 1) go to 1
  2   w = b * (u1/u2) * (u1/u2)
      if ( (c*u2-d+w+1./w) .le. 0.) go to 4
  3   if ( (c*alog(u2)-alog(w)+w-1.) .ge. 0.) go to 1
  4   gt = a * w
      return
      end
c
c
c
      function rnd(idum)
c
c  Generates random uniform numbers, use 0=idum
c
      save p,q,m,j,nn

      integer p,q,m(98)
      logical inuse
      data j/40/,nn/2147483647/,p/98/,q/27/
      data (m(k),k=1,50)/
     *1387256442, 539505633,7126687,2115653676,480642437,
     *1403109719,898019591,1609472695,742049136,964528840,
     *1774590149,531014893,1478060509,224730595,1413365137,
     *1415397063,370513614,1981855272,1672294721,
     *1559669404,1992066581,440083042,1552169384,949029171,
     *1848294689,1014369863,1226252978,199445637,552539314,
     *101995811,1795618857,1468200845,403608434,466262418,
     *1783034892,2125486341,1437171068,839437811,685760609,
     *311739045,1876584692, 223544964, 667792106,1829604735,
     *887026472,688815796,1153871680,1135467106,1975710098,
     *1393037901/
      data (m(k),k=51,98)/
     *330755675,804762632,393596594,1695657725,50479950,
     *1039358666,1885424316,400881551,142829986,187416368,
     *821029919,1292641081,415120294,1104581275,
     *1258423968,304285054, 400491932,2014625087,1619263031,
     *750624285,1996732699,97476312,1250544934,
     *2145510054,1510875684,262891578, 616032534,
     *1316668730,1500747974,2138561534,809719156,
     *1605036043,510086967,317411066,54278455,2052774305,
     *439191668,1881943474,1397167115,2046084812,
     *644321591,328615697,1004646018,1110120728,2007784487,
     *992677826,1756605308,796797739/
      if(idum)200,100,300
 100  j=j+1
      if(j.gt.p)j=1
      k=j+q
      if(k.gt.p)k=k-p
      m(j)=m(k).xor.m(j)
      rnd=float(m(j))/nn
      return
```

```
  200 iunit=100
  201 iunit=iunit-1
      inquire(unit=iunit,opened=inuse)
      if(inuse)goto 201
      open(unit=iunit,file='rnd.str',status='unknown')
      read(iunit,*)m,j
      close(unit=iunit)
      return
  300 iunit=100
  301 iunit=iunit-1
      inquire(unit=iunit,opened=inuse)
      if(inuse)goto 301
      open(unit=iunit,file='rnd.str',status='unknown')
      write(iunit,*)m,j
      close(unit=iunit)
      return
      end
c
c
c
      double precision function dbti (x,a,b)
c
c  Incomplete beta function value for x with a, b parameters
c   calls betac and gamln  -- double precision function
c
      real*8 betacf,gamln,a,b,x,bt,one,zero,two
      data one,zero,two/1.d0,0.d0,2.d0/
      if(x.lt.zero.or.x.gt.one) print *,'bad argument x in
     *betai'
      if(x.lt.zero) x=zero
      if(x.gt.one) x=one
      if(x.eq.zero.or.x.eq.one) then
        bt=zero
      else
        bt=exp(gamln(a+b)-gamln(a)-gamln(b)
     *        +a*dlog(x)+b*dlog(one-x))
      endif
      if(x.lt.(a+one)/(a+b+two))then
       dbti=bt*betacf(a,b,x)/a
       return
      else
       dbti=one-bt*betacf(b,a,one-x)/b
       return
      endif
      end
c
c
c
      double precision function betacf(a,b,x)
      parameter(itmax=100,eps=3.d-7)
      implicit real*8 (a-h,o-z)
      data fone/1.d0/
      am=fone
      bm=fone
      az=fone
      qab=a+b
      qap=a+fone
      qam=a-fone
      bz=fone-qab*x/qap
      do 11 m=1,itmax
```

```
        em=m
        tem=em+em
        d=em*(b-m)*x/((qam+tem)*(a+tem))
        ap=az+d*am
        bp=bz+d*bm
        d=-(a+em)*(qab+em)*x/((a+tem)*(qap+tem))
        app=ap+d*az
        bpp=bp+d*bz
        aold=az
        am=ap/bpp
        bm=bp/bpp
        az=app/bpp
        bz=fone
        if(abs(az-aold).lt.eps*abs(az)) go to 1
   11 continue
      pause 'a or b too big, or itmax too small; Hit CR'
    1 betacf=az
      return
      end
c
c
c
      double precision function gamln(xx)
      save cof,stp,half,fone,fpf,x,tmp,ser
       real*8 cof(6),stp,half,fone,fpf,x,tmp,ser
      data cof,stp/76.18009173d0,-86.50532033d0,
     *    24.01409822d0,
     *    -1.231739516d0,.120858003d-2,-.536382d-5,
     *    2.50662827465d0/
      data half,fone,fpf/0.5d0,1.0d0,5.5d0/
      x=xx-fone
      tmp=x+fpf
      tmp=(x+half)*log(tmp)-tmp
      ser=fone
      do 11 j=1,6
       x=x+fone
       ser=ser+cof(j)/x
   11 continue
      gamln=tmp+log(stp*ser)
      return
      end
c
c
c
      subroutine sort (ra,n)
c
c   Sorts an array RA of length N into ascending numerial
c   order using the Heapsort algorithm.  N is input;  RA is
c   replaced by its sorted rearrangement.
c
      dimension ra(n)
      l = n/2 + 1
      ir = n
   10 continuc
      if (l .gt. 1) then
         l = l - 1
        rra = ra(l)
        else
        rra = ra(ir)
        ra(ir) = ra(1)
```

```
         ir = ir - 1
         if (ir .eq. 1) then
            ra(1) = rra
            return
            endif
         endif
       i = l

       j = l + l
   20 continue
       if (j .le. ir) then
          if (j .lt. ir) then
            if (ra(j) .lt. ra(j+1)) j=j+1
            endif
         if (rra .lt. ra(j)) then
            ra(i) = ra(j)
            i = j
            j = j + j
            else
            j = ir + 1
            endif
         go to 20
         endif
       ra(i) = rra
       go to 10
       end
c
c sorts and outputs Monte Carlo results
c

       subroutine finish (pval,nn)
       dimension pval(10000),per(13)
       fnn=nn
       i01=fnn*.010001
       i05=fnn*.050001
       i10=fnn*.100001
       i20=fnn*.200001
       i30=fnn*.300001
       i40=fnn*.400001
       i50=fnn*.500001
       i60=fnn*.600001
       i70=fnn*.700001
       i80=fnn*.800001
       i90=fnn*.900001
       i95=fnn*.950001
       i99=fnn*.990001
       ss=0.0
       sum=0.0
       do 412 i=1,nn
       sum=sum+pval(i)
       ss=ss+pval(i)**2
  412 continue
       avg=sum/fnn
       var=(ss-sum**2/fnn)/(fnn-1.)
       stdev=sqrt(var)
       call sort (pval,nn)
       fmin=pval(1)
       fmax=pval(nn)
       per(1)=pval(i01)
       per(2)=pval(i05)
```

```
      per(3)=pval(i10)
      per(4)=pval(i20)
      per(5)=pval(i30)
      per(6)=pval(i40)
      per(7)=pval(i50)
      per(8)=pval(i60)
      per(9)=pval(i70)
      per(10)=pval(i80)
      per(11)=pval(i90)
      per(12)=pval(i95)
      per(13)=pval(i99)
      i51=i50+1
      fmedian=(pval(i50)+pval(i51))/2.
      write(22,41) nn
   41 format(/,'Monte Carlo results for ',i6,' samples:',/)
      write(22,42) fmin,fmax
   42 format(' minimum value = ',e12.6,/,' maximum value =
     *',e12.6)
      write(22,43) avg,var,stdev
   43 format(' mean = ',e12.6,/,' variance = ',e12.6,/,
     *' standard deviation = ',e12.6)
      write(22,44) fmedian,(per(k), k=1,13)
   44 format(' median = ',e12.6,/,/,'  1st percentile = ',
     *e12.6,/,'  5th percentile = ',e12.6,/,' 10th
     *percentile = ',e12.6,/,' 20th percentile = ',e12.6,/,'
     *30th percentile = ',e12.6,/,' 40th percentile = ',
     *e12.6,/,' 50th percentile = ',e12.6,/,
     *' 60th percentile = ',e12.6,/,' 70th percentile = ',
     *e12.6,/,' 80th percentile = ',e12.6,/,' 90th
     *percentile = ',e12.6,/,' 95th percentile = ',e12.6,/,
     *'99th percentile = ',e12.6)
      return
      end
```

Appendix C—EMPIRICAL

```
      program empirical
c
c
c forms empirical distribution functions for a given set
c of percentiles for multiple experts.
c
c uses simulation to combine weighted aggregations of these
c distributions according to a specified weighting function
c
c empirical cumulative distribution functions (for each
c expert) are sampled in the simulation using lines
c connecting the individual points of the distribution.  The
c more percentiles provided by the experts, the less
c influence this linear approximation has on the results
c A step function version of this code is available from the
c authors.
c
c inputs can be made directly from the terminal or through a
c file called emp.in (on unit 52).
c outputs are on a file called emp.out (on unit 59).
c
c emp.in file has the following lines and formats:
c  line 1 idim = no. of experts (distributions) - free
c     format
c  line 2 iper = no. of percentiles for each distribution -
c     free format for each expert & DM do lines 3,4,5,6
c  line 3 pe array = estimates of the iper percentiles -
c     free format
c  line 4 pl array = levels (e.g. 0.95) for the percentiles
c     - free format
c  line 5 pmin = minimum value for the estimates - free
c     format
c  line 6 pmax = maximum value for the estimates - free
c     format
c     last 3 lines are:
c  line 7 nn = number of simulations (e.g. 1000) - free
c     format
c  line 8 ifun = 1 for equal weights, = 2 for unequal - free
c     format
c  line 9 wt array = weights for experts & DM - free format
c
c
      dimension pe(20,20), pl(20,20), fx(21), fy(21)
      dimension val(20), fxa(20,21), fya(20,21),plx(20,21)
      dimension pmin(20),pmax(20),pval(10000),per(13),wt(20)
      open (unit=59,file='emp.out',status='new')
      open (unit=52,file='emp.in',status='old')
```

```
      write(*,'("To use input file EMP.IN, type 1; else type
     *0: ",\)')
      read(*,*) ifile
      if(ifile.eq.1) go to 50

      write(*,'("Enter the sum of the experts and DM: ",\)')
      read(*,*) idim
      write( *,'("Enter the # of percentiles for an
     * expert/DM: ",\)')
      read(*,*) iper
      do 10 i=1,idim
      write (*,11) iper, i
   11 format ('Enter the ',i3,' percentile estimates for
     *  expert ',i3)
      read(*,*) (pe(i,j), j=1,iper)
      write (*,12) iper, i
   12 format (' Enter the ',i3,' percentile levels ',
     *'(e.g. .95 for 95th percentile) for expert ',i3)
      read(*,*) (pl(i,j), j=1,iper)
      write(*,'("Enter the absolute minimum value possible:
     *",\)')
      read(*,*) pmin(i)
      write(*,'("Enter the absolute maximum value possible:
     *",\)')
      read(*,*) pmax(i)
   10 continue
      write( *,'("Enter the number of samples for the
     *simulation ",\)')
      read(*,*) nn
      write(*,48)
  48  format(/,'Specify the aggregation function to be
     *used:')
      write(*,'("Enter 1 for equal weights, else enter 2:
     *",\)')
      read(*,*)ifun
      if(ifun.eq.2) then
      write(*,'("Enter the weights for experts & DM,
     *including 0s: ")')
      do 8 i=1,idim
      write(*,7) i
    7 format('  weight for person ',i2,' = ',\)
      read(*,*) wt(i)
    8 continue
      endif
      sumw=0.0
      do 6 i=1,idim
    6 sumw=sumw+wt(i)
      epsilon=0.0001
      if(abs(sumw-1.0).gt.epsilon) then
      write(*,'("The weights will be normalized to 1.0")')
      do 5 i=1,idim
    5 wt(i)=wt(i)/sumw
      endif
      write(*,'("Output is on file EMP.OUT")')
      go to 40
```

```
   50 continue
      read(52,*) idim
      read(52,*) iper
      do 19 i=1,idim
      read(52,*) (pe(i,j), j=1,iper)
      read(52,*) (pl(i,j), j=1,iper)
      read(52,*) pmin(i)
      read(52,*) pmax(i)
   19 continue
      read(52,*) nn
      read(52,*) ifun
      read(52,*) (wt(i), i=1,idim)
      sumw=0.0
      epsilon = 0.0001
      do 4 i=1,idim
    4 sumw=sumw+wt(i)
      if(abs(sum-1.0).gt.epsilon) then
      write(59,'("The weights are normalized to 1.0")')
      do 3 i=1,idim
    3 wt(i)=wt(i)/sumw
      endif
   40 continue
      write(59,13) idim,iper,nn
   13 format ('Number of experts=',i5,/,'Number of
     *percentiles=',i5,/,'Number of samples=',i5)
      if(ifun.eq.1) write(59,49)
   49 format('The aggregation function uses equal weights.')
      if(ifun.eq.2) then
      write(59,47)
   47 format('The aggregation function has weights as:')
      write(59,*) (wt(i), i=1,idim)
      endif
      do 14 i=1,idim
      write(59,16) i
   16 format(/,'Estimates for expert ',i3,':')
      write(59,15) pmin(i),pmax(i),(pl(i,j),j=1,iper)
   15 format(' min & max=',2f6.3,/,' levels ',5f7.4,/,
     *'           ', 6f7.4)
      write(59,9)(pe(i,j),j=1,iper)
   9  format(' estimates',5f7.4,/,'           ',6f7.4)
   14 continue
c
c
      do 20 i=1,idim
      call distmake (i,pe,pl,iper,pmin(i),pmax(i),fx,fy)
      iper1=iper+1
      do 30 j=1,iper1
      fxa(i,j)=fx(j)
      fya(i,j)=fy(j)
   30 continue
   20 continue
c
c  forms the nn samples of the product distribution
c
      call monte (fxa,fya,idim,iper,nn,pval,plx,pl,ifun,
     * wt,pmax,pmin)
```

```
c
c  calculates stats for pval
c
      call calc(nn,pval,avg,fmedian,var,stdev,fmin,fmax,per)
c
c  print results
c
      write(59,41) nn
   41 format(/,'Monte Carlo results for ',i5,' samples:')
      write(59,42) fmin,fmax
   42 format(' minimum value = ',e12.6,/,' maximum value =
     *',e12.6)
      write(59,43) avg,var,stdev
   43 format(' mean = ',e12.6,/,' variance = ',e12.6,/,
     *' standard deviation = ',e12.6)
      write(59,44) fmedian,(per(k), k=1,13)
   44 format(' median = ',e12.6,/,'  1st percentile =
     *',e12.6,
     */,'  5th percentile = ',e12.6,/,' 10th percentile = ',
     *e12.6,/,' 20th percentile = ',e12.6,/,' 30th
     *percentile = ',e12.6,/,' 40th percentile = ',
     *e12.6,/,' 50th percentile = ',e12.6,/,
     *' 60th percentile = ',e12.6,/,' 70th percentile = ',
     *e12.6,/,' 80th percentile = ',e12.6,/,' 90th
     *percentile = ',e12.6,/,' 95th percentile = ',e12.6,/,
     *' 99th percentile = ',e12.6)
      close (unit=59)
      close (unit=52)
      end
c
c  Forms empirical distribution function for each
c  expert & DM: fx and fy.
c
      subroutine distmake (ie,pe,pl,n,pmn,pmx,fx,fy)
      dimension pe(20,20), pl(20,20), fx(21), fy(21)
      ibin=n+1
      fy(1) = pl(ie,1)/(pe(ie,1)-pmn)
      fy(ibin) =(1.0-pl(ie,n))/(pmx-pe(ie,n))
      do 100 i=2,n
      j=i-1
      fy(i)=(pl(ie,i)-pl(ie,j))/(pe(ie,i)-pe(ie,j))
  100 continue
      do 110 i=1,n
      fx(i)=pe(ie,i)
  110 continue
      fx(ibin)=1.
      write(59,113) ie
  113 format(/,'Empirical distribution; fx,fy for
     *person:',i4)
      do 114 j=1,ibin
  114 write(59,*) fx(j),fy(j)
      return
      end
```

```
c
c  Performs the monte carlo simulation of the supplied
c  functions
c
      subroutine monte (fxa,fya,idim,iper,nn,pval,plx,pl,
     *ifun,wt, pmax,pmin)
      dimension fxa(20,21),fya(20,21),val(20),plx(20,21),
      wt(20)
      dimension pval(10000),pl(20,20),pmax(20),pmin(20)
      n=iper+1
      do 223 i=1,idim
      do 222 j=1,iper
      plx(i,j)=pl(i,j)
  222 continue
      plx(i,n)=1.0
  253 format(' plx=',6e12.6)
  223 continue
      do 200 isamp=1,nn
      do 201 i=1,idim
      n=iper+1
      idum = 1
      val(i)=0.0
      if (isamp.eq.1.and.i.eq.1) idum = -iabs(487320587)
      t=ran3(idum)
      if(t.eq.0.0) val(i)=0.0
      if(t.le.plx(i,1).and.t.gt.0.0) then
        b1 = plx(i,1)/(fxa(i,1)-pmin(i))
         b0 = plx(i,1)-b1*fxa(i,1)
         val(i)=(t-b0)/b1
      endif
      if(t.gt.plx(i,iper)) then
        b1 = (1.0-plx(i,iper))/(pmax(i)-fxa(i,iper))
         b0 = 1.0-b1*pmax(i)
         val(i) = (t-b0)/b1
      endif
      do 202 k=2,iper
      km1=k-1
      if(t.le.plx(i,k).and.t.gt.plx(i,km1)) then
        b1 = (plx(i,k)-plx(i,km1))/(fxa(i,k)-fxa(i,km1))
         b0 = plx(i,k)-b1*fxa(i,k)
         val(i) = (t-b0)/b1
      endif
  202 continue
  213 format(' isamp',i4,' expert',i2,' rand#',e12.6,'
     *value',e12.6)
  201 continue
      dim=idim
      if(ifun.eq.1) go to 275
      if(ifun.eq.2) go to 265
  265 continue
      sumy = 0.0
      do 266 i=1,idim
      sumy = sumy + val(i)*wt(i)
```

```
  266 continue
      pval(isamp)=sumy
      go to 270
  275 sumy=0.0
      do 276 i=1,idim
      fi=idim
      sumy=sumy+val(i)/fi
  276 continue
      pval(isamp)=sumy
      go to 270
  270 continue
  214 format('isamp',i4,' pval',e12.6)
  200 continue
      return
      end

      function ran3 (idum)
c
c         Returns a uniform random deviate between 0.0 and
c         1.0. Set IDUM to any negative value to initialize
c         or reinitialize the sequence.
c
      parameter (mbig=1000000000,mseed=161803398,mz=0,
     *fac=1./mbig)
c
c         According to Knuth, any large MBIG, and any
c         smaller (but still large) MSEED can be substituted
c         for the above values.
c
      save inext, inextp, ma
      dimension  ma(55)
      data iff /0/
      if (idum.lt.0 .or. iff.eq.0) then
        iff = 1
        mj = mseed - iabs(idum)
        mj = mod(mj,mbig)
        ma(55) = mj
        mk = 1s
        do 11 i = 1, 54
          ii = mod(21*i,55)
          ma(ii) = mk
          mk = mj - mk
          if (mk .lt. mz) mk = mk + mbig
          mj = ma(ii)
   11   continue
        do 13 k=1, 4
          do 12 i=1, 55
            ma(i) = ma(i) - ma(1+mod(i+30,55))
            if (ma(i) .lt. mz) ma(i) = ma(i) + mbig
   12     continue
   13     continue
        inext = 0
        inextp = 31
        idum = 1
      endif
```

```
      inext = inext + 1
      if (inext .eq. 56) inext = 1
      inextp = inextp + 1
      if (inextp .eq. 56) inextp = 1
      mj = ma(inext) - ma(inextp)
      if (mj .lt. mz) mj = mj + mbig
      ma(inext) = mj
      ran3 = mj*fac
      return
      end
c
c
c
      subroutine calc(nn,pval,avg,fmedian,var,stdev,fmin,
     *fmax,per)
      dimension pval(10000),per(13)
      fnn=nn
      i01=fnn*.010001
      i05=fnn*.050001
      i10=fnn*.100001
      i20=fnn*.200001
      i30=fnn*.300001
      i40=fnn*.400001
      i50=fnn*.500001
      i60=fnn*.600001
      i70=fnn*.700001
      i80=fnn*.800001
      i90=fnn*.900001
      i95=fnn*.950001
      i99=fnn*.990001
  401 format(' 1,5,10,50,90,95,99',7i4)
      scale=0.0
      ss=0.0
      sum=0.0
      do 400 i=1,nn
      scale=scale+pval(i)
  400 continue
      scale=1.0
      do 411 i=1,nn
      pval(i)=pval(i)/scale
  411 continue
  422 format('scale',e12.6)
      do 412 i=1,nn
      sum=sum+pval(i)
      ss=ss+pval(i)**2
  433 format ('pval =' ,e12.6)
  412 continue
      avg=sum/fnn
      var=(ss-sum**2/fnn)/(fnn-1.)
      stdev=sqrt(var)
      call sort (nn,pval)
      fmin=pval(1)
      fmax=pval(nn)
```

```
      per(1)=pval(i01)
      per(2)=pval(i05)
      per(3)=pval(i10)
      per(4)=pval(i20)
      per(5)=pval(i30)
      per(6)=pval(i40)
      per(7)=pval(i50)
      per(8)=pval(i60)
      per(9)=pval(i70)
      per(10)=pval(i80)
      per(11)=pval(i90)
      per(12)=pval(i95)
      per(13)=pval(i99)
      i51=i50+1
      fmedian=(pval(i50)+pval(i51))/2.
      return
      end


      subroutine sort (n,ra)
c
c         Sorts an array RA of length N into ascending
c         numerial order using the Heapsort algorithm.
c         N is input;  RA is replaced by its sorted
c         rearrangement.
c
      dimension ra(n)
      l = n/2 + 1
      ir = n
   10 continue
      if (l .gt. 1) then
         l = l - 1
          rra = ra(l)
          else
          rra = ra(ir)
          ra(ir) = ra(1)
          ir = ir - 1
          if (ir .eq. 1) then
           ra(1) = rra
           return
           endif
       endif
      i = l
      j = l + l
   20 continue
      if (j .le. ir) then
         if (j .lt. ir) then
           if (ra(j) .lt. ra(j+1)) j=j+1
          endif
```

```
  if (rra .lt. ra(j)) then
     ra(i) = ra(j)
     i = j
     j = j + j
     else
     j = ir + 1
     endif
  go to 20
  endif
ra(i) = rra
go to 10
end
```

Appendix D—BOOT

```
      program boot
c
c  Constructs bootstrap samples from the original sample of
c  size n.
c
c  Each sample is randomly formed on its own.
c
c  Input is done on the terminal or file BOOT.IN
c  BOOT.IN has the following lines in free format:
c     line 1 iabb = 0 if all simulated values are printed,
c                 =1 if not
c     line 2 irange = 1 if the sample of n experts each
c     supplies two ranges values (3 values per person),
c                     = 0 if not **
c     line 3 n = number of values supplied = number of
c     persons or 3 times that
c     line 4 m = number of simulations
c     line 5 x array = the n values supplied
c                       (a space between each value)
c
c  Output is sent to file BOOT.OUT
c
c  ** if each expert provides a best estimate and an upper &
c     lower range value, then irange = 1 and there are 3
c     times as many values as experts.  for this case, the
c     sample size used in the program is changed to n/3 =
c     number of experts, not the number of values.
c
c
      character title*75
      dimension x(65),xsamp(65),xbig(20000),xnew(65)
      dimension xmed(1000)
      open (unit=11,file='boot.in',status='old')
      open (unit=22,file='boot.out',status='new')
      write(*,'("to use input file boot.in, type 1 else type
     *0: ",\)')
      read(*,*) infile
      if(infile.eq.1) go to 61

      write(*,31)
   31 format('to print all simulated values, type 0; else
     *type 1: ',\)
      read(*,*) iabb
      write(*,32)
   32 format('if range values are included, type 1; else
     *type 0: ',\)
      read(*,*) irange
      if(irange.eq.1) write(22,33)
      if(irange.eq.0) write(22,34)
```

```
   33  format(' irange=1, ranges with sample values are
      *assumed')
   34  format(' no range values included')
       write(*,'("enter the sample size: ",\)')
       read(*,*) n
       write(*,'("enter the number of simulations
      *(e.g.1000):",\)')
       read(*,*) m
       fn=n
       write(*,39)
   39  format('enter the sample values with a space between
      *each:')
       read(*,*) (x(i),i=1,n)
       go to 65

   61  continue
       read(11,*) iabb
       read(11,*) irange
       if(irange.eq.1) write(22,63)
       if(irange.eq.0) write(22,64)
   63  format(' irange=1, ranges with sample values are
      *assumed')
   64  format(' no range values included')
       read(11,*) n
       read(11,*) m
       fn=n
       read(11,*) (x(i),i=1,n)

   65  continue
       write(22,66) m
   66  format(' the number of simulations is ',i5)
       write(22,37) n, (x(i),i=1,n)
   37  format(' the ',i2,' sample values are: ',/,10e9.2)
c
c  Calculate simple statistics for the sample only
c
       slog=0.0
       sum=0.0
       ss=0.0
       do 51 i=1,n
       sum=sum+x(i)
       if(x(i).eq.0.0) go to 51
       slog=slog+alog10(x(i))
       ss=ss+x(i)**2
   51  continue
       xbar=sum/fn
       xgbar=10.0**(slog/fn)
       xvar=(ss-sum**2/fn)/(fn-1.)
       xstd=sqrt(xvar)
       do 52 i=1,n
       xnew(i)=x(i)
```

```
   52  continue
       call sort(xnew,n)
       k=n/2
       k1=k+1
       kind=mod(n,2)
       if(kind.eq.0) xtild=(xnew(k)+xnew(k1))/2.0
       if(kind.eq.1) xtild=xnew(k1)
       write(22,53)xbar,xtild,xgbar,xvar,xstd,xnew(1),xnew(n)
   53  format(/,'simple statistics for the original
      *sample:',/, ' sample mean = ',e9.2,/,' sample median
      *= ',e9.2,/, ' sample geometric mean = ',e9.2,/,'
      *sample variance = ',e9.2,/,' sample standard deviation
      *= ',e9.2,/, ' sample minimum = ',e9.2,/,' sample
      *maximum = ',e9.2,/)
c
c   Begin bootstrap sampling of x
c   All bootstrap samples are randomly formed and stored in
c   xbig
c
       do 50 i=1,m
       do 50 j=1,n
       k=(i-1)*n+j
       idum=1
       if(i.eq.1.and.j.eq.1) idum = -iabs(5739784770)
   12  t1=rand(idum)
       temp=t1*fn
       do 13 kk=1,n
       kk1=kk-1
       if(temp.eq.0.0) xbig(k)=x(1)
       if(temp.le.kk.and.temp.gt.kk1) xbig(k)=x(kk)
       if(temp.gt.n) go to 12
   13  continue
   50  continue
       if(irange.eq.1) n=fn/3.0
c
c  Loop for the 3 estimators: median, mean, geomean
c
       iwrite=0
       do 100 ii=1,3
       do 40 i=1,m
       do 41 j=1,n
       k=(i-1)*n+j
       xsamp(j)=xbig(k)
   41  continue
       call sort(xsamp,n)
c
c  Median calculations
c
       if (ii.eq.1) then
         k=n/2
         k1=k+1
         kind=mod(n,2)
         if(kind.eq.0) xmed(i)=(xsamp(k)+xsamp(k1))/2.0
         if(kind.eq.1) xmed(i)=xsamp(k1)
       endif
```

```
c
c  Mean calculations
c
      if (ii.eq.2)then
        xmean=0.0
        do 101 ij=1,n
 101    xmean=xmean+xsamp(ij)/float(n)
        xmed(i)=xmean
      endif
c
c  Geometric mean calculations
c
      if (ii.eq.3)then
        xgeo=0.0
        do 102 ij=1,n
        if(xsamp(ij).eq.0.0) go to 102
        xgeo=xgeo+alog10(xsamp(ij))
 102    continue
        if(xgeo.eq.0.0) then
          iwrite=iwrite+1
         xmed(i)=0.0
          go to 107
        endif
        xmed(i)=10.0**(xgeo/float(n))
 107    continue
      endif
  40  continue
      if(iwrite.gt.0.0) write(22,104)iwrite
 104  format(/,/,'warning: ',i4,' samples had no geometric
     *mean,',' results may be biased low')
c
c  End sampling, sort resulting estimators
c
      call sort(xmed,m)
c
c  Calculate percentiles, estimators, variances
c   Print results
c
      if(ii.eq.1) write(22,27)
      if(ii.eq.2) write(22,28)
      if(ii.eq.3) write(22,29)
  27  format(' ',//,'medians')
  28  format(' ',//,'means')
  29  format(' ',//,'geomeans')
      if(iabb.ne.1) write(22,43)
  43  format('the sample estimator values from the
     *simulation')
      if(iabb.ne.1) write(22,21) (xmed(i),i=1,m)
  21  format(' ', 6f10.6)
      smed=0.0
      ssmed=0.0
      slmed=0.0
      do 45 i=1,m
      smed=smed+xmed(i)
      ssmed=ssmed+xmed(i)**2
      if(xmed(i).eq.0.0) go to 45
      slmed=slmed+alog10(xmed(i))
```

```
   45 continue
      amed=smed/float(m)
      vmed=(ssmed-(smed**2)/float(m))/float(m-1)
      sdmed=sqrt(vmed)
      gmed=10.0**(slmed/float(m))
      delta=float(m)/100.0
      m99=99.0*delta
      m95=95.0*delta
      m90=90.0*delta
      m50=50.0*delta
      m20=20.0*delta
      m30=30.0*delta
      m40=40.0*delta
      m60=60.0*delta
      m70=70.0*delta
      m80=80.0*delta
      m1=1.0*delta
      m5=5.0*delta
      m10=10.0*delta
      write(22,23) xmed(1),xmed(m)
   23 format(' minimum value = ',f10.6,/,' maximum value
     *= ',f10.6)
      write(22,26) xmed(m1),xmed(m5),xmed(m10),xmed(m20),
     *xmed(m30),xmed(m40),xmed(m50),
     *xmed(m60),xmed(m70),xmed(m80),xmed(m90),
     *xmed(m95),xmed(m99)
   26 format(' percentiles: ',/,11x,'  1:',f10.6,/,11x,'
     *5:',f10.6,/,
     * 11x,' 10:',f10.6,/,11x,' 20:',f10.6,/,
     * 11x,' 30:',f10.6,/,11x,' 40:',f10.6,/,
     * 11x,' 50:',f10.6,/,11x,' 60:',f10.6,/,
     * 11x,' 70:',f10.6,/,11x,' 80:',f10.6,/,
     * 11x,' 90:',f10.6,/,11x,' 95:',f10.6,/, 11x,'
     *99:',f10.6)
      write(22,22) amed,vmed,sdmed,gmed
   22 format(' mean = ',f10.6,/,' variance',f10.6,/,
     *' standard deviation = ',f10.6,/,
     *' geometric mean = ',f10.6)
  100 continue
      stop
      end

      subroutine sort (ra,n)
c
c   Sorts an array ra of length n into ascending numerial
c   order using the heapsort algorithm.  n is input;  ra is
c   replaced by its sorted rearrangement.

      dimension ra(n)
      l = n/2 + 1
      ir = n
   10 continue
      if (l .gt. 1) then
         l = l - 1
```

```
        rra = ra(l)
        else
        rra = ra(ir)
        ra(ir) = ra(1)
        ir = ir - 1
        if (ir .eq. 1) then
           ra(1) = rra
           return
        endif
       endif
      i = l
      j = l + l
   20 continue
      if (j .le. ir) then
         if (j .lt. ir) then
          if (ra(j) .lt. ra(j+1)) j=j+1
         endif
         if (rra .lt. ra(j)) then
          ra(i) = ra(j)
          i = j
          j = j + j
         else
          j = ir + 1
         endif
       go to 20
      endif
      ra(i) = rra
      go to 10
      end


      function rand (idum)
c
c   Returns a uniform random deviate between 0.0 and 1.0.
c   set idum to any negative value to initialize or
c   reinitialize the sequence.
c
      parameter (mbig=1000000000,mseed=161803398,mz=0,fac=1
     */mbig)
c
c   According to Knuth, any large mbig, and any smaller (but
c   still large) mseed can be substituted for the above
c   values.
c
      save     inext,     inextp,      ma
      dimension      ma(55)
      data           iff /0/
      if (idum.lt.0 .or. iff.eq.0) then
         iff = 1
        mj = mseed - iabs(idum)
        mj = mod(mj,mbig)
        ma(55) = mj
        mk = 1
```

```
      do 11 i = 1, 54
        ii = mod(21*i,55)
        ma(ii) = mk
        mk = mj - mk
        if (mk .lt. mz) mk = mk + mbig
        mj = ma(ii)
11     continue
      do 13 k=1, 4
        do 12 i=1, 55
           ma(i) = ma(i) - ma(1+mod(i+30,55))
           if (ma(i) .lt. mz) ma(i) = ma(i) + mbig
12      continue
13    continue
      inext = 0
      inextp = 31
      idum = 1
    endif
    inext = inext + 1
    if (inext .eq. 56) inext = 1
    inextp = inextp + 1
    if (inextp .eq. 56) inextp = 1
    mj = ma(inext) - ma(inextp)
    if (mj .lt. mz) mj = mj + mbig
    ma(inext) = mj
    rand = mj*fac
    return
    end
```

Glossary of Expert Judgment Terms

ADVISORY EXPERT: The in-house employee or consultant considered an expert in the subject matter. This person assists the project personnel in developing the questions that the external experts will answer at a later session.

AGGREGATION: See Behavioral or Mathematical Aggregation.

ALPHA LEVEL OR α LEVEL: See Significance Level.

ANALYSIS OF VARIANCE: A statistical technique for testing the equivalence of mean values from several different data groups or classes. The test is done by comparing variation between the groups to the variation within the groups. The within-group variation represents the background-, error-, or noise-level variation. Groups are determined before the study so that the test can be made from a minimal number of measurements. Such efficient planning is called experimental design.

ANALYST: A member of the project personnel who analyzes the expert data and judgment. The analyst may have other roles (e.g., interviewer).

ANCHORING AND ADJUSTMENT HEURISTIC: This effect occurs when individuals reach a final answer by starting from an initial value and adjusting from it. In some cases, the questions supply the initial values while in others the experts derive the values through their impressions or computations. The use of this heuristic usually skews the answer toward the initial value. See also Heuristic.

ANCHORING BIAS: An individual's failure to sufficiently adjust from his or her first impression in solving a problem—the individual *anchors* to the first impression. Sometimes this bias is explained in terms of Bayes theorem as the failure to adjust a judgment in light of new information as much as it would be adjusted in terms of Bayes mathematical formula.

ANCILLARY DATA/INFORMATION: Any information or data gathered as part of the elicitation that does not consist of the expert's answer. For example, information on the expert's background and problem-solving processes (expert data) are considered ancillary information. Because this information may relate to the answers, it can form conditional variables.

ANSWER–ONLY DOCUMENTATION: A written record that exclusively documents the experts' answers.

ANSWERS OR EXPERT ANSWERS: The expert's final response to a technical question. This term includes responses given in quantitative (estimates) or qualitative (solution) form.

ANSWER PLUS PROBLEM-SOLVING DOCUMENTATION: A written record of the experts' answers and their problem-solving processes.

AVAILABILITY BIAS: How easily an individual can retrieve particular events from long-term memory. Data involving catastrophic, familiar, concrete, or recent events may be easier to recall. Availability bias affects an individual's ability to accurately estimate frequencies and to recall other aspects of the event.

BACKGROUND INFORMATION: Information that the expert needs to interpret the question or problem. Background information includes the sequence of events leading up to the event in question, pictorial representations of the question (e.g., flow charts), and decompositions of the question.

BAYESIAN METHODS/APPROACH: A technique that combines information of various types or from various sources. The combining calculations are based on Bayes theorem, which defines the probability distribution function of the data as conditioned on its parameters. These parameters are also assumed to have a probability distribution called a prior distribution. Combining the data and prior distributions produces a posterior distribution. The expected value of this posterior distribution is the desired final estimate. The philosophy of this method is to use and combine all available information to form the final estimate rather than to rely exclusively on the data from a single study or experiment.

BEHAVIORAL AGGREGATION: A means of obtaining one answer from multiple experts through the use of behavioral techniques that encourage consensus. For example, group-think bias can be fostered to create pressures toward unanimity.

BETA PROBABILITY DISTRIBUTION: A very useful probability distribution function (pdf) for probability estimates because the range of the distribution is between 0 and 1. It is also a very versatile distribution that can generate many different shapes simply by changing the 2 parameters. The two parameters are functions of the number of successes and the number of trials. In a Bayesian application, a beta prior distribution easily combines with a binomial process. The resulting distribution is also a beta.

BIAS: Bias occurs when (1) expressions of the experts' thinking do not match their actual thinking at the time of the elicitation, and (2) the experts' estimates do not follow normative statistical or logical rules. An example of the first would be if the expert judged a particular event to be extremely rare but had to select from response options that did not extend as far as his or her judgment. An example of the second would be if the expert claimed that A was better than B, B better than C, and C better than A. Sources of bias can be a person's needs (*motivational bias*) or thought processes (*cognitive bias*).

BOOTSTRAP: A data-based simulation technique useful for finding estimates and distributions of estimates when statistical distribution theory does not apply. This technique is based on forming multiple (e.g., 1000) random samples with replacement from the original sample data, calculating the estimator of interest in each sample, and forming the distribution of that estimator from these calculated 1000 values.

CENTRAL LIMIT THEOREM (CLT): Theorem used for identifying the theoretical distribution for the sample mean. The sample mean is distributed approximately normal, with a mean equal to the population mean and a variance equal to the population variance divided by the sample size. The theorem works best for large sample sizes (> 30) and for symmetrically distributed populations.

CLUSTER ANALYSIS: Techniques for identifying how values or variables are grouped numerically or how variables themselves are grouped according to shared information. For example, the data set of values (0.2, 0.25, 0.30, 0.77, 0.80, 0.95) would cluster into two groups. Various grouping methods determine group size and membership. Most are based on some measure of distance between values and groups of values.

COGNITION: The mental processing of information that humans do when they solve problems.

COGNITIVE BIAS: Biases that arise from the limitations of the human mind. Anchoring bias is one example.

COGNITIVE DISSONANCE: Cognitive dissonance occurs when individuals find a discrepancy either between their beliefs or between their beliefs and their actions (Festinger, 1957). For example, individuals may find that they hold an opinion that conflicts with that of the other group members, and if they have a high opinion of the intelligence of the group, they may resolve the discrepancy by unconsciously changing their judgment to agree with that of thc group.

CONDITIONAL VARIABLE: An ancillary quantity (variable) influential or important for determining the experts' answers. More than one analysis technique helps make this determination.

CONDITIONALITY: The description of the phenomenon in which one variable has an influence on or a significant relationship to another variable. Conditionality also includes the case in which several variables influence several other variables.

CORRELATION: The Pearson correlation coefficient is the statistic calculated from the data set which measures the strength of the linear relationship between two variables. A perfect line has correlation 1.0. No correlation between the variables equals 0.0. See also Dependence.

COVARIANCE: Approximately the average of the product of the deviations from their means of two random variables. Covariance is the numerator of the correlation coefficient. If the two random variables are identical, their covariance is the variance.

CRITICAL VALUE: A term used in significance testing representing the value in the tails of the test distribution that delineates the areas according to the specified significance level. It is the value compared to the estimated value from the sample to determine if the estimated value belongs to the test distribution.

CUMULATIVE DISTRIBUTION FUNCTION (CDF): The function resulting from the accumulation of area under the probability distribution function (pdf); in other words, the integral of the pdf. The function value, $F(x)$, is the probability that the random variable takes on values less than or equal to the value at x.

DATA GATHERER: See Interviewer or Knowledge Engineer.

DECISION ANALYSIS: A structured means used for conceptualizing and resolving complex problems. This structure involves breaking the problem into parts to make it more tractable. The decision structure often includes the initial options or acts, the possible consequences of these values, and uncertainty measures. A variety of structural forms are used, ranging from event trees to hierarchies of factors. Decision analysis is applied mainly to business problems such as plant siting, procurement, and portfolio analysis. This analysis has also been used in crisis management, international negotiations, intelligence analyses, and labor-management negotiations (Peaslee, 1981).

DECOMPOSITION: The division of a problem or question into its component parts to make it easier to solve. This technique has been shown to increase accuracy.

DEGREES OF FREEDOM: Parameter of some pdfs (e.g., the F distribution). It is a function of the sample size (usually n-1) and represents the amount of information available in the sample after estimating certain parameters.

DELPHI SITUATION: An elicitation method developed by Rand Corporation to limit the biasing effects of interaction. In a true Delphi, the experts do not interact with one another and only interact with the moderator in a limited way. The experts, in isolation from one another, give their judgments and, in some cases, their reasons for making these judgments. The moderator collects these judgments, makes the judgments anonymous, distributes these judgments to the individual experts, and allows each of them to revise their previous judgments. This process can be repeated for as many times as desired (e.g., until consensus is achieved).

DEPENDENCE: Experts' estimates may be conditioned on some factor. Their estimates are therefore affected by this conditioning. In this book, *dependence* is used interchangeably with *correlation*. Dependence can refer to the data from different experts (between-expert dependence) or it can refer to dependencies of estimates given by the same expert (within-expert dependence).

DEPENDENT VARIABLE: The quantity of interest that contains the expert answers. These answers are usually conditioned and therefore dependent on other ancillary variables or information.

DETAILED STRUCTURED DOCUMENTATION: See Structured Documentation.

DETAILED VERBATIM DOCUMENTATION: See Verbatim Documentation.

DISAGGREGATION: See Decomposition.

DISCRIMINANT ANALYSIS: A multivariate statistical technique for determining how well a chosen variable discriminates or classifies each datum from a data set into specified groups.

DISPERSION MEASURE: An estimate of how much spread or dispersion appears in the sample data. Dispersion measures take the form of percentiles, variances, ranges, or error bars.

DISTRIBUTION: See Probability Distribution Function.

ELICITATION: Process of gathering expert judgment in a specially designed manner.

ELICITATION DESIGN: Planning the method in terms of (1) the project's constraints, such as time, budget, and personnel; (2) goals, such as for obtaining particular data; and (3) additional considerations, such as the logistics and cost of different elicitation situations and models of communication, the structuring of the elicitations, the treatment of bias, the presentation of the problems, and the documenting of the elicitation. The reason for designing the elicitation is to create the optimal combination of techniques for a particular situation. Different techniques possess differing advantages and abilities to control for particular factors, such as those which would introduce bias. One of the main techniques used in designing the data gathering is *structuring*, or placing controls on the elicitation (see Structuring). An example of structuring applied to interactive groups is to have the natural leader present his or her views last, so as to prevent the follow-the-leader effect (see Group Think Bias).

ELICITATION SITUATION: The setting in which the expert's judgment is elicited. Experts' judgments can be elicited in private, in a group setting, or when the experts are alone but receiving information on the other experts' judgments. (See Delphi Situation, Individual Interview, and Interactive Group Situation.)

ELICITATION TECHNIQUE: The means by which the experts are led to describe aspects of their knowledge (e.g., their definitions for terms and the processes and procedures by which they solve problems). See Verbal Report, Verbal Probe, Ethnographic Technique, and Problem-Solving Elicitation.

ESTIMATE: The expert's answer encoded in the response mode. Estimates specifically refer to answers given in numerical form, such as probabilities or ratings. It also is the value of an estimator from a calculation, measurement, or observation.

ESTIMATOR: The formula or procedure used for calculating or determining the value that estimates a population parameter such as the mean or variance.

ETHNOGRAPHIC TECHNIQUE: An elicitation technique from cultural anthropology that involves restating the subject's words into questions. This method avoids the danger of having the interviewer bias the subject's account by using the subject's own words. This technique is used to pursue in greater depth information that the subject has mentioned.

EXPECTED VALUE: The first moment of a pdf which corresponds to its mean value. The expected value of a random variable y is calculated by •$yf(y)dy$ integrated over the range from -• to +•.

EXPERT: A person who has a background in the subject matter at the desired level of detail and who is recognized by his or her peers or those conducting the study as being qualified to solve the questions. In this book, we sometimes refer to the experts as the *external experts* to distinguish them from the *advisory experts* who assist in selecting and refining the questions. In knowledge acquisition applications, the external experts are often referred to as the *source experts*.

EXPERT DATA: See Problem-Solving Data.

EXPERT JUDGMENT: Judgments of those with expertise or knowledge in the area. Expert judgment is usually elicited when experimental data is sparse or lacking. In this book, expert judgment refers to a combination of the experts' answers, their data on how these answers were reached (e.g., definitions, assumptions, and algorithms), and ancillary information on the experts themselves (e.g., educational background and work experience).

EXTERNAL EXPERT: See Expert.

F DISTRIBUTION: A pdf not symmetric about the mode but skewed in shape such that the mode is to the left of center and the tail on the right side is elongated. The two parameters of this distribution are called the numerator degrees of freedom and the denominator degrees of freedom. The F distribution is used to test the equivalence of 2 variances according to their ratio.

FACTOR ANALYSIS: A multivariate statistical technique for determining how a set of variables shares common information. The original variables are transformed to a set of new variables called factors. Some factors, called common factors, are composed of shared information of more than one original variable, while other factors, called unique factors, are formed from information composed of only one variable.

GENERAL LINEAR MODEL (GLM): See Regression.

GEOMETRIC MEAN: A mean value based upon a log scale interpretation. It is the product of n values raised to the $1/n$th power.

GRANULARITY: Level of detail used in gathering, examining, or analyzing data. For example, data on an expert's problem solving could be viewed at a coarse granularity, such as the of type of heuristic used, or at a fine granularity, such as the actual calculations performed as a part of each heuristic.

GROUP THINK BIAS: The tendency to modify a judgment so that it agrees with that of the group or of the group leader. The individuals are generally unaware that they have modified their judgment to be in agreement. This bias is classified as a motivational bias because it stems from the human need to be accepted and respected by others. Individuals are more prone to group think if they have a strong desire to remain a member, if they are satisfied with the group, if the group is cohesive, and if they are not a natural leader in the group.

HEURISTIC: A faster and easier method used to reduce the mental effort of solving a complex problem. A common heuristic is anchoring and adjustment. Instead of performing many detailed calculations, individuals adjust in small increments from their initial impression of the answer.

IMPRESSION MANAGEMENT BIAS: A type of social pressure bias that occurs when the subject responds to the reactions of those not physically present. For example, on a survey the subject tries to answer in such a way as to bring the most approbation (e.g., from society in the abstract or from the question writer in particular).

INCONSISTENCY BIAS: Inability to maintain the same problem-solving heuristic, definitions, or assumptions through time because of the limited information-processing capacity of the human mind.

INDEPENDENT VARIABLE: A quantity that is fixed, measured, recorded, or determined before or during the study. This quantity is possibly related to the answers given. (See also Ancillary Information and Conditional Variable.)

INDIVIDUAL INTERVIEW: One of the three basic situations for eliciting expert judgment One individual is interviewed at a time, usually in a face-to-face situation. The interviewer can structure the elicitation to any degree. An unstructured interview resembles a conversation; a structured one an interview driven by prepared questions. Because the separate responses are often mathematically combined in some way, this interview type is also called *staticized* or *nominal group*.

INFERENCE (GENERAL): The process of drawing conclusions from information for interpretation on a general or universal scale.

INFERENCE (STATISTICAL): The process of drawing conclusions about the population of interest or study from the results of statistically valid sampling and analysis.

INTENSIVE PILOT TEST: A type of pilot testing that combines structured interviews and observations. The intensive pilot test provides two kinds of feedback: (1) how the experts progress through the elicitation, their general impressions, and when and why they decide to respond to particular questions; and (2) how the experts specifically interpret each direction, statement of the question, or response mode option.

INTERACTIVE GROUP SITUATION: One of the three basic elicitation situations. In the interactive group, the members participate in a face-to-face situation with one another and a session moderator. The participants' interactions can be structured to any degree. A totally unstructured group resembles a typical meeting; a highly structured group is carefully choreographed as to when the participants present their views and when open discussion takes place.

INTERVIEWER: The person who elicits the expert judgment. (See also Knowledge Engineer.)

KNOWLEDGE ACQUISITION: Part of artificial intelligence connected with "extracting, structuring, and organizing knowledge from some source, usually human experts, so it can be used in a program" (Waterman, 1986, p. 392).

KNOWLEDGE-BASED COGNITION/BEHAVIOR: The level of thinking that most of expert judgment involves. It is an interpretive, analytical, high-level, conscious activity caused by thinking about rare or uncertain phenomena.

KNOWLEDGE ENGINEER: The knowledge engineer, like the interviewer, elicits information from the expert. However, the term *knowledge engineer* refers to someone who, in addition to interviewing, represents and enters the expert knowledge into a computer system with the goal of creating an expert or knowledge-based system.

LIKELIHOOD: The probability distribution form assumed for and formed from the data. Likelihood refers to the chance that a particular sample is chosen from the underlying population.

LIMITED INTERACTIVE GROUP: See Delphi Situation.

LIMITED PILOT TEST: A type of pilot testing done with a very small sample of experts. The limited pilot test is done after the intensive pilot test to provide a time estimate of the duration of each part of the elicitation.

LINEAR SCALE: A continuous line of numbers on a scale linear in structure; that is, each number on the line is separated from its neighbor by a value equal to the difference between it and its neighbor. For example, 4.0 would be twice as far from 0.0 as 2.0.

LOGNORMAL DISTRIBUTION: A pdf for a random variable whose log is normally distributed. In reliability applications, the lognormal distribution is commonly used as the distribution for failure rates.

LONG-TERM MEMORY (LTM): Memory of large capacity and relatively permanent duration. A portion of what is processed in the individual's short-term memory is stored in this type of memory.

MATHEMATICAL AGGREGATION: The use of mathematical methods to combine multiple experts' answers into one answer, usually when a single estimate is needed. Multiple experts' answers or distributions can also be combined into a single distribution. Some mathematical methods weight the experts' answers equally, such as the mean; others use more complex weighting schemes.

MAXIMUM LIKELIHOOD: A method used to determine estimators such as means or model coefficients by finding the maximum of the likelihood distribution function and solving for the desired estimator.

MEAN: The numeric average of a set of values (sample) calculated by summing the values and dividing by how many there are in the set. The mean of a population is the theoretically derived expected value.

MEAN SQUARE: In analysis of variance, mean square represents the variance between the groups or the variance for the error. It is calculated by the ratio of the sum of squares and df.

MEDIAN: The middle value of a sample or a distribution. The median is the 50th percentile of a distribution. It is the value such that half of the sample or the distribution is larger, and half of the sample or distribution is smaller.

MISINTERPRETATION: The altering of the expert's thoughts as a result of the methods of elicitation and documentation. See Training Bias.

MISREPRESENTATION: The altering of the expert's thoughts and answers as a result of modeling or analyzing them. See Tool Bias.

MODE: The most frequent value in a data set (sample) or distribution. The mode is the hump of the distribution. A distribution or data set with more than one hump is bimodal (for 2), trimodal (for 3), or multimodal (more than 3).

MODES OF COMMUNICATION: Communicating with the expert in person, by mail (electronic or postal), or by telephone. Each mode has its advantages and disadvantages. For example, the mail mode is the least expensive but has the lowest response rate.

MONTE CARLO SIMULATION: A computational technique for investigating the properties and behavior of a variable by repeated random sampling from a known or assumed distribution (e.g., normal) representing the variable.

MOTIVATIONAL BIAS: Biases whose source involves the emotional needs and wishes of the subject. Group-think bias is one example.

MULTIVARIATE ANALYSIS: A statistical analysis technique that allows two or more variables of interest to be considered simultaneously.

NOMINAL GROUP METHOD: See Individual Interview.

NONPARAMETRIC TECHNIQUE: Statistical analysis techniques that do not require assuming that the sample or population follow a particular distribution (e.g., the normal distribution). These are sometimes referred to as distribution-free techniques.

NORMAL DISTRIBUTION: A probability distribution function with a bell-like shape, symmetrical about the mean, and with identical values for mean, median, and mode. It is also known as the Gaussian distribution or a bell-shaped curve.

NORMATIVE EXPERTISE: Expertise in the statistical or mathematical principles of the response mode.

PAIRWISE COMPARISONS: Establishing the relative ratings of a set of objects, events, or criteria by comparing them two at time. For a set of n things, $n(n-1)/2$ comparisons are required to make all possible pairwise comparisons. Comparing object A to object B is the reciprocal of comparing object B to object A.

PARAMETER: The characteristic of a probability distribution. It is also the characteristic of the population (which is represented by a probability distribution).

PERCENTILE: The value from a distribution of a random variable that divides the area under the distribution curve into specified percentages. For example, the 5th percentile is the value of the distribution such that 5% of the distribution is smaller and 95% of the distribution is larger.

PILOT TESTING: A type of practice that involves taking a sample of the larger expert population, presenting these experts with an aspect of the elicitation, obtaining their feedback, and revising the elicitation accordingly. (See also Intensive and Limited Pilot Test.)

POPULATION: The entire existing or theoretical set of items under study. Examples include all the people on the earth—past, present, and future; all earthquakes on the earth; all earthquakes in a particular location on the earth; and failures and successes of all components of a certain type. The population is often represented as a probability distribution for the random variable that describes the items under study.

POSTERIOR DISTRIBUTION/DENSITY FUNCTION: The distribution or density function that results from a Bayesian analysis. This distribution combines the prior information and the data.

PRACTICING THE ELICITATION: Rehearsing the elicitation to detect any problems in its design before its actual use. One type of practice is pilot testing. (See Pilot Test.)

PRIOR DISTRIBUTION/DENSITY FUNCTION: The distribution or density for the parameters of interest. This represents the information known before the gathering of the data takes place. It is combined with the data to form the posterior.

PROBABILITY: Refers to the chance of something occurring. One important property of probabilities is that they are values from 0.0 to 1.0. A probability equal to 0.0 means that the event never happens. A probability equal to 1.0 means that the event always happens. A probability equal to 0.5 means that the event happens half of the time. Another important property is that the probabilities of all exhaustive (all events in a set), mutually exclusive events (nonoverlapping events) must sum to the value 1.0.

PROBABILITY DISTRIBUTION FUNCTION (PDF) OR PROBABILITY DISTRIBUTION: The mapping of a random variable onto a functional representation of probabilities for the possible values of that random variable. The nonnegative function (f) is mathematically represented by an equation in terms of the values of the random variable (x). The area under the entire curve resulting from this equation equals 1.0. Sectional areas under the curve correspond to the probabilities of the corresponding x values. The values of the equation (f) are probabilities per unit interval (dx).

PROBLEM-SOLVING ELICITATION: Techniques used to elicit how the subject solved the problem. These elicitation techniques can be used to deliver data of differing levels of detail. For example, in a risk analysis study, a few sentences on the expert's reasoning may be enough. An artificial intelligence project, however, may require more detailed information to help model each step of the expert's thinking.

PROBLEM-SOLVING METHOD OR PROCESS: The method the expert uses to solve a problem. These methods could include the expert's interpretation of the problem, assumptions, definitions, and algorithms.

PROJECT PERSONNEL: Persons working on the project, such as in-house managers, data gatherers (interviewers or knowledge engineers), and analysts.

PUTATIVE INTERVAL: An interval estimate calculated from a simulated or computed distribution, usually from a bootstrap or Monte Carlo simulation. For example, the 5th and 95th percentiles of the simulated distribution of the median for a sample would form the central 90% putative interval for that distribution.

QUALITATIVE DATA/INFORMATION: Any nonnumeric data (e.g., verbal descriptions, classifications, categories, or preferences).

QUANTIFICATION: The process of transforming qualitative information into quantitative or numerical forms. In addition, quantification also refers to the process of transforming raw data (in any original form) to a desired numerical form. This transformation can change the granularity of the data from coarse to fine.

QUANTITATIVE DATA/INFORMATION: Any numerical data (e.g., integers, ranks, or values) on the real number line.

QUESTION: The concrete, detailed points within the question areas to which the expert is asked to respond. Questions are also called *problems*. An example of a question or problem is "what is the leak rate in gallons per minute as a function of time to seal failure due to loss of cooling to the pump shaft?" The test for whether a query qualifies as a question is whether the expert finds it sufficiently specific to provide an answer.

QUESTION AREA: The specific issue for investigation or the general area in which the experts will be questioned. Question areas are developed by considering such information as the goal of the project and the client's directives. For example, for an application whose goal was the provision of likelihoods and consequences of severe accidents in light-water reactors, separate question areas existed for front- and back-end phenomena, the different plant manufacturers, and so on.

QUESTION PHRASING: The wording of the question and the response mode done to maximize the chances that the expert understands it and is not unduly influenced by the wording. Payne (1951) has shown that different word choices and orderings can change the answer reached by 4% to 15%.

RANDOM VARIABLE: The quantity of interest that can take on any of a set of possible values or outcomes of an experiment or observation. The symbol for a random variable is X; the symbol for a generic value of a random variable is x.

RANGE: The difference between the maximum and minimum possible values of a sample or pdf. (See Volunteered Dispersion Measure.)

RANKS: A set of numeric or descriptive values assigned to an original set of values or descriptions. Ranks are usually cardinal (i.e., composed of integer values, in ascending or descending order, and equally spaced 1, 2, 3, etc.). Ranks can also be ordinal, or descriptive in nature (worst, better, best). Ratings are usually assigned numbers from a chosen scale (e.g., from 1 to 10). The analyst assigns the numbers according to some criteria.

RATINGS: See Ranks.

REGRESSION: The analysis that finds the best fit line for a dependent variable (y) in terms of the independent variables (x_i). The form of the model is $y = b_0 + b_1x_1 + b_2x_2 + \ldots + e$, where the b_0 is the intercept term, the other bs represent slopes, and e is the residual or remaining error not accounted for in the model. The regression line is fit to minimize the squared distances between the data points and the line. Regression is a subset of the analyses known as general linear models (GLMs), where various techniques determine linear relationships.

RELIABILITY ANALYSES: Studies of process or equipment failure or operability. For example, a reliability study would analyze "how frequently a chemical reactor might overheat due to malfunctioning pumps, heat exchangers, human operators, control systems, and other plant equipment . . ." (Henley & Kumamoto, 1981, p. 8).

RESIDUAL: The error term in a linear model representing the difference between the dependent variable's value and its value as predicted by the model.

RESPONSE MODE: The form used to ask the subjects to give their judgments. Some numeric response modes that are commonly used include probabilities, odds, intervals, ratings, logs, and pairwise comparisons. Qualitative response modes include verbal or written descriptions, classifications, categories, or preferences.

RISK ANALYSIS: Risk analysis includes the data and techniques used to quantify risk. These analyses usually include a model of events leading to risks and their consequences. A risk analysis of a nuclear plant could describe major factors relating to accident events, frequency and uncertainty ranges of accident events, major factors and accident phenomena leading to these damage events, and consequences and risks of these to the public.

SAMPLE: A subset of a population of items chosen for examination. A statistically valid sample is chosen by using sampling techniques designed for representativeness of the population and for random selection of the items.

SHORT-TERM MEMORY: A memory of limited capacity and intermediate duration. Ericsson and Simon (1980) depict short-term memory as the place where information is processed in problem solving.

SIGNIFICANCE OR SIGNIFICANCE LEVEL: The result of a statistical test or technique is significant if the conclusions indicate a difference between the data and the assumed normal state of the world. For example, a test concludes that two experts are correlated where the assumed state is that experts are not correlated. Because nothing is 100% certain, a chance exists that the conclusion drawn from a test is incorrect. The probability of a significant result being incorrect is the significance level. This level, which analysts choose before conducting the test, indicates the chance that they are willing to take that the conclusion is incorrect. This level is commonly 5% or less, but the choice is always indicated in the statement of the conclusion. For example, "the positive correlation between expert 1 and 2 is significant at the 5% level," means that experts 1 and 2 are positively correlated, and a 5% chance exists that they are not. This 5% is sometimes called the alpha level or the type I error.

SIMULATION: See specific simulation techniques: Bootstrap and Monte Carlo.

SOCIAL PRESSURE: An effect that induces individuals to slant their responses or to silently acquiesce to the views that they believe the interviewer, their group, supervisors, organization, or society in general will accept. This altering of an individual's thoughts can take place consciously or unconsciously. The social pressure can come from those physically present or from the subject's internal evaluation of how others would interpret his or her responses. People's need to be loved, respected, and recognized induce them to behave in a manner that will bring affirmation.

SOLUTION: Expert judgments given as descriptive text or diagrams, as opposed to numerical estimates.

SOURCE EXPERT: See Expert.

SPONSOR: The person who requests the gathering of expert judgment. In other words, the sponsor is the one whose needs the project will serve. The sponsor is often the person funding the study. Whether the sponsor funds the study or not, the sponsor can usually say what the purpose and goals of the project are, what information is needed from the experts, and what resources will be available for the project.

STANDARD DEVIATION: The square root of the variance.

STATICIZED GROUP: See Individual Interview.

STATISTIC: A characteristic of the sample derived from the data set (e.g., the mean of the data).

STRUCTURED DOCUMENTATION: A detailed type of record of the expert's answers and problem-solving processes. The person asked to perform the documentation is usually provided with a format of what should be recorded. The format lists those aspects deemed the most important (e.g., answers and uncertainty levels, assumptions, and algorithms) and the level of detail at which the information is desired.

STRUCTURING ELICITATION: The amount of controls placed on the elicitation process. The interaction between the experts is one aspect of an elicitation method that is typically structured. Varying degrees of structure can be imposed, ranging from none to a high degree of structuring. No structuring would allow spontaneous interaction between the experts; a high degree would produce carefully choreographed communications. (See the first section in chapter 8, *Considerations in Designing the Elicitation*, for a description of the larger process of which structuring forms a part.)

SUBSTANTIVE EXPERTISE: Expertise stemming from the expert's experience in the field in question (e.g., the rupture rate of Westinghouse pipes).

SUM OF SQUARES: The sum of the squared deviations of each datum from the appropriate mean of the data. It is the numerator term of variances.

SUMMARY DOCUMENTATION: A type of record of the experts' answers and problem-solving processes. Typically, it provides a few sentences or paragraphs on the experts' thinking, such as the sources of information that they used, their major assumptions, and their reasons for giving particular answers.

THINK ALOUD METHOD: See Verbal Report.

TOOL BIAS: The misrepresentation of the expert's data as a result of forcing these to fit the tools selected for analysis. Analysts tend to use those models or methods with which they are most comfortable. As a result, they are often unable to objectively judge whether they have used the tool appropriately (e.g., the model required that the data have a normal distribution, and the data may not have).

TRAINING BIAS: The tendency of the data gatherer to introduce bias into the expert's data by misinterpreting it. It is an unconscious human tendency to interpret incoming information in terms of what is already believed, such as what has been learned through professional training. For example, it is common for a data gatherer to define a term using those definitions that he or she has learned rather than to elicit the expert's definitions.

TYPE-I ERROR: See Significance Level.

UNDERESTIMATION OF UNCERTAINTY BIAS: The tendency to underestimate the true amount of uncertainty in giving an answer. For example, people are frequently asked to estimate ranges around their answers to reflect their uncertainty. If people are requested to put a range around their answers such that they are 90% sure that the range encompasses the correct answer, they will underestimate the uncertainty by providing a range that is two to three times too narrow. In other words, their range will only have a 30% to 60%, not a 90%, chance of covering the correct answer (Capen, 1975).

UNIFORM DISTRIBUTION: A pdf with parameters that specify the minimum (a) and maximum (b) values of the distribution function. The pdf has a value of $1/(b - a)$ for every value of the random variable.

VARIABLE: See Random Variable.

VARIANCE: A measure of dispersion based on the squared differences between individual values and their mean or expected value. The standard deviation is the square root of the variance.

VARIANCE/COVARIANCE MATRIX: A matrix with entries corresponding to the variances and covariances of several variables (multivariate). The diagonals of the matrix are the variances and the off-diagonals are the covariances.

VERBAL PROBE: An elicitation technique from educational psychology used to obtain information on the subject's problem solving. Different types of verbal probes vary in when and how they are asked. In this book, we use verbal probe to refer to a question which has a nonleading wording that is asked while the expert is still attending to the subject of the question.

VERBAL REPORT: An elicitation technique from educational psychology (often called *protocol analysis*) requiring the subject to *think aloud* as he or she works through the problem. The verbal report is used in face-to-face interviews.

VERBATIM DOCUMENTATION: A record of the expert's answers and problem-solving processes. Obtaining a verbatim account is usually done by mechanically recording the expert's elicitation sessions and then transcribing them. This type of documentation occurs more often in artificial intelligence than in traditional expert judgment applications.

VOLUNTEERED RANGES: A type of dispersion measure that the experts volunteer without being asked. This dispersion marks a spread of values around the expert's best estimate.

WISHFUL THINKING BIAS: A tendency that occurs when individuals' hopes influence their judgments. For example, people typically overestimate what they can produce in a given amount of time. In general, the greater the subjects' involvement and the more they stand to gain from their answers, the greater this bias. Also called *conflict of interest* bias.

References

ADAMS-WEBER, J.R. (1981). George A. Kelly--An appreciation. In M. L. G. SHAW Ed., *Recent Advances in Personal Construct Technology*, London, England: Academic Press, pp. 1-7

AHRENS, J. H. & DIETER, U. (1974). Computer methods for sampling from gamma, beta, Poisson, and binomial distributions. *Computing*, **12**, 223-246.

AMOS, C. N., BENJAMIN, A. S., BOYD, G. J., KUNSMAN, D. M., LEWIS, S. R., SMITH, L. N. & WILLIAMS, D. C. (1987). Evaluation of severe accident risks and the potential for risk reduction: Grand Gulf, Unit 1, Vol. 4. Sandia National Laboratories Report SAND86-1309, Albuquerque, New Mexico and Nuclear Regulatory Commission Report NUREG/CR-4551, Washington, D.C.

ARMSTRONG, J. S. (1985). *Long-Range Forecasting: From Crystal Ball to Computer.* New York, NewYork: Wiley-Interscience.

ARMSTRONG, J. S., DENNISTON, W. B., JR. & GORDON, M. M. (1975). Use of the decomposition principle in making judgments. *Organizational Behavior and Human Performance,* **14**, 257-263.

ASCHER, W. (1978). *Forecasting: an Appraisal for Policymakers and Planners.* Baltimore, Maryland: Johns Hopkins University Press.

BAECHER, G. B. (1979). Correlations among experts' opinions. Department of Civil Engineering, Massachusetts Institute of Technology. Boston, Massachusetts.

BARCLAY, S., BROWN, R. V., KELLY, C. W., III, PETERSON, C. R., PHILLIPS, L. D. & SELVIDGE, J. (1977). *Handbook for Decision Analysis.* McLean, Virginia: Decisions and Designs, Inc.

BARON, R. A. & BRYNE, D. (1981). Understanding human interaction. *Social Psychology.* Boston, Massachusetts: Allyn and Bacon Inc.

BATESON, G. (1979). *Mind and Nature: A Necessary Unity.* New York, New York: Bantam Books.

BENJAMIN, A. S., KUNSMAN, D. M., WILLIAMS, D. C., BOYD, G. J. & MURFIN, W. B. (1987). Evaluation of severe accident risks and the potential for risk reduction: Surry Power Station, Unit 1, Vol. 1. Sandia National Laboratories Report SAND86-1309, Albuquerque, New Mexico and Nuclear Regulatory Commission Report NUREG/CR-4551, Washington, D. C.

BERNREUTER, D. C., SAVY, J. B., MENSING, R. W., CHEN, J. C. & DAVIS, B. C. (1985). Seismic hazard characterization of the Eastern United States. Lawrence Livermore National Laboratory Report UCID- 20421, Livermore, California.

BOOKER, J. M. (1978). Mean wheat yield prediction when predictor variables are subject to error. Ph.D. dissertation. Texas A & M University, College Station, Texas.

BOOKER, J. M. & BRYSON, M. C. (1985). Decision analysis in project management: an overview. *IEEE Transactions in Engineering Management,* **32**, 3-9.

BOOKER, J. M., BRYSON, M. C. & MCWILLIAMS, T. P. (1984). Decision analysis package for R & D project selection and evaluation. Los Alamos National Laboratory Report LA-UR-84-41, Los Alamos, New Mexico.

BOOKER, J. M. & MEYER, M. A. (1985). Sources and effects of correlation of expert opinion. Los Alamos National Laboratory Report LA-UR-85-1879, Los Alamos, New Mexico.

_______ (1988*a*). Sources and effects of interexpert correlation: an empirical study. *IEEE Transactions on Systems, Man, and Cybernetics,* **18**, 135-142.

_______ (1988*b*). Determining the independence of experts using simulation. Los Alamos National Laboratory Report LA-UR-88-301, Los Alamos, New Mexico.

BOOSE, J. H. & GAINES, B. R. (1988). Knowledge acquisition for knowledge-based systems. *AAAI-88 Tutorial Notes, Seventh National Conference on Artificial Intelligence, S*t. Paul, Minnesota.

BOOSE, J. H. & SHAW, M. (1989). Knowledge acquisition for knowledge-based systems. *AAAI-89 Tutorial Notes, Eleventh International Conference on Artificial Intelligence,* Detroit, Michigan.

BRACHMAN, R. J. & LEVESQUE, H. J. (1985). *Readings in Knowledge Representation.* Los Altos, California: Morgan Kaufman Publishers.

BRADSHAW, J. M. & BOOSE, J. H. (1990). Decision analysis techniques for knowledge acquisition: Combining information and preferences in AQUINAS and AXOTL. *International Journal of Man-Machine Studies,* **32**, 121-186.

BREUKER, J. & WIELINGA, B. (1988). Models of expertise in knowledge acquisition. In G. GUIDA & C. TASSO Eds., *Topics in Expert System Design: Methodologies and Tools.* Amsterdam, The Netherlands: North Holland Publishing Co., pp. 265-296.

BRUCKNER, L. A. & MARTZ, H. F. (1987). Beta product. Statistics Group draft document, available from the Statistics Group, MS F600, Los Alamos National Laboratory, Los Alamos, New Mexico.

CAPEN, E. C. (1975). The difficulty of assessing uncertainty. *Society of Petroleum Engineers and the American Institute of Mining, Metallurgy, and Petroleum Engineers 50th Annual Fall Conference and Exhibit,* Dallas, Texas (September 28-October 1).

CATTELL, R. B. (1963). The personality and motivation of the researcher from measurements of contemporaries and from biographies. In C. W. TAYLOR & F. BARRON Eds., *Scientific Creativity: Its Recognition and Development,* pp. 119-138. New York, New York: John Wiley.

CHENG, R. C. H & FEAST, G. M. (1979). Some simple gamma variate generators. *Applied Statistics*, **28**, 290-295.

CLEAVES, D. A. (1986). Cognitive biases and corrective techniques: proposals for improving elicitation procedures for knowledge-based systems. *Proceedings from the AAAI sponsored 2nd Annual Knowledge Acquisition for Knowledge-Based Systems Workshop,* Banff, Canada, pp. 9-0 to 9-11.

CLEMEN, R. T. (1986). Calibration and the aggregation of probabilities. *Management Science,* **32**, 312-314.

CLEMEN, R. T. & WINKLER, R. L. (1985). Limits for the precision and value of information from dependent sources. *Operations Research,* **33**, 427-442.

CLUB OF ROME (1974). *Limits To Growth: A Report for the Club of Rome's Project on the Predicament of Mankind.* New York, New York: Universe Books.

COCHRAN, W. (1963). *Sampling Techniques.* New York, New York: John Wiley & Sons.

COMER, M. K., SEAVER, D. A., STILLWELL, W. G. & GADDY, C. D. (1984). Generating human reliability estimates using expert judgments, Vol. 4. Sandia National Laboratories Report SAND84-7115, Albuquerque, New Mexico and Nuclear Regulatory Commission Report NUREG/CR-3688, Washington, D.C.

CONOVER, W. J. (1971). *Practical Nonparametric Statistics.* New York, New York: John Wiley & Sons.

COOK, I. & UNWIN, S. D. (1986). Controlling principles for prior probability assignments in nuclear risk assessment. Sandia National Laboratories Report SAND85-1323, Albuquerque, New Mexico and Nuclear Regulatory Commission Report NUREG/CR-4514, Washington, D.C.

DALKEY, N. C. (1969). An experimental study of group opinion: the Delphi method. *Futures,* **1**, 403-406.

DAWES, R. M., FAUST, D. & MEEHL, P. E. (1989). Clinical versus actuarial judgment. *Science,* **243**, 1668-1673.

DENNING, P. J. (1986). The science of computing--will machines ever think? *American Scientist,* **74**, 344-46.

_______ (1988). The science of computing--blindness in designing intelligent systems. *American Scientist,* **76**, 118-120.

DHALIWAL, J. S. & BENBASAT, I. (1989). A framework for the comparative evaluation of knowledge acquisition tools and techniques. *Proceedings of the AAAI-sponsored 4th Annual Knowledge Acquisition for Knowledge-Based Systems Workshop,* Banff, Canada, pp. 9-1 to 9-24.

DOUGHERTY, E. M., JR., FRAGOLA, J. R. & COLLINS, E. P. (1986). Human reliability analysis. Science Applications International Corporation Report SAIC/NY-86-1-OR, Oak Ridge, Tennessee.

DOUGLAS, M. (1970). *Purity and Danger: An Analysis of Concepts of Pollution and Taboo.* London, England: Routledge and Kegan Paul.

DREYFUS, H. L. & DREYFUS, S. E. (1986). *Mind Over Machine.* New York, New York: Free Press.

DURAN, B. S. & ODELL, P. L. (1974). *Cluster Analysis: A Survey.* New York: Springer-Verlag.

EFRON, B. (1979). Bootstrap methods: another look at the jackknife. *Annals of Statistics,* **7**, 1-26.

EFRON, B. & GONG, G. (1983). A leisurely look at the bootstrap, the jackknife, and cross validation. *The American Statistician,* **37**, 36-48.

EGAN, D. E. & SCHWARTZ, B. J. (1979). Chunking in recall of symbolic drawings. *Memory and Cognition,* **7**, 149-158.

ELSTON, A., EMERSON, J. D., MEYER, M. A., OSBORN, M. M. & STODDARD, M. L. (1986). Report of the Fort Knox AOBC Prototype I data collection. Military Systems Group Report S-6:86-U-429, Los Alamos National Laboratory, Los Alamos, New Mexico.

ERICSSON, K. A. & SIMON, H. A. (1980). Verbal reports as data. *Psychological Review,* **87**, 215-250.

_______ (1984). *Protocol Analysis: Verbal Reports as Data.* Cambridge, Massachusetts: MIT Press.

FAUST, D. & ZISKIN, J. (1988). The expert witness in psychology and psychiatry. *Science,* **241**, 31-35.

FELDMAN, B. Z., COMPTON, P. & SMYTHE, G. A. (1989). Hypothesis testing: An appropriate task for knowledge-based systems. *Proceedings of the 4th Knowledge Acquisition for Knowledge-Based Systems Workshop*, Banff, Canada, pp. 10-1 to 10-20.

FELLER, W. (1957). *An Introduction to Probability Theory and Its Applications*, Vol. 1. New York, New York: John Wiley & Sons.

FESTINGER, L. (1957). *A Theory of Cognitive Dissonance.* Palo Alto, California: Stanford University Press.

FOGEL, L. J. (1967). *Human Information Processing.* Englewood Cliffs, New Jersey: Prentice-Hall.

FRENCH, S. (1986). Calibration and the expert problem. *Management Science,* **32**, 315-321.

GAINES, B. R. (1989). Social and cognitive processes in knowledge acquisition. *Knowledge Acquisition*, **1**, 39-58.

GAINES, B. R. & BOOSE, J. Eds. (1988). *Knowledge Acquisition for Knowledge-based Systems.* London, England: Academic Press.

GAINES, B. R. & SHAW, M. L. (1989). Comparing the conceptual systems of experts. *Proceedings of the Eleventh International Joint Conference on Artificial Intelligence,* Detroit, Michigan, pp. 633-638.

GENEST, C. & ZIDEK, J. V. (1986). Combining probability distributions: a critique and an annotated bibliography. *Statistical Science,* **1**, 114, 148.

GOFFMAN, E. (1959). *The Presentation of Self in Everyday Life.* Garden City, New York: Doubleday Anchor Books.

GOLD, P. E. (1987). Sweet memories. *American Scientist,* March-April, 151-155.

GORDEN, R. L. (1980). *Interviewing: Strategy, Techniques, and Tactics.* Homewood, Illinois: Dorsey Press.

GRIZZLE, J. E., STARMER, C. F. & KOCH, G. G. (1969). Analysis of categorical data by linear models. *Biometics,* **25**, 489-504.

GUSTAFSON, D., SHUKLA, R., DELBECQ, A. & WALSTER, G. (1973). A comparative study of differences in subjective likelihood estimates made by individuals, interacting groups, Delphi groups, and nominal groups. *Organizational Behavior and Human Performance,* **9**, 280-291.

HARRISON, J. M. (1977). Independence and calibration in decision analysis. *Management Science,* **24**, 320-328.

HAYES-ROTH, B. (1980). Estimation of time requirements during planning: interactions between motivation and cognition. Rand Corporation Report N-1581-ONR, Santa Monica, California.

HENLEY, E. J. & KUMAMOTO, H. (1981). *Reliability Engineering and Risk Assessment.* Englewood Cliffs, New Jersey: Prentice-Hall.

HENRION, M. & COOLEY, D. R. (1987). An experimental comparison of knowledge engineering for expert systems and for decision analysis. *Proceedings of the 6th National American Association for Artificial Intelligence,* Seattle, Washington, pp. 471-476.

HIRELY, W. (1989). Survey probes tensions between science and democracy. The Science Observer, *American Scientist,* **77**, 24-27.

HOFFMAN, R. R. (1987). The problem of extracting the knowledge of experts from the perspective of experimental psychology. *Artificial Intelligence Magazine,* Summer, 53-67.

HOGARTH, R. (1975). Cognitive processes and the assessment of subjective probability distributions. *Journal of the American Statistical Association,* **70**, 271-291.

_______ (1980). *Judgement and Choice: The Psychology of Decisions.* Chicago, Illinois: Wiley- Interscience.

JANIS, I. C. (1972). *Victims of Group Think: A Psychological Study of Foreign Policy Decisions and Fiascos.* Boston, Massachusetts: Houghton Mifflin.

JOHNSON, M. E. (1987). *Multivariate Statistical Simulation.* New York, New York: John Wiley & Sons.

KAHNEMAN, D., SLOVIC, P. & TVERSKY, A. Eds. (1982). *Judgment Under Uncertainty: Heuristics and Biases.* Cambridge, Massachusetts: Cambridge University Press.

KAHNEMAN, D. & TVERSKY, A. (1982). Subjective probability: A judgment of representativeness. In KAHNEMAN, D., SLOVIC, P. & TVERSKY, A. Eds., *Judgment Under Uncertainty: Heuristics and Biases,* pp. 32-47. Cambridge, Massachusetts: Cambridge University Press.

KEENEY, R. L. & VON WINTERFELDT, D. (1989). On the uses of expert judgment on complex technical problems. *IEEE Transactions on Engineering Management,* **36**, 83-86.

KNAPP, R. H. (1963). *Demographic, cultural, and personality attributes of scientists.* In C. W. TAYLOR & F. BARRON Eds., *Scientific Creativity: Its Recognition and Development,* pp. 205-216. New York, New York: John Wiley.

KOUTS, H., CORNELL, A., FARMER, R., HANAUER, S. & RASMUSSEN, N. (1987). Methodology for uncertainty estimation in NUREG-1150 (draft): conclusions of a review panel. Department of Nuclear Energy, Brookhaven National Laboratory Report BWL-NUREG-52119, Upton, New York and Nuclear Regulatory Commission Report NUREG/CR-5000, Washington, D.C.

KRUPKA, M. C., PEASLEE, A. T., JR. & LAQUER, H. L. (1983). Gaseous fuel safety assessment for light-duty automotive vehicles. Los Alamos National Laboratory Report LA-9829-MS, Los Alamos, New Mexico.

KSHIRSAGAR, A. M. (1972). *Multivariate Analysis.* New York, New York: Marcel Dekker.

LAFRANCE, M. (1988). The knowledge acquisition grid: A method for training knowledge engineers. In B. R. GAINES & J. H. BOOSE Eds., *Knowledge Acquisition for Knowledge-Based Systems,* Vol. 1, pp. 81-91. London, England: Academic Press.

LANGER, E. J. (1989). *Mindfullness.* Reading, Massachusetts: Addison-Wesley.

LICHTENSTEIN, S. & FISCHHOFF, B. (1977). Do those who know more also know more about how much they know? *Organizational Behavior and Human Performance,* **20**, 159-183.

LICHTENSTEIN, S., FISCHHOFF, B. & PHILLIPS, L. D. (1982). Calibration of probabilities: the state of the art to 1980. In D. KAHNEMAN, P. SLOVIC & A. TVERSKY Eds., *Judgment Under Uncertainty: Heuristics and Biases,* pp. 306-334. Cambridge, Massachusetts: Cambridge University Press.

LINDLEY, D. V. & SINGPURWALLA, N. D. (1984). Reliability and fault tree analysis using expert opinions. George Washington University Report GWU/IRRA/TR-84/10, Washington, D.C.

MAHONEY, M. (1976). *The Scientist as Subject: The Psychological Imperative.* Cambridge, Massachusetts: Ballinger Publishing Co.

MARTZ, H. F., BECKMAN, R. J., CAMPBELL, K., WHITEMAN, D. E. & BOOKER, J. M. (1983). A comparison of methods for uncertainty analysis of nuclear power plant safety system fault tree models. Los Alamos National Laboratory Report LA-9729-MS, Los Alamos, New Mexico.

MARTZ, H. F., BRYSON, M. C. & WALLER, R. A. (1985). Eliciting and aggregating subjective judgments--some experimental results. Los Alamos National Laboratory Report LAUR-85-800, Los Alamos, New Mexico.

MARTZ, H. F. & WALLER, R. A. (1982). *Bayesian reliability analysis.* New York, New York: John Wiley & Sons.

MATHIOWETZ, N. A. (1987). Response error: correlation between estimation and episodic recall tasks. *Survey Research Methods Proceedings of the Joint Statistical Meetings,* San Francisco, California, pp. 430-435.

MCGRAW, K. L. & HARBISON-BRIGGS, K. (1989). *Knowledge Acquisition: Principles and Guidelines.* Englewood Cliffs, New Jersey: Prentice Hall.

MEYER, M. A. (1986). Human factors affecting subjective judgments. *Proceedings of the Thirty-First Conference on the Design of Experiments in Army Research and Development,* Madison, Wisconsin and Los Alamos National Laboratory Report LA-UR-84-3176, Los Alamos, New Mexico.

_______ (1987). Eliciting data on subject's problem solving of computerized exercises: a cognitive interview. Statistics Group Internal Report A-1:87-U-601, Los Alamos National Laboratory, Los Alamos, New Mexico.

MEYER, M. A. & BOOKER, J. M. (1987*a*). Problems with expert data in studies of inter-expert correlation. *Social Statistics Proceedings of the Joint Statistical Meetings,* San Francisco, California, pp. 88-97.

_______ (1987*b*). Sources of correlation between experts: empirical results from two extremes. Los Alamos National Laboratory Report LA-10918-MS, Los Alamos, New Mexico and Nuclear Regulatory Commission Report NUREG/CR-4814, Washington, D.C.

_______ (1989). A practical program for handling bias in knowledge acquisition. *Proceedings of the AAAI-sponsored 4th Annual Knowledge Acquisition for Knowledge-Based Systems Workshop,* Banff, Canada, pp. 23-1 to 23-20.

MEYER, M. A., BOOKER, J. M., CULLINGFORD, H. S. & PEASLEE, A. T., JR. (1981). A data-gathering method for use in modeling energy research, development and demonstration programs. Technology Assessment Group Report S-4/81-11, Los Alamos National Laboratory, Los Alamos, New Mexico.

______ (1982). A data-gathering method for use in modeling energy research, development and demonstration programs. In T. N. VESIROGLU Ed., *Energy Programs, Policy, and Economics: Alternative Energy Sources IV.* pp. 421-430. Florida: Butterworth Publishers.

MEYER, M. A. & JOHNSON, E. R. (1985). Proceedings from the technology control panel workshop. Los Alamos National Laboratory Report LA-UR-85-4078, Los Alamos, New Mexico.

MEYER, M. A., MNISZEWSKI, S. M. & PEASLEE, A. T., JR. (1989). Using three minimally biasing elicitation techniques for knowledge acquisition. *Knowledge Acquisition,* **1**, 59-72.

MEYER, M. A., PEASLEE, A. T., JR. & BOOKER, J. M. (1982). Group consensus methods and results. Los Alamos National Laboratory Report LA-9584-MS, Los Alamos, New Mexico.

MILLER, G. A. (1956). The magical number seven, plus or minus two: some limits on our capacity for processing information. *Psychological Review,* **63,** 81-97.

MORRIS, P. A. (1977). Combining expert judgments: a Bayesian approach. *Management Science,* **23**, 679-693.

______ (1986). Observations on expert aggregation. *Management Science,* **32,** 321-328.

MUMPOWER, J. L., PHILLIPS, C. D., RENN, O. & UPPULURI, V. R. R. (1987). *Expert Judgment and Expert Systems. NATO ASI series F,* Computer and Systems Sciences, Vol. 35. New York: Springer-Verlag.

ORTIZ, N. R., WHEELER, T. A., MEYER, M. A. & KEENEY, R. L. (1988). The use of expert judgment in NUREG-1150. *Proceedings of the 16th Water Reactor Safety Information Meeting.* Gaithersburg, Maryland. In Nuclear Regulatory Commission Report NUREG/CP-0097, Vol. 5, pp. 1-25.

PAYNE, S. (1951). *The Art of Asking Questions.* Princeton, New Jersey: Princeton University Press.

PERT COORDINATING GROUP (1963). PERT guide for management use. Department of Defense, Washington, D.C.

RAIFFA, H. (1970). *Decision Analysis: Introductory Lectures on Choices Under Uncertainty.* Reading, Massachusetts: Addison-Wesley.

REGOCZEI, S. & HIRST, G. (1989). Sortal analysis with Sortal, a software assistant for knowledge acquisition. *Proceedings of the AAAI-sponsored 4th Annual Knowledge Acquisition for Knowledge-Based Systems Workshop,* Banff, Canada, pp. 25-1 to 25-18.

RIPLEY, B. A. (1987). *Stochastic Simulation.* New York, New York: John Wiley & Sons.

ROE, A. (1952). *The Making of a Scientist.* Westport, Connecticut: Greenwood Press.

______ (1963). Personal problems and science. In C. W. TAYLOR & F. BARRON Eds., *Scientific Creativity: Its Recognition and Development*, pp. 132-138. New York, New York: John Wiley.

ROSENFIELD, I. (1988). *The Invention of Memory: A New View of the Brain.* New York: Basic Books.

SAATY, T. L. (1980). *The Analytic Hierarchy Process: Planning, Priority Setting, and Resource Allocation.* New York: McGraw-Hill.

______ (1982). *Decision Making for Leaders.* Belmont, California: Lifetime Learning Publications.

SCHERVISH, M. J. (1986). Comments on some axioms for combining expert judgements. *Management Science,* **32**, 306-312.

SEAVER, D. A. (1976). Assessments of group preferences and group uncertainty for decision making. Technical Report 76-4. Social Science Research Institute, University of Southern California, Los Angeles, California.

______ (1978). Assessing probability with multiple individuals: group interaction versus mathematical aggregation. SSRI Research Report 78-3, Social Science Research Institute, University of Southern California, Los Angeles, California.

SEAVER, D. A. & STILLWELL, W. G. (1983). Procedures for using expert judgment to estimate human error probabilities in nuclear power plant operations. Sandia National Laboratories Report SAND82-7054, Albuquerque, New Mexico and Nuclear Regulatory Commission Report NUREG/CR-2743, Washington, D.C.

SHAW, M. L. G. & WOODWARD, J. B. (1989). Mental modes in the knowledge acquisition process. *Proceedings of the AAAI-sponsored 4th Annual Knowledge Acquisition for Knowledge-Based Systems Workshop,* Banff, Canada, pp. 29-1 to 29-24.

SKUCE, D. & SOWA, J. F. (1988). Knowledge representation: design issues. *AAAI Tutorial Notes, Seventh National Conference on Artificial Intelligence,* St. Paul, Minnesota.

SNEDECOR, G. W. & COCHRAN, W. G. (1978). *Statistical Methods.* Ames, Iowa: Iowa State University Press.

SOWA, J. F. (1984). *Conceptual Structures: Information Processing in Mind and Machine.* Reading, Massachusetts: Addison-Wesley.

SPETZLER, C. S. & STAEL VON HOLSTEIN, C. A. (1975). Probability encoding in decision analysis. *Management Science,* **22**, 340-352.

SPRADLEY, J. P. (1979). *The Ethnographic Interview.* New York: Holt, Rhinehart, and Winston.

STAEL VON HOLSTEIN, C. A. (1971). Two techniques for assessment of subjective probability distributions: an experimental study. *Acta Psychologia,* **35**, 378-394.

STILLWELL, W. G., SEAVER, D. A. & SCHWARTZ, J. P. (1982). Expert estimation of human error probabilities in nuclear power plant operations: a review of probability assessment and scaling. Sandia National Laboratories Report SAND81-7140, Albuquerque, New Mexico and Nuclear Regulatory Commission Report NUREG/CR-2255, Washington, D.C.

STOTO, M. A. (1988). Dealing with uncertainty: statistics for an aging population. *American Statistician,* **42**, 103-110.

STROUD, M. (1980). *Professional writing: Strategies to improve the transfer of knowledge.* Communications Strategies of Advanced Technologies, Albuquerque, New Mexico.

TIETJEN, G. L. (1986). *A Topical Dictionary of Statistics.* New York: New York: Chapman and Hall.

TVERSKY, A. & KAHNEMAN, D. (1974). Judgments under uncertainty: Heuristics and Biases. *Science,* **185**, 1124-1131.

______ (1981). Framing of decisions and the psychology of choice. *Science.* **211**, 453-458.

U.S. NUCLEAR REGULATORY COMMISSION (NRC), Office of Nuclear Regulatory Research (1983). PRA procedures guide: a guide to the performance of probabilistic risk assessments for nuclear power plants, Vol. 1-2, prepared under the auspices of the American Nuclear Society and the Institute of Electrical and Electronic Engineers under a grant from the Nuclear Regulatory Commission, Nuclear Regulatory Commission Report NUREG/CR-2300, Washington, D.C.

U.S. NUCLEAR REGULATORY COMMISSION (NRC), Office of Nuclear Regulatory Research (1989). Severe accident risks: an assessment for five U.S. Nuclear Power Plants, (formerly entitled Reactor risk reference document), Vol. 1-2, second draft for peer review, Nuclear Regulatory Commission Report NUREG-1150, Washington, D.C.

VAN DE VEN, A. & DELBECQ, A. (1974). The effectiveness of nominal, Delphi, and interacting group decision making processes. *Academy of Management Journal,* **17**, 605-621.

VEDDER, R. G. & MASON, R. O. (1987). An expert system application for decision support in law enforcement. *Decision Science ,* **18**, 400-414.

WAERN, Y. (1987). Mental models in learning computerized tasks. In M. FRESE, E. ULICH & W. DZIDA Eds., *Psychological Issues of Human-computer Interactions in the Work Place,* pp. 275-294. Amsterdam: North-Holland Press.

WATERMAN, D. A. (1986). *A Guide to Expert Systems.* Reading, Massachusetts: Addison-Wesley Publishing.

WEISSENBERG, P. (1971). *Introduction to Organizational Behavior: A Behavioral Science Approach to Understanding Organizations.* Scranton, Ohio: Intext Educational Publishers.

WELBANK, M. (1983). *A Review of Knowledge Acquisition Techniques for Expert Systems.* British Telecom Research Laboratories Report, Martlesham Heath. Ipswich, England: Martlesham Consultancy Services.

WHEELER, T. A., HORA, S. C., CRAMOND, W. R. & UNWIN, S. D. (1989). Analysis of core damage frequency from internal events: expert judgment elicitation, Vol. 2. Sandia National Laboratories Report SAND86-2084, Albuquerque, New Mexico and Nuclear Regulatory Commission Report NUREG/CR-4550, Washington, D.C.

WIELINGA, B., AKKERMANS, H., SCHREIBER, G. & BALDER, J. (1989). A knowledge acquisition perspective on knowledge-level models. *Proceedings of the AAAI-sponsored 4th Annual Knowledge Acquisition for Knowledge-Based Systems Workshop,* Banff, Canada, pp. 36-1 to 36-22.

WINKLER, R. L. (1968). The consensus of subjective probability distributions. *Management Science,* **15**, B-61 to B-75.

___________ (1981). Combining probability distributions from dependent information sources. *Management Science,* **27**, 479-488.

WINKLER, R. L. (1986). Expert resolution. *Management Science,* **32**, 298-303.

ZIMBARDO, P. G. (1983). To control a mind. *Stanford Magazine,* Winter, 59-64.

Index

C

F

G

S